# THE GOLDEN FRONTIER

*The recollections of*

*Herman Francis Reinhart*

1851–1869

*Personal Narratives of the West*


J. Frank Dobie, General Editor

Herman Francis Reinhart

# THE GOLDEN FRONTIER

*The recollections of
Herman Francis Reinhart
1851-1869*

Edited by DOYCE B. NUNIS, JR.
Foreword by NORA B. CUNNINGHAM

UNIVERSITY OF TEXAS PRESS · AUSTIN

Library of Congress Catalog Card No. 62-9786
Copyright © Nora B. Cunningham, 1962
All rights reserved
ISBN 978-0-292-74158-4
First paperback printing, 2012

*To the memory of*

Rena Reinhart

# FOREWORD

The bundle, I remember, made a sizable armful. It was wrapped in light quilted material, time-yellowed, fastened by large safety pins. Opened, it revealed five long narrow ledgers with red-marbled hard backs, such ledgers as were once used by country storekeepers.

All the ledgers were filled with writing in black ink, not much faded. It was a firm and flowing hand, but the style was old-fashioned, with "f"s for "s's," and with many peculiarities of spelling and capitalization. The writer was economical of space; he left no margins. At the top of the first page of the top ledger was written "The Life and Recollections of Herman Francis Reinhart."

This was my first sight of the Reinhart manuscript, and perhaps this was the first time it had been seen by anyone since it was put away after the death of the writer more than sixty years before.

I had had a casual acquaintance with the Reinhart sisters, both of whom worked in a downtown dry-goods store, for many years, but it was not until after their retirement, when they came to live in a little house up our street that I became better acquainted with them. When the death of her sister left Miss Rena alone, my sister and I tried to show our sympathy by neighborly attentions, and sometimes took her riding.

We happened to be talking of the West as we returned from a ride one day, and Miss Rena remarked that her father had spent most of his youth in the West. She added that he had written some of his reminiscences. I said politely that those should be interesting, and she said she would show them to me sometime.

I had forgotten the conversation when she phoned a few days later to say she had decided to give me her father's writings. I was taken aback.

I knew nothing of her father except that he was one of the town's earliest citizens, but I felt sure he was not a writer, and did not imagine that his reminiscences would be of much value to anyone but his family. I would just borrow them to read, I told her. No, she said, she wanted to give them to me. She was an old woman and was clearing out her possessions. Her brother and sister were dead. She and her sister had talked about it, and had decided that the writings should be burned. So, if I did not want them —Hastily, I told her I would be right up to get them. (But I wondered a little, just how much there was of the writings, and in what form.)

Her beautiful old walnut bureau stood in the "front room" of her little house, and it was from the bottom drawer of this bureau that she took the bundle I have described. She gave it to me quite matter-of-factly and cut short my thanks by beginning to talk of something else.

I carried the bundle home before opening it, and having seen what was in it, I put it away. The summer was a busy one, and I had no leisure to pore over old-fashioned handwriting. There came a day, however, when I began the story, and once begun, I could not stop. For some reason, I thought of the narrator as "Franz"—he was christened "Herman Franz" but probably he thought the name "Francis" sounded more "American." The German name seemed to fit him best, however.

It was slow work reading the old-fashioned handwriting, so I began typing it off, concentrating on the words, and later enjoying the story in type. I talked so much of "Franz" that Elsie (my sister) became interested, and took a hand at the copying. As she is a proficient typist, the work went more rapidly, but even so, it took most of our spare time for almost a year.

We grew fond of Franz as we followed his adventures. He was so frank, so honest, so good-humored; he had such a zest for life and enjoyment of new experiences. His brother Charley was a year older, but much more reckless and hot-headed. Franz was always having to get him out of scrapes. But he didn't seem to mind it. He was fond of Charley.

Franz loved fine clothes, and there was a certain fancy buckskin coat, bought from the Indians, that caused a stampede one night on the trip across the plains. Later, the wagon train was robbed by the Indians, and Franz lamented the loss of his coat, but he wrote that he "had to laugh to think what a swell the Indians would cut in our clothes, white shirts and broadcloth and my Fancy Buckskin coat, too."

In his writing he seemed concerned only about setting down the way things happened, whether they showed him in the best light or not. Yet he had an innocent vanity, shown by his habit of saying, when he had met an old acquaintance, "They was glad to see me," instead of the more usual observation that he was glad to see them.

He had a good deal of humor, and for an uneducated man—he had

*Foreword* xiii

only about four years schooling, and "was never a good scholar"—he showed an unexpected tendency to philosophize and speculate about his experiences. But he was a man of action, and it mattered little what sort of action was required. Once, for instance, after a hazardous trip up the Klamath River and across the mountains, he reached his brother's bakery and, finding him short-handed, at once set to work making pies.

As Elsie and I followed the story with our typing, two questions haunted us: how could Franz have written all this from memory alone? And why did he want to write it? He says that he wrote from memory, yet how could he have kept the chronological sequence and all the mass of detail in his head? On the other hand, how could he have kept diaries, or even notebooks, during those ever-active, ever-changing Western years? I questioned Miss Rena but she could tell me nothing. She was fourteen when her father died, and she remembered him in those last years as a semi-invalid, sitting about the house writing in his ledgers. Did he have anything to refer to as he wrote, any notebooks, any diaries? She did not think so. But she was only a child and did not pay much attention.

And why did he write it? Was it merely a way of passing the time for a sick man? Did he hope it might be published? He knew nothing of the writing or the publishing of books. He liked to read, and sometimes records his buying books in frontier towns of the west, but the world of writers was outside his ken; perhaps no one could have been farther removed from "literary atmosphere." Yet if it was written for his family, it would seem that he might have glossed over his failures and blunders and unethical actions, such as playing with marked cards. Then too, if he wrote the story for his family, he must have known that they were not readers. They cherished the written ledgers as "keepsakes," wrapping them carefully and putting them away in the darkness of a bureau drawer for sixty years.

We shared the story with some of our friends, and my dear poet-friend, Billy B. Cooper of Neodesha, Kansas, suggested that I offer some of Franz' adventures to Mr. Henniker-Heaton, Home Forum editor of the *Christian Science Monitor*. These were found acceptable, and in 1953–1954, over a period of several months, six "Franz" stories were published on the Home Forum Page.

These brought a gratifying response, particularly from Western readers familiar with the locale that Franz had known in the 1850's. Three historical societies, of California and Oregon, sought to acquire the ledger-manuscripts for their archives. But the ledgers had already gone, as soon as the typing was completed, to the Kansas State Historical Society at Topeka. Special thanks are due Mr. Nyle H. Miller, secretary of the Society, for his courtesy and cooperation in the publishing of this portion of the ledger-contents.

In our Memorial Building here in Chanute, Kansas, is one room devoted to local history. Here Franz and Maggie, his wife, look down from their tintypes on the wall, among pictures of their contemporaries. In Elmwood, the Chanute cemetery, is a row of headstones bearing the names of Franz and Maggie and their children—of whom Miss Rena, who died in the spring of 1961, was the last. But the spirit of Franz lives on, ever youthful, in this journal which he left.

The story the journal tells has great historical value, I know, but for me the human interest is transcendent. Franz was a common man, with little schooling, perhaps a rather typical American of his time. He shows that time in all its lusty vigor, and in so doing leaves the reader with an unforgettable impression of his own personality.

NORA B. CUNNINGHAM

# ACKNOWLEDGMENTS

"Seek, and ye shall find!" This ancient admonition is most applicable to the historian engaged in editing a manuscript for publication. The success of the search for information, in many instances, rests exclusively on the help rendered by others. By such generous and selfless acts, editorial puzzles find solutions: small or large, that assistance adds its measure to the finished work. I wish to express to the following my debt of gratitude and appreciation for their efforts.

For special searches, I am grateful to Willard E. Ireland, Provincial Librarian and Archivist, Victoria, British Columbia, who furnished me with several answers and pertinent extracts from the Victoria *Gazette*. Miss Mary K. Dempsey, Assistant Librarian, Historical Society of Montana, not only solved three dilemmas for me, but also helped procure appropriate scenes portraying Montana during Reinhart's labors there. Mr. J. M. Brown of *The Oregonian*, Portland, gave ready response to my requests for photostats from his newspaper files (where needed), as did Robert D. Young of *The Oregon Journal*. Miss Priscilla Knuth, Research Associate, Oregon Historical Society, graciously replied to a query with the available information.

Mr. Herman J. Hogensen, City Recorder, Salt Lake City, solved research conflicts by his ready and full response to a query concerning 1868 law enforcement officials in his city; as did Lauritz G. Petersen, Assistant Librarian, Church of Jesus Christ of Latter-Day Saints, Office of the Church Historian, on several biographical matters. Mr. Guy L. Robins, Millard County Clerk, likewise readily gave attention to my quest for the name of that County's 1868 Sheriff. Dr. A. R. Mortensen, formerly Director of the Utah Historical Society (now Director of the University of

Utah Press), and John James, Jr., the Society Librarian, graciously provided information that made possible a more articulate editing of Reinhart's stay in Utah, 1867–1868.

Miss Hazel B. Rider, Secretary, Siskiyou County Historical Society, rendered service to two queries placed in her hands, besides making available an 1851 Yreka reproduction. Mr. F. Hal Higgins, Davis, California, responded quickly with information pertinent to a singular editorial problem. And, Mr. Ernest P. Mauk, Los Angeles Valley College, shared his theater library and knowledge with me in solving several identifications for theater personalities mentioned by Reinhart.

To these thoughtful and cooperative people, my thanks.

A special appreciative debt is owed to the Henry E. Huntington Library staff for their never failing smiles, quiet suggestions, and assistance during my work there these past months. The facilities of the Bancroft Library, University of California, Berkeley, provided several solutions on a brief visit, made more profitable and enjoyable by their quick response to my needs. Particularly, I wish to express my long standing debt to Dale L. Morgan of that staff for his never failing kindness, encouragement, and rich knowledge of the West. He provided answers to several difficult problems which confronted me.

James DeT. Abajian, Librarian, California Historical Society, rendered immeasurable assistance by compiling index cards from the Reinhart manuscript to search through his extensive and valuable information-card file, one of the real treasures of the Society's library. For that effort, above the call of our friendship, I am deeply grateful.

Lastly, I wish to thank Mrs. Laurel Allen for providing me with a final typed draft, and to Mrs. Elizabeth I. Dixon, Librarian, University of California at Los Angeles, for her generous assistance in helping to proofread the galleys.

DOYCE B. NUNIS, JR.<br>
*University of California*<br>
*Los Angeles*

We wish to express our gratitude to the Kansas State Historical Society and to the *Christian Science Monitor* for their cooperation in this publication of the recollections of Herman Francis Reinhart.

THE UNIVERSITY OF TEXAS PRESS

## CONTENTS

| | | |
|---|---|---|
| Foreword | | xi |
| Acknowledgments | | xv |
| Prologue | | xix |
| Chapter 1 | Overland to California | 1 |
| Chapter 2 | Mining Apprenticeship: Siskiyou County | 26 |
| Chapter 3 | Gold Rush Days in Southern Oregon | 49 |
| Chapter 4 | A Pacific Rancher on the Oregon Coast | 77 |
| Chapter 5 | Return to Siskiyou: Indian Wars and Mining | 90 |
| Chapter 6 | To the Fraser and Return | 108 |
| Chapter 7 | Oregon Interlude | 146 |
| Chapter 8 | People, Places, and Things | 162 |
| Chapter 9 | Winter in Walla Walla | 181 |
| Chapter 10 | Settling Down on Dry Creek | 202 |
| Chapter 11 | Hauling to the Boise Basin | 218 |
| Chapter 12 | Badmen, Gunmen, and Vigilantes | 237 |
| Chapter 13 | Teamster in Montana | 247 |
| Chapter 14 | Utah: Mormons and Lost Horses | 271 |
| Chapter 15 | Hauling for the Union Pacific | 293 |
| Epilogue | | 307 |
| Appendix | | 308 |
| Bibliography | | 311 |
| Index | | 319 |

## ILLUSTRATIONS

Herman Francis Reinhart                            *Frontispiece*

1. Yreka, California, in 1851                 Following page 116
2. Jacksonville, Oregon, in 1854
3. Browntown, Oregon, (c. 1852–1853)
4. A Whipsaw-mill near the Oregon Mines
5. An Early Gold Mine
6. Fort Yale, British Columbia, in 1858
7. Fort Langley, British Columbia, in 1858
8. Map of the Gold Regions of British Columbia (1862)
9. Freighting to the Idaho Mines
10. Helena, Montana, in 1866
11. Virginia City, Montana (c. 1866)
12. Fort Benton, Montana (c. 1867–1868)
13. Salt Lake City, Utah, in 1869
14. Bear River City, Wyoming
15. Supply trains in 1869

## MAPS

Map 1. Reinhart's Pacific Northwest . . . . . . . xxiv, xxv
Map 2. The Oregon Trail . . . . . . . . . 12, 13
Map 3. Western Oregon and Northern California . . . . 38
Map 4. The Fraser River Trip . . . . . . . . 110
Map 5. The Eastern Mines . . . . . . . . . 198
Map 6. Utah and Wyoming . . . . . . . . . 275

# PROLOGUE

Herman Francis Reinhart. This was not an uncommon name, but *he was* a *common man*. His life story shows what might be called a typical pattern—perhaps just a bit more, for he was born in Prussia and died in Kansas. He was one of those immigrant thousands who sought the American dream in a land ripe with promise. If his life be measured in terms of material achievement, then by most standards he was a failure. Yet, the very foundation of the nation rests on such men as Reinhart, for each in his own small way helped to build and stimulate the growth of their homeland.

Reinhart was born on New Year's Eve, 1832. His parents, Carl Heinrich and Maria Concordia, were living in Jena, in Prussia, having only the year before moved from Gabestadt. Both in their early thirties, they had four children, two daughters and two sons, Herman Francis being the youngest. Reinhart's father was engaged, at the time, in operating a bakery-confectionery shop and a dance pavilion, and, in fact, a large ball was in progress at the pavilion at the time of the boy's birth.

From the very beginning, Herman Francis took after his father, a man passionately devoted to horses, a wagon and carriage maker, apparently an accomplished baker and confectioner, a keeper of hotels who was fond of providing recreational facilities for his community and guests. These were the trades young Reinhart learned in his youth, both in Prussia and later when the family immigrated to the United States. Primarily educated as a tradesman with no more than a few years of formal schooling, two years in Prussia, two in New York City, he turned his manual skills to good advantage in after years: his vocational training provided him the basic ingredients for livelihood in America.

When he was eight Reinhart's family took steerage passage and sailed for New York City. There Reinhart continued to master the various trades he had been weaned on at his father's side. After eight years in the city the family moved to Lake County, Illinois, to a newly acquired farm. Here the young man became a skilled thresher, adding another asset to his trades' repertoire, one that was to prove an advantage when he emigrated to the Oregon Territory.

The death of their parents, coupled with growing indebtedness incurred in operating the family farm, drove Reinhart and his brother, Charles, infected by "gold fever," to head West in 1851 for Eldorado. Their hope was to strike it rich and return East with enough capital to build their farm into a profitable operation. Their hope was vain.

For almost twenty eventful years, Reinhart sought fortune in gold mining or what have you. Jack-of-several-trades, he was better equipped for the disappointments which came to him on the mining frontier than many of his compatriots. He at least made a tolerable living, for he was skilled and enterprising. Although defeated in his pursuit of illusive fortune, he left, ironically, a more enduring record. He took the time, for reasons unknown, to write the recollections of his American odyssey. That odyssey is neither spectacular nor impressive—it is quite ordinary; it is human.

The recollections which follow are the memoirs of an ordinary man, a *common man*—indeed, *everyman* of like bent—who was lured West by the glittering promise of wealth to be found in the gold fields. It is a personal record of experience on the Pacific Northwest frontier, 1851–1869. With the exception of the 1849 rush to California, and the later rushes to Colorado and the Black Hills, Reinhart was a participant in practically all the gold-mining discoveries of the two decades—from Yreka to Jacksonville, from Althouse Creek to the Fraser, from Gold Beach to Indian Creek, Clearwater, Orofino, Boise, and the Salmon, from Helena to the Sweetwater. His account is the story of the expanding mining frontier, of the plain men who sought the earth's golden treasure, and of the events that shaped and colored their quest for fortune. I know of no other miner's recollections encompassing such a continuous time span, embracing such a broad geographic region, and furnishing glimpses of so many major events, as those related by Reinhart.

Certainly the limitations of Reinhart are obvious. He was neither a good writer nor a thoughtful witness. His language has many shortcomings, and his storyteller style falls far short of many an available recollection. He realizes almost nothing of the national or regional import of much that he records. Even the most local and personal happenings are invariably judged according to their effect upon his own well-being and enterprise—this despite the fact that he relates his tale years

*Prologue*                                                                 xxi

later when even a moderately thoughtful man should have begun to see such experiences in a broader light. One would wish more information on the Plummer Gang, the opening of various mining enterprises, the death of Bozeman, a chance meeting with Brigham Young. No. Instead one reads about business, cronies, horses, and money. Reinhart's limited awareness, his lack of recognition of significant events, his apparent inability to delineate cause and effect, outside of his immediate experience and personal objectives, frustrate the reader.

Yet, these seeming defects in textual exposition cast the recollections in a mold that is unique. Never for a moment does one question the fact that Reinhart is recalling just what happened, or what he believed happened in his twenty years on the gold-mining frontier: he gives to every incident the emphasis that it actually had for him in his memory's eye. His lack of vision, proportion, relevance, and value—all are clearly those of Reinhart. He reveals himself with startling directness and naïveté. He is, above all, himself.

His account, in substance, is that rarest of all things—the *common man's* view of life around him. His narrative reflects the way things happened to the vast majority of those men who took part in the gold rush. Through such men and their admittedly selfish quest, the West was nurtured: they blazed the way for a more enduring accomplishment.

Reinhart's recollection mirrors that very fact. One sees the beginning of settlements that are to grow into towns and cities. Frontier initiative in business, both retail and recreational, is illustrated by Reinhart's varied enterprises. The foundations for future developments in lumbering, transportation, communication are woven into his story, for he played an initial part in such ventures.

Law and order, politics and elections are treated in more than a casual fashion. What Reinhart has to relate on these matters is rather detailed in comparison to his reports on other events. Particularly, the tragedy of the Civil War is reflected in fairly extensive firsthand accounts of the explosive and divided opinion of Westerners at that time.

And more. Reinhart tells the story of ordinary folk, of their struggle for dignity and security, economic and personal, on the western frontier. One sees only too clearly the need for adaptation and the response to the challenge of life in a primitive environment. What Reinhart presents is an intimate and simply etched portrait of the *common man's* West.

In preparing Reinhart's recollections for publication, care has been taken to preserve the flavor of his remembered experiences as he records them. Occasional paragraphing, additional punctuation for clarity, correction of misleading spelling, have been the editorial guides. Certain portions of the text have been deleted when repetitious, or when the related events were overly drawn out. The primary objective has been to present the reader with a faithful rendering of Reinhart's original text without excessive editorial embellishments.

In a postscript to his narrative, Reinhart wrote: "1887. I have written this out in haste from memory without notes and in fear that I might not complete it." He wrote a decade and a half after his western adventures and his memory, though accurate in many respects, was not infallible. For this reason, annotations of the text have been reserved for correction of erroneously recorded events and added clarifications for the reader's convenience.

During his twenty years on the mining frontier, Reinhart was practically everywhere in the Pacific Northwest. Because of the variety of his friends and associations, and the lack of available sources to check his remembrances against, errors may still exist which have escaped the editor. It is hoped that if such be the case there will be few of them.

Lastly, it should be noted that several portions of the Reinhart recollections have been previously published in an unedited form in the *Christian Science Monitor*. These excerpts were prepared by Miss Nora B. Cunningham. They were published under the following titles and dates: "Covered Wagon Days," December 8, 1953; "Gold Miner to Pieman," January 26; "Franz at Cow Creek," March 30; "Franz Moves Forward," May 19; "Franz Remains Unwed," September 15; "Franz Decides," October 14, 1954. "Franz," of course, is Herman Francis Reinhart.

What follows is his story.

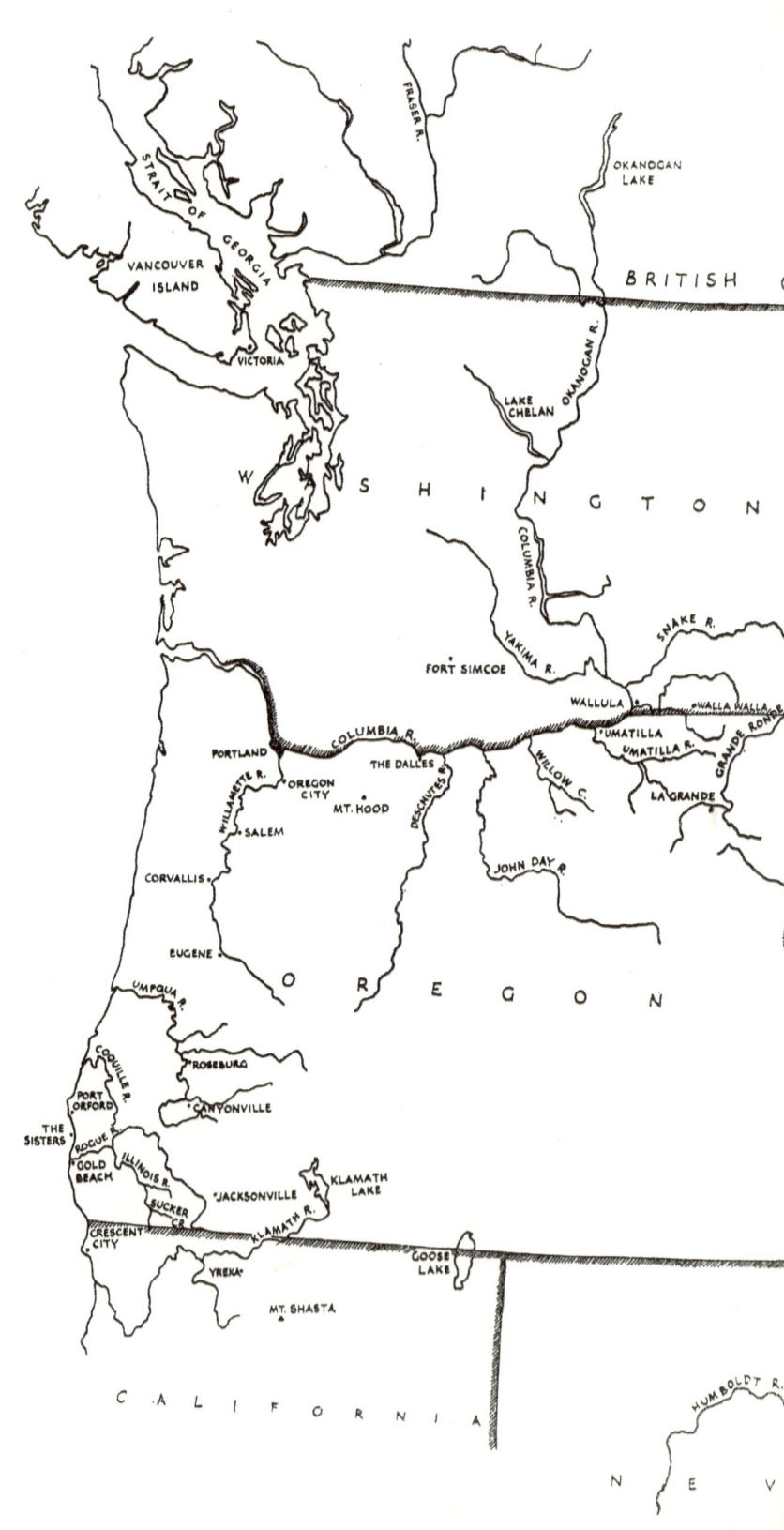

# Map 1. Reinhart's Pacific Northwest

# THE GOLDEN FRONTIER

*The recollections of*

*Herman Francis Reinhart*

**1851–1869**

# 1

## OVERLAND TO CALIFORNIA

(1851)

After our coming to the conclusion to go to California, we made our preparation to trade. Charley commenced and got a light wagon for the heavy [one], and a yoke of cattle for a horse, and got some boot. I traded Billy the bay horse and got two yoke of steers [oxen], unbroken, 3 and 2-year-olds, and Charley got one yoke steers [oxen] and some money. We had 4 yoke and the light wagon and cover and some little money and feed and provisions to start. So we bid our friends . . . goodbye and started some time in April [1851] for California or Oregon.

Some of our strongest reasons for going was Charley had got his money on 80 acres of land for four months, expecting to make it threshing, and there was a Payment to be made on the Threshing Machine besides, and no way to meet the payments, and our farms at home were rented for two years. We thought by going out to California or Oregon would give us a chance to make something and meet our promises by the time I, the youngest, would become of age, before the balance of the land could be divided.

After the first day a man by the name of Pease caught up with us. He was a schoolteacher in Wisconsin and thought of going to California. He traveled with us a few days, but before we got to Cedar Rapids [Iowa] he turned back and concluded not to go for a while. He had a young wife at home and he got homesick. Another young man wanted to go with us, named Parker. He was a Blacksmith, but we shipped him in a few days; he was on the sponge too much and had no means of travel.

Our young cattle had to be broke in; if we put the old broke one on the lead, the others would not hold back, so we had to put the Old Big Yoke on the tong[ue] to steady the others, then one of us would have to be on each side to head them around when we wanted them, for they would not mind our call, Haw or Gee. We just had to run around them and head them to stop. In going down hill we had to catch on to the wagon to keep up. We could not stop them by calling Hoh, and the old one could not hold back enough to stop them. We used to go thirty miles a day and leave all ox teams behind. They called us the Tellagraph team. Our wagon was a light one-horse, iron-axled wagon. Going through the timber on Skunk River, Iowa,[1] we caught one axle on to a tree and sprung the axle, and came near breaking up several times when our four yoke of steers became unmanageable, till we got them broke. We had several offers of Girls that wanted to go with us to California, for the fun of it. We had a nice little outfit.

We had a hard time to get flour or wheat bread in most all of Iowa, excepting few of the larger towns. We would ask them what they eat—they said, "Corn bread and potatoes." We had to do the best we could.

One morning between Des Moin[es] River and Skunk River, Charley was driving up the cattle with the whip and he called to me to look what he had found. He was driving a nice white and black spotted little thing like a cat with a bushy tail. He kept it close to the cattle and was coming right into camp with it. I never saw anything like it, nor had he. When one of the boys saw it, he called out to let it go as it was a Skunk or Poll Catt. Charley had been trying to get up to it several times, so he let it go, and the boys made fun of Charley's Pet that he had found.

We had to inquire our way a great deal in Iowa for it was not much settled in them parts. Fort Des Moines was but a small place. Iowa City was the Capitoll.[2] We found a lot of Mormons were settled in about Council Bluffs or [Kane]sville. This last place was a large town seven or eight miles northwest of the river from Council Bluffs, and where all went to fit out Crossing the Plains.[3]

---

[1] The Skunk lies northeast of and flows somewhat parallel to the Des Moines River. It empties into the Mississippi seven miles south of Burlington, Iowa.

[2] Governor Robert Lucas founded Iowa City as the territorial capital, May 4, 1839. Under the 1846 Iowa Constitution, provision was made for the possible selection of another site. After considerable acrimonious debate in the state legislature, Des Moines was officially proclaimed the new capital site by Governor James W. Grimes, October 19, 1857.

Des Moines, situated at the confluence of the Racoon and Des Moines rivers, was established as a military fort by Captain James Allen, November 12, 1842, being evacuated in early March, 1846. By legislative act, January 28, 1857, the word "fort" was deleted from its name.

The Reverend John A. Nash, a New York missionary sent west by the American Baptist Home Mission Society, arrived in Des Moines, January 3, 1851, to take up his "call" and found the settlement of around 550 inhabitants "anything but inspiring." Dan E. Clark (ed.), "Some Episodes in the Early History of Des Moines [Selections from the Autobiography of John A. Nash]," *Iowa Journal of History and Politics*, XIII (1915), 175-237. The 1850 census, taken in June, listed the population as 494.

[3] When the vanguard of the Mormon emigration to Utah reached this jumping-off-place, June 14, 1846, having been expelled from Nauvoo, Illinois, the names Hart's

We found a lot of teams on the banks of the Missouri River between Council Bluffs and [Kane]sville, some bound for Oregon, some for California, and lots for Salt Lake City. We had not enough money to buy our outfit for six or eight months, and we found three yoke of cattle would be plenty to haul our little wagon, and we got a chance to sell one yoke. So Charley sold one yoke for $55 and laid in our provisions and ammunition, and we moved our team down to the ferry where we had to cross the Missouri from Council Bluffs to the other shore, where Omaha now is, but then nothing but an Indian Agency.[4] We saw several Indians, Sa[cs] and Foxes and Omahas. We crossed and went out a mile or two and camped. We found a large lot of wagons that were going to start for Oregon in a few days. We concluded to join them and go with them as far as we saw fit to travel their road. There were 21 men and boys.

We crossed the Missouri, May 15, 1851. Our train was composed of Capt. Hill[s] and family (two brothers, Eli[jah] and Erastus, single) and Mr. Briggs and family, father-in-law of Capt. Hill[s]'s; Wm. R. Riddle of Sangam[on] County, Illinois, Springfield, and family; Charles W. Beckw[or]th from Wisconsin and his family; Col. Helms and family from Missouri; J[ohn] B. Welch and brother Jerry Welch (single) from Rochester, New York; Bowen Lindley, Elish[a] Hammer from Iowa; two brothers, Douglas and James Downer from Janesville, Wisconsin and Lester Clark, the same; myself and Brother Charles. Nineteen wagons, twenty-one boys and men.

Some of the teams had extra teamsters or ox drivers. Helms had about

---

Bluffs (after an 1820's French fur trader) or Miller's Hollow were in vogue. On April 8, 1848, the Mormons renamed the settlement Kanesville in honor of Colonel Thomas Leiper Kane who had befriended them in their hour of persecution. It was not until 1852 that the community was christened Council Bluffs. Charles H. Babbitt, *Early Days at Council Bluffs*, pp. 9–16.

[4] Omaha as a settled community was not established until 1854. It served as the territorial capital from 1855 to 1867. In 1851, it had an Indian school, taught by Samuel Allis and his wife, and functioned as a subagency to the Council Bluffs Indian Agency, headed by Major John E. Barrow. The school was abandoned that very year. Barrow, in reporting to his chief, Colonel D. D. Mitchell, St. Louis, October 1, 1851, noted a problem which was to plague Reinhart's emigrant train. He wrote:

> The complaints of the Indians [Ottoe, Missouris, and Omaha] against the large emigration through their country are becoming very general, and I think some compensation justly due them for this purpose would satisfy them, and prevent many robberies from being committed upon those of our people who journey to California and Oregon.

The Indian school was located at Bellevue, a few miles south of the present city of Omaha. *U. S. Senate, Executive Documents*, 32 Congress, 1 Session, Document 1, III, 357, 359–361.

At this time, 1851, two men are known to have operated ferries across the river: William D. Brown, a California-bound emigrant who saw golden opportunity in transportation and who was one of Omaha's founders; Peter A. Sarpy, a former factor for the American Fur Company. Walker D. Wyman, "Omaha: Frontier Depot and Prodigy

three; John Welch had one, John Lockridge, a gunsmith, the best short-distance marksman I ever saw. Two Blacksmiths had a team: J[ames] Douglas and Clark Richardson, both from Janesville, Wisconsin. The men Helms had were Mr. Paddy and one Lingenfelter. One of Riddle's was Shaffer, and Jack Middleton drove a team of Riddle's for his daughter, a young widow named Art[i]n[eci]a Chapman.[5] Mr. Jerry Welch

---

of Council Bluffs," *Nebraska History Magazine*, XVII (1936), 143-144; Babbitt, *Early Days*, p. 86.

[5] Like so many emigrants who came west with emigrant trains, Reinhart does not give an accurate description of the members of his party. To reconstruct the names of the party who made the trek west in 1851 and to give a broader and more accurate account, a number of sources have had to be tapped. Those book sources which have been found most useful are: George W. Riddle, *History of Early Days in Oregon*; A[lbert] G. Walling, *Illustrated History of Lane County, Oregon*; A[lbert] G. Walling, *History of Southern Oregon* . . .; Joseph Gaston, *The Centennial History of Oregon, 1811-1912*; Hubert H. Bancroft, *History of Oregon*; Lewis A. McArthur, *Oregon Geographic Names*; [n.a.], *Portrait and Biographical Record of Western Oregon*; and various articles found scattered in the *Oregon Historical Quarterly*.

In identifying the party, I have accepted Reinhart's basic list, with such corrections as follows.

Charles W. Beckworth and family from Wisconsin. Reinhart, along with several other sources, sometimes spells the last name "Beckwith." On arrival in Oregon, Beckworth settled in the vicinity of Canyonville.

Samuel B. Briggs and family from Iowa. Briggs was born in Massachusetts in 1812; later moved to Ohio, and in 1841 to Iowa. His wife was Susanna Phillips, a native of Maine. His daughter, Sophronia married the captain of the 1851 overland train, C. J. Hills. (A son of Hills, Elijah C. Hills, states in an interview in *The Oregon Journal*, December 31, 1932, that his maternal grandparents were David and Almira Briggs. Other sources are contradictory on these names.) The Briggs family had six children. William, Sophronia, and Almira apparently made the trip west, possibly the others as well. The family settled two and a half miles northwest of Canyonville.

William H. Riddle and family from Sangamon County, Illinois, near Springfield. Born, December 14, 1809, in Bourbon County, Kentucky. Riddle was a descendant of the founder of that state's famed Riddle Station. After living in Ohio, he moved to Illinois in 1838. He married Maximilla Bouseman. Those of his family who made the trek west in 1851 were his sons (William H., age 13—who died in 1857; George W., 11; Abner, 9; John Bouseman, 8; T. Stilley, 2); his daughters (Isabella—or Isabell—18; the recently widowed Artinecia Chapman—Riddle's eldest daughter—with her infant son, John); Lucinda McGill, about 45, Mrs. Riddle's half sister-in-law (who later married Campbell Crismas in Lane County); Anna Hall, a cousin, age 11. Employed by Riddle to assist in driving three wagons, each drawn by three yoke of oxen, and to herd forty head of cattle, were Newtown and George Bramson, brothers, and John (Jack) Middleton (whom Reinhart lists). *The Oregonian*, March 29, 1925. The Riddle family settled in an area in Douglas County which has been named for them. The sons, Abner, Tobias Stilley, George, and John, have received separate biographical sketches in various Oregon historical publications mentioned above.

Accompanying the Riddles were the following families: Stephen Hussy and family; Samuel Yokum and family, and "Sandy" Yokum, a bachelor. The Yokums are not named by Reinhart, although later in his narrative, he implies that they were members of the 1851 train. (See Chapter III, *note 21, post*.)

James W. Lingenfelter settled near Jacksonville. During the Civil War he was made captain of a volunteer company and was killed at Fortress Monroe, October 8, 1861. Robert T. Platt, "Oregon and Its Share in the Civil War," *Oregon Historical Quarterly*, IV (1903), 104.

A sketch of Cornelius J. Hills is found in *note 14, post*.

had been sick with the fever for some time, and he died at this place where we first got to them. We helped burry him that afternoon.[6]

Next day we all started on. When we got to Elk Horn and Loupe Fork,[7] the streams were very high and we had to use all the tight wagon boxes to ferry ourselfs and our goods over. Some places a mile wide the cattle had to swim. One man in another train was killed by lightning which was fearfull on the Platte River. Some few men were drowned out of other trains. All the little streams and ravines, at other times very dry, were now very high and full of water. Charley in helping to drive the cattle to swim over could not swim himself, but caught an ox by the tail and held on until he pulled him over all safe.

We used to drive in turn; one day one team would be ahead, and the next day that team dropt back and the next team would be ahead, and so on through in turn. Charley and I took our turns driving. One would drive the afternoon and next morning till noon, and then the other would drive the same, and so on. The one not driving could walk, hunt or ride or sleep in the wagon.

One day I was driving and my team was on the lead. We ran over a rough hill and the first thing I knew the wagon turned over, and Charley was in a sleep at the time. We had some flour loose in a box and it turned over Charley, and when I turned back the wagon you should have seen Charley, all covered with flour, and everything tumbled around him! He was not hurt but he was awful angry at me for upsetting him in the wagon but I could not help it, and we both upset more than once after that.

Another day I was driving and Charley was laying in the wagon. We were ahead again. We got to a small hill at one side; I called to Charley to take the whip and drive until I could go around the hill out of sight and I jumped out. He, it seems, was hardly wide awake. When I got to the other side of the hill there was a creek or "dry run" as it was called, but awful high at the time. When I got near, Charley started his cattle right into the creek to cross over—he thought it all right and supposed it was fordable, but his little leading steers commenced swimming, and he

---

[6] It should be noted from the language Reinhart uses in reference to the death and burial of Jerry Welch, "he died at this place where we first got to them," that Reinhart, in company with the Riddle family and their party members, did not join the Hills party until some 200 miles west of the Missouri River near the banks of the Loup River. Riddle, *Early Days*, p. 10, records in his recollections of the 1851 westward trip, that his family overtook the Hills company there and states: "They were being ferried across the river and while engaged in this work death claimed a member of their party, a young man named Welch, who, from accounts, had been very sick for several days." Thus it would appear that Reinhart and his brother may have joined the Riddles and their friends at Council Bluffs and became members of the Hills company near the banks of the Loup for the remainder of the westbound journey.

[7] This is the Elkhorn River and the Loup Fork. The latter is formed by the joining of the North and Middle Loup near St. Paul, Nebraska. The Loup empties sixty-eight miles east into the Platte at present-day Columbus.

urged the rear cattle on to cross. It would not have been safe to turn back, so Charley whipt up the cattle and they all had to swim, and our little wagon being very low and light, swam too, and the wagon bed came off and floated down stream with Charley in it. The cattle went on through with the running gear of the wagon, and stopt on the other side. Charley caught on to the wagon box to hold it—he could not swim—and our boxes and tin ware commenced swimming off down stream. A lot of our men saw Charley's danger and four or five jumped in and got him and the wagon box and what things they could ketch in the water, and took him over to the other side (They could all swim). So we had to all lay there and prop up our large wagons above the boldsters out of reach of the water to cross over without wetting their goods. We had to do the same at many places before and after. Some four teams got across and I with them. Fixt up Charley's team and wagon again and put in our effects all wet. Charley was awful mad at me for leaving him to drive and he not half awake.

While the balance of teams were crossing, some four or five teams of us went on to a large river called Wood River, which we had to ferry.[8] We wished to cross and camp on the other side that night. Two of our men that could swim went over with a rope and our little wagon being lightest was unloaded and to be pulled over first, for they wanted more help to pull over the larger wagons. So I was to get in our wagon. We had tied it down to the running gear so that it could not come off again. There was a cover and bows on our wagon, and I would make three over to pull the heavier wagons. Now when my little wagon got into deep water, it filled quite fast and sunk down to the bottom and I had all I could do to force my head out between the water and the wagon top. I was nearly drowned, but I managed to get on the front gate with my feet and held to the top and bows, and when the men on the other side that were pulling our wagon over saw the wagon sink, they knew I could not swim and they both let go the rope to come to my rescue by swimming, but the wagon washed down a little and got on the bottom, and I was safe on the top. They then run and got hold of the rope again before it got away, and pulled until I could jump in and help get the wagon out. They had a big laugh at me that time, and we both came near being drowned in that half day. It was a narrow escape for both of us, but we went right to work and by the time the balance of the train drove up, we had our wagons all over and had about ferried our goods over, and we made up big fires to dry our clothes, bedding, and what provision was not spoiled. All the balance of the train got over that evening, some after dark.

We had some awful storms on the Platte River and several wagons got upset. We had to stake our wagons down to keep them from blowing

---

[8] The Wood River is in south-central Nebraska and flows 110 miles, altering its course from southeast to east-northeast where it empties into the Platte near Central City.

away. By the time we got to Fort Lar[a]m[ie],[9] our provision got quite short, having lost a lot by the water damaging it. I went over to the Fort to try and buy some flour and sugar, but could not get any, they being short themselves.

All they could spare was some hard bread, for which I paid twelve cents per pound. Our train was made up of ox teams; all but one of Capt. Hill[s]'s. He had horses and a mule. We seemed to get along very slow, and some of the men got impatient and wanted to go faster. We had to stand guard every night for fear of the Indians and thieves.

The day that I stopt back at Fort Lar[a]m[ie], Charley and I had dropt behind, intending after getting what we wanted at the Fort, to catch up to the train that evening, when they would camp. We were four or five miles behind, but being so close to a United States fort, thought no danger to be feared from the Sioux Indians. The Indians were just then gathering together for a conference with the government, and were camped not far away on the Lar[a]m[ie] River.[10] As we were crossing some low hills six or eight miles from the Fort, we met a lot of Indians, some afoot, and some on ponies. They acted friendly, and called to us "Howtodo" and shook hands and offered to sell us a pony. We had some ten or fifteen dollars in half-dollars, and we had been expecting a chance to buy a good cheap pony, so we offered them so many half-dollars. But they saw that we were alone and rather green, and they tried their best to keep back some of our half-dollars. They would climb into our wagon and try to handle our guns and commenced to help themselves to crackers, and got quite saucy. I soon saw that we might be robbed, so took my rifle and Charley his shotgun (double-barrelled) and made them get off the wagon and started our cattle and they had to give us up. I am satisfied if we had not been so close to the Fort they would have taken our team and all our effects and maybe killed us both for what we had.

We caught up to our train at night and camped with them. We had to take our turns standing guard at night. Some of the families had a lot of loose stock, driving along some work cows in yokes, same as oxen, and some would turn out some of the work oxen when the roads were good

---

[9] Fort Laramie, located at the mouth of Laramie's Fork where it empties into the Platte River, was originally erected as a trader-fort by William Sublette, construction beginning, June 1, 1834. First dubbed Fort William, it subsequently took its geographic place name and served as a major way station for the westbound emigrants. In 1848 the government acquired the property for a military post. Albert J. Partoll (ed.), "Anderson's Narrative of a Ride to the Rocky Mountains in 1834," Historical Reprints No. 27, *Sources of Northwest History*; LeRoy R. Hafen and Francis M. Young, *Fort Laramie and the Pageant of the West, 1834–1890*, pp. 18–38.

[10] At the insistent urging of Thomas Fitzpatrick, U. S. Indian agent for the Upper Platte and Arkansas Agency, a general peace council of the Plains Indians was called to meet at Fort Laramie in the late summer of 1851. The conference did not officially open until September 8, adjourning after successful negotiations thirteen days later. LeRoy R. Hafen and W[illiam] J. Ghent, *Broken Hand, The Life Story of Thomas Fitzpatrick, Chief of the Mountain Men*, pp. 228–245.

and some got lame. It took herders to take the cattle as soon as they were turned out and herd them until night, when all our wagons would be formed into a large corrall by chaining the tongues of the wagons, each to the hind wheel of the wagon ahead of it. The twenty-one wagons made a large corrall and all the cattle were drove into it every night as soon as they had eat enough grass, and then a guard of three or four men walked around the camp and corrall until midnight, when another guard would stand until daylight, when the herders would take them to grass again.

Charley did not like to stand guard and kept telling me he wanted to leave the train and take up with some other and go to California. I was not in favor of leaving that train because we had got acquainted with them. If we should get out of provisions, we might expect some help from some of them that we could not expect from strangers in another train. But Charley got dissatisfied and tried all kinds of pretexts to leave the train.

One day, it [was] our turn to go ahead. Our young steers not well broke were quite wild and unmanageable, and our wagon light and nothing of a load at all. So our cattle kept going faster until our train was miles behind. We caught up to several other trains and our young leaders would horn the back of the wagons ahead of us, so that one of us would jump out of the wagon and turn our leaders to one side and go right by the train. After passing several trains, we stopt at noon to feed our cattle and let our train catch up with us.

But it seems our train stopt before, and nooned, and we waited, and then hitched up and started on, expecting our train to overtake us at night. But they did not. Charley still urged me to take up with some other train, but I would not, so next morning we started on still alone. We stopt at noon and I was in favor of staying there till our own train would come up.

A Californian with several wagons and a large flock of sheep came along. Charley had used up all our matches for smoking—we had lost a lot by getting wet and we happened not to have a match to light a fire. So I got up a lot of grass and weeds to make a fire and Charley shot off his double-barreled shotgun to get the wad and kindle before the wad went out. Charley saw the wad was about to go out in a hurry, so took the powder can between his thumb and finger, and commenced to pour the powder in a stream onto the burning wad. The Californian and his wife were both just then on horseback speaking to us, when the can with powder exploded out of Charley's hand and scared the horses so that they came near throwing their riders off.

The empty can just grazed Charley's head and charred his hand and nearly blowed off his thumb and finger. It made the fire burn but it was an awful careless trick, the Californian and his wife thought. Charley wanted to go on right then and not wait any more for the other train, that they were too slow, and we could get into a [faster] train and beat them in several weeks. We hitched up, but I would not let the team go until our train should come up (I could see them coming at a distance). Charley tried to take the whip away from me to drive on, but I would not give it up. He got a holt to take it, and I clinched him and threw him down. He gave it up then, for the train was right there and saw it all.

So we drove on the balance of the day with our train. At night, Charley being still of the will to leave the train, I told him he would have to take his yoke of Cattle and Wagon, and I would stay with my two yoke of steers. So we divided up and I sold my two yoke of steers to Col. Helms for $80, and Charley put his old yoke on to one of W[illiam] Riddle's wagons and drove for him. We left the wagon and some one else took it and left a heavier one. There was wagons left every day by the different trains. Some burnt them up to cook by. The wood was very scarce on Platte River, and for over five hundred miles we had to pick up weeds and dried Buffalo chips to cook by.

I had no provisions, so I bought a passage from there to Oregon to be boarded and my clothes or bedding for fifty dollars. That left me some thirty dollars from my cattle. At Soda Springs [in Idaho] I bought me a buckskin coat, the fanciest I ever saw, all embroidered and finished with ribbons of silk, a large cape and nice brass buttons. I got it very cheap ($10) from a half-breed Frenchman's squaw. She could do beautiful embroidering and needlework.

Our train did not go by Salt Lake City. We took the Landers & Headspath Cutoff and after a few days after the roads came together our train parted. Col. Helms took the Northern route to Oregon by Fort Hall.[11] He had some four wagons and 3 or 4 drivers. They laid by at the junction of the road for more company as it was not safe to travel without going in large companys of men and teams, for the Indians had committed some depredations on Emigrants all ready on both routes to California and the Government had some soldiers placed at Bear River near the Fork of the two roads to delay small partys from travelling by themselfs.[12] (So we

[11] The Lander Cutoff was not so named until 1858. In 1851 it was part of the "Sublette Cut-off" at least to the Green River. That route, along the northwest base of the Wind River Mountains and on the upper Snake below Jackson Lake, had been pioneered by William Sublette in 1832. The Hudspeth Cutoff was blazed by Benoni M. Hudspeth, guided by John J. Myers, in 1849. Dale L. Morgan (ed.), *The Overland Diary of James A. Pritchard*, pp. 144, 159, 175.

Reinhart's party most likely took the Sublette Cutoff. Lander Cutoff begins east of South Pass. The company probably split in the vicinity of Soda Springs, Idaho, with Reinhart's party setting forth on the Hudspeth Cutoff. A glance at the map in Irene Paden, *The Wake of the Prairie Schooner*, p. 264, coupled with subsequent references by Reinhart would seem to sustain this surmise. Riddle, *Early Days*, pp. 14–23, sketches the route more carefully and it is almost the standard Oregon Trail route into southern Oregon.

[12] The road juncture was probably between Soda Springs and Alexander, Idaho.

parted with Col. Helms. He had two sons in his family, one named Thomas Benton Helms. I saw them some 8 or 9 years after in the Willamette Valley, Linn County near Albany, where they had a very fine farm.)

We kept around North of Salt Lake City and the road that went by Salt Lake City struck ours again on the Humboldt River, about 100 miles west of the City.[13] Our Captain, Col. Cornelius Hill[s], had went to California in 1849 from Oregon. He went to Oregon [in] 1847.[14] He came back the Applegate Road from Oregon to the Lawson Road to Feather River and from there to where he struck the Humboldt, 80 miles east of the sink of Humboldt River. The route had not been travelled since 1847, but it was the nearest and best for Northern California and Southern Oregon (but when we took the road we expected that there would be considerable travel).[15] We took in to our train a few wagons, about the same number of teams and men as when Helms was with us. We were quite cautious about Indians and watched and guarded our camp very close. [We] had several stampedes of our cattle by the storms. The wind, rain, thunder and lightning were fearfull. Some nights when the cattle would start in

---

Troops were stationed at Cantonment Loring near Fort Hall in 1849–1850, but apparently not in 1851.

[13] The road west from Salt Lake City (the Hasting Cutoff) joined Reinhart's trail at the confluence of the South Fork of the Humboldt River, not far from present Elko, Nevada.

[14] Cornelius Joel Hills was born in New York State in 1818. He emigrated to Oregon in 1847, taking up a 640 acre donation claim at Jasper, Lane County. (The name Jasper was for Hills' son who was born there in 1859.) In the fall of 1848 he went to the California gold fields, but returned to Oregon from San Francisco on board the ninety-ton schooner, *W. L. Hackstaff*, under the command of Captain William White, July 22, 1849. Although shipwrecked at the mouth of Rogue River, Hills made his way safely back to Cow Creek.

In 1850, accompanied by Isaac Constant, he returned east with a saddle and packhorse train to persuade his two brothers, Erastus and Elijah, to return to Oregon with him, which they did. On the 1851 return trip, he married Sophronia P. Briggs in Lee County, Iowa, February 19, 1851. The newlyweds set out for Oregon on April 16, reaching Lane County, September 7. They settled in Lane County to raise their children, descendants of whom still reside in the region. Hills Creek in that county bears their name. Fred Lockley, "Impressions and Observations of the Journal Man," *The Oregon Journal*, December 30, 31, 1932; *The Oregonian*, March 29, 1925; Riddle, *Early Days*, pp. 4, 9; McArthur, *Oregon Names*, pp. 300, 324.

[15] The description of the route is garbled. Hills probably took the Applegate Road from Oregon to the Lassen Road and then struck west, hitting the Humboldt River about 80 miles north of Humboldt Sink. This route would take him north of the Feather River. Had he actually descended to the Feather River, this would place him considerably south and east of the Sink and Humboldt River. The Lassen Road was not established until 1848, thus Hills in returning east probably followed some variation of that trail. (Reinhart consistently uses Lawson for Lassen.) Peter H. Burnett, *Recollections and Opinions of an Old Pioneer*, pp. 256–270.

Reinhart's statement, "The route had not been traveled since 1847," is hardly correct. The Applegate Cutoff was traversed eastbound by Isaac Pettijohn in 1848. In 1850 the route was traveled westbound by George Keller. Henry R. Wagner, *The Plains and the Rockies*, revised by Charles L. Camp, pp. 265–266.

*Overland to California*

the dark and rain, the guard would try to head them around. You could only see them when the lightning would flash and no telling if the Indians were not running them off, maybe never see them again and [us] so far away from any settlements—it was awful to think of. Several familys and their wagons [left] together; the men nearly all [would] run out to help stop the cattle. Sometimes they would run 10 or 15 miles and not get them till the next day.

We had several men in our train that thought they were great hunters and must go out to kill some Buffaloes, deer and antelopes. They were afraid they would never have another chance to kill some to brag about. Three men started off in one party and some in others, but the balance got in the same night. But 3 did not come up to the train. We had drove on right along the road and the men had been cautioned not to go too far into the hills and to try and keep our train in sight. At noon or at night they could catch up, but they found that cattle travelling right along whilst they took to the hills back and forth and the excitement of hunting made them walk a great deal more than they had any idea of. So next day after breakfast two men started back to look for them and we were to travel on slow until they could catch up to us again. But at noon, then the second night, none caught up. We got uneasy and thought maybe the Indians had got them. In the night one of the men caught up on a horse and told the Partners of the teams to stop their teams till the balance could catch. They had Killt a Buffalo and cut off some meat to carry and catch up but after carrying the meat till they give out they threw it away. One only kept the ears for fear we would not believe he had killed the Buffalo. At last suffering for the want of water, they nearly gave out the first night; they must have been 40 miles behind us the 2nd day. We had to lay over nearly 2 days for all to come up, awfully used up and footsore. One took so sick from over exertion he came nearly dying for 2 weeks. That stopt hunting awhile for quite a time. . . .

Some places we found grass very short and had to drive our cattle acrost the streams or rivers and swim them over and back. Wood was very scarce and we had to pick up weeds and willows and grass to cook by. We frequently heard of Indian troubles, shooting into camps or shooting and driving off stock, that we were kept in a continual flutter and uneasyness.

Capt. Co[rnelius] Hill[s] with his horse team and young wife would go ahead of our train. When they got to a good place to noon, [they] would stop until the balance of us would come on . . . and [then] go ahead again until toward evening [when] he would hunt up a good camping place for us all and wait till we came up and we would all camp. Several of the Familys had some loose young cattle, cows, calves and (oxen to rest) that would be drove behind the train by men having to take turns driving. Some of the stock would be lame and footsore and would be so

Map 2. The Oregon Trail

slow that they were often 2 or 3 miles behind. When the cattle got so they could not keep up, we would have to leave them (and when several would have to be left by lameness or drinking alkaly water or die). [We] would have to leave a wagon once in a while or burnt it up for firewood. We found lots of wagons left and in one place found 10 or 15 wagons, buggies or carriages and trunks and boxes of books all strewn around, with all kind of tools for mining and cooking utensils thrown away and left for the want of teams or the wagons too heavy for what teams they had. Some left their wagons and packed and rode their horses or mules and some even packed oxen.

The year before, 1850, the Cholera got on the Plains and many Emigrants died and were buried close by the road, some with headboard and some with none, and Coyotes or Wolves had dug up many graves.[16]

Before we got to the Salt Lake road (west), one night I caused a big stampede in our train and the cause of it none ever suspected and I did not until it took place. We had camped and our cattle had been drove into the Correll for the night. I came on guard with C. W. Beckworth and one other. I had a Rifle and had on my new Buckskin coat that I had bought a few days before. The cattle were quiet and all laid down and I lay down in a hollow near the mouth of the Correll where I could watch the cattle and have a good view of the hills around for the approach of anyone if they should come from that side. I was one side of the mouth of the correll and Beckworth and the other man on the other side. After

[16] Bancroft, *Oregon*, II, 174, blames the crowded road conditions as one of the causes for the pestilence. Some thirty to forty thousand emigrants were on the trail to Oregon and California in 1850.

I had laid a long time I concluded to get up and walk around to see that all was quiet, as I got up the cattle all jumped up and ran for the other side of the correll and made a break and stampeded right over Beckworth and the other guard. They hollowed and tried to stop the cattle and all the people in camp got out and thought the Indians had attacked us and stampeded the cattle and got their guns and called to find out what was up. After I took a second thought I accounted for the scare. The cattle had forgot when I lay down in the hollow and did not notice me until I got up. When I lay down it was dark; now the moon shone bright and my buckskin coat showed white and seeing something raise just like it raised out of the ground scared them and they all jumped like wild away from me to get out of the correll and some [that] did not know what the others had seen, run with them snorting and bellowing like mad. After I got over my scare and began to know that I was the cause of it I just ran over and tried to head around the runaways and we got them back to camp in 2 or 3 hours and the camp was wide awake and they were expecting to be taken by the Indians. I never let on what was the cause of it. Old C. W. Beckworth swore the cattle ran right over him and knockt him down. Next day a mail driver past us and said the Indians had attackt and ran off some stock back of us and he said the Indians had been following some trains several days, and one party had seen some Indians go ahead to ambush a train and the Capt. saw them. He got some 7 men to get on horses and ride around the Indians and got into an ambush before the Indians got there, and when the Indians were preparing to ambush the Emigrant train they themselfs run into the old Capt's ambush and he killed all but one out of 8 or 9.

So that was a damper on the Indians. Next morning they fired acrosst the river at a white man and he fell as if dead. He had on a double breasted coat, vest, and woollen over and undershirt and the bullet went through coat, vest and overshirt and stopt in the undershirt right over his heart. It stunned him for a while and he thought he was surely killed but it was a spent ball from a long distance but well aimed. We were a little scared about the Indians for a couple of days but got as careless as ever again.

One day I was helping drive loose cattle. I was afoot. A man named Lindley was helping but he rode a little mule, [and] there was some other boy helping drive. Neither of us had a gun. Lindley had a 6-shooter of the pepperbox kind. Capt. Hill[s] had gone on ahead in the morning and stopt at noon at a place called Thousand Spring Valley.[17] [As] we passed by he was still nooning. We 3 drivers had gone about 1 or 2 miles and I happened to look back and saw something at a distance coming towards us. I told Lindley and he and I stopt to see what it was. When it got close I saw it was a woman on horseback riding the horse as fast as it could run, and motioning for us and as she got closer we recognized her as Mrs. C. Hill[s]. [As] she tried to turn on her horse (for she was riding crosslegged), she fell off. I run back towards her and when she got up she said her husband was killed by the Indians. I caught her horse and called to the hind drivers of the teams (they happened to be Erastus and Eli[j]a[h] Hill[s]) that the Indians were killing [their] brother. They got their guns and called to others and stopt the train ahead and came back. I got on the horse Mrs. Hill[s[ rode and Lindley on his mule.

We both started back towards the place. [I rode] ahead of Lindley's mule [because] I could not hold the horse back (he had left his mate with Capt. Hill[s]), but I had no gun or pistol. When I got back to Capt. Hill[s]'s wagon the other horse was tied to the wagon and Capt. Hill[s] was all right. He had his gun in his hand and told us that the Indians had run and retreated up the hill when Mrs. Hill[s] came to us for help. The Indians were among the rocks in the hills and we could not see them even when they fired off their guns at us. We could not see the smoke to locate them by it, the sun being so hot and the air sort of hazy. Capt. Hill[s] kept walking back and forth to draw their fire so that we could see them and the balls kept striking around us and in the wagon, so that the men that had come back with their guns could not get at the Indians. The two Hill[s] brothers proposed to ride up the hill among the rock after the Indians, but all concluded that would be foolhardy and dangerous, for the Indians were as good as fortified among the rock and would see us and have the first shots. [Since] so many of us had come back and left the wagons and all the women and children with not three

---

[17] Thousand Springs Valley is situated between Goose Creek and the head of the Humboldt River in northeastern Nevada.

persons to protect them, we hitched up the horses onto the wagon and started back towards the train. After we got about a half mile the Indians jumped on rocks and hollowed after us to come on, that we were cowards and could not fight. They kept shouting at us while Lindley and I harnessed the horses to the wagon and several bullets spatted the ground around us, mostly spent balls. When the Indians jumped up and called us names old Beckworth fired off his rifle at them over half a mile off. We then commenced to feel uneasy that we had been decoyed back from the train so as to attack it, but we found all safe, more by good luck than by good management. For if there had been 8 or 10 Indians, they could have killed all with the wagons and could have drove off every hoof before we could have got back to them. It gave Mr. Capt. Hill[s] a lesson, and he told us how it had began.

He had just gone out to get the horses that were grazing a short distance off when he happened to see some 6 or 8 Indians coming towards him. When within a hundred yards, Capt. Hill[s] [saw] them and motioned them off with his hands. They called out, "How, how, how do!" and kept trying to come closer. Hill[s] drew up his rifle to keep them off. One [Indian] threw down his gun on the ground to show that he was friendly and would come unarmed. But Capt. Hill[s] was not to be fooled that way. [He] drew up his gun to shoot, only to scare them and keep them back, for if he had fired, he could not have had time to load again before they would have killed him, and then his wife and horses and wagon would have been at their mercy. So Capt. Hill[s] saw he was in a bad fix. [The] only way to get out was to catch a horse and put his wife on and let her ride for life to us for help, [while] he would try and keep them off. When the Indians saw what he was about, they give their leader a rifle. He aimed at Hill[s] but missed him. Hill[s] kept his gun pointed at [the Indian leader] so he could not get good aim, [while] he and his wife got to the horse. He just put her on without saddle or bridle, with only a halter, and she came for us.

The same Indian that first shot at Capt. Hill[s] must have been their best shot, for as fast as he shot one gun they would hand him another, and Hill[s] kept jumping and threatening with his gun, so that the last shot at Hill[s] was not over fifty or sixty yards off, and just grazed the leg of Capt. Hill[s]'s buckskin pants. But when Mrs. Hill[s] got off on the horse safe, they seemed to get demoralized and commenced to retreat towards the hills, still firing as they went. They did not seem to be very anxious to kill Hill[s], or were very cowardly, for a white man could have captured Capt. Hill[s] easily.

We traveled down the Humboldt River a long ways without any more trouble, until Capt. Hill[s] got just as careless as before. Some of the ox teams would go ahead a mile or two, or lag behind the same distance. We ought to have been more cautious, for the Indians were very troublesome and had been for several years on the Humboldt River and had killed and murdered whole trains of emigrants. In 1849 and 1850 they had been espeshly bad.

But one day I was again driving loose cattle and I called the attention of some of our men driving with me to what seemed to be five or six persons at quite a distance across the Humboldt River. They seemed to be carrying something on their backs. I thought they could not be white men, for they were off the road and going southeast from us, say one or two miles off. After a little while, some parties of our train came back and told us that the wagon I was with (J. B. Welch's) had been robbed by the Indians, and John Welch had got shot. The front of the train had stopt and when we caught up was in great confusion. Our wagon was sacked and shot into several places, nearly everything gone. John Welch had his arm broke with the shot and two fine blooded colts were cut loose from the wagon, and the cattle scattered all around. Capt. Hill[s] that afternoon had gone ahead again with his wife; as usual he was about two miles and a half ahead of the main train. John Welch and John Lock[ridge] (with the wagon I was traveling with) drove with four yoke of cattle ahead of the main train about one and a half miles and were within a half mile of Capt. Hill[s] team. John Lockridge (the gunsmith from Janesville, Wis.) was driving. John Welch (owner of the team) was sleeping in the wagon. The balance of the train was strung along for two miles out of sight behind.

John Lockridge happened to look back the road to see how far the teams were behind him, when he saw three persons coming along the road, one with gun; only at first he thought that they were some of our men who often went ahead to hunt and look for camp grounds, until they got closer, when they looked like Indians to him. He waked up John Welch and told him to look and they saw that they were Indians, two with bows and arrows. They came within a few hundred yards. [Ahead] there was a bend in the river all grown up with young timber at edge [and] the road took a turn around the bend. [T]hen John Lockridge noticed the Indians cut acrost the bend to strike the road ahead where Capt. Hill[s] was about a half mile ahead. Lockridge told Welch he believed the Indians were trying to cut off Capt. Hill[s], who had pulled out for a camping place. So John Welch took the whip and told John Lockridge to take his rifle and run ahead and warn Capt. Hill[s] of the Indians.

Lockridge jumped out of the wagon with his gun and got about 60 yards when he heard a shot fired and saw that John Welch had dropt his whip and a lot of Indians, about 20, just coming up out of the banks of the river. The lead yoke of cattle at the crack of the gun broke their yoke, and the next yoke, a pair of cows, swung the wagon into the bend out of sight of the road or train behind. John Welch had raised the whip to start up his leaders when the shot broke his arm; he then ran towards

Lockridge (and Capt. Hill[s]). Lockridge came back to meet Welch to lead him and he saw the Indian that shot Welch and raised his rifle and took good aim, not over 30 yards (he, Lockridge said) and he missed the Indian (and Lockridge was the best rifle shot I ever saw) so they both had to run towards Capt. Hill[s] and the Indians took possession of the wagon and contents.

I had a large wagon sheet, a featherbed and feather pillows and my nice buckskin coat and another coat and some gold coins, all I had, all tied up in the wagon sheet, a bundle I could hardly carry, just rolled in the back of the wagon. John Welch had Feather bed and lots of blankets and quilts, and several of the other boys of other wagons had sacks of fine clothes in our wagon. We could not keep them in boxes, they were too heavy to haul. The Indians took all they could carry, and besides flour, crackers, sugar, coffee, meat, everything but a keg of powder they did not find and a large mattress was all that was left in our wagon, and part of a barrell of hard bread. They cut loose two fine blooded colts tied on behind but when cut loose the colts run off back towards the train and they could not get them. After taking out all they could carry out of our wagon, some 2 or 3 put the muzzles of their guns against the wagon and fired them off, expecting to burn up the wagon. They then crosst the river and struck Southeast and it was then I saw them at a distance with my big bundle containing my Fancy Buckskin Coat and all my bedding and what little money in gold I had left, all but 7 or 8 Dollars in silver (50-cent pieces) I had tied in a rag and it was in our grub box which they did not find. They took all they could carry in a hurry to get away before the rest of the train should come up.

So when I got up to our robbed wagon and found my things all gone, I looked and felt quite Blue and discouraged, still glad no one was killed.

John Welch and John Lockridge got close enough to Hill[s] to make him hear and the shots made him look back and came back to our wagon. The whole train stopped and held council what to do. If we had 20 or 25 men to spare we could have overtaken the Indians loaded with our plunder and all afoot, but we were so few that we could not afford to divide our men to follow and maybe run into an Indian ambush. So, it was thought best to let the Indians go with what they had than to some of us lose our lives in trying to get the things back and we might still lose more. James Downer and Clark and Douglas and others had lost some fine broadcloth suits and they as well as I felt bad over it, but it made us all laugh to think what a swell the Indians would cut in our nice clothes, white shirts and broadcloth, and my Fancy Buckskin Coat too. It was laughable but still we had to grin and bear it. We blamed Hill[s] and John Welch, but Welch got his punishment in a bad wound and broken arm. He could not ride in the big wagon with us any more for it jolted him too much, and so he got a chance to ride in one of Riddle's spring wagons, and he suffered dreadful with his arm for fear he would have to lose it.

We saw some fine sights in our travels and objects lookt close by that were a long way off. Some of us thought Chimney Rock on the Platte River was but a few miles off, when some tried to go to them and went 12 or 15 miles and found themselfs not much closer. And Scott[s] Bluff the same way. We could see Fremont's Peak and other Peaks a hundred and fifty miles off from a hill on a clear day.

We got to Fremont's Peak the 3d of July and several of us climbed to the top and cut our names and date and train on the rocks. On the Fourth we had camps at Devil's Gate, a great canyon 500 or 600 feet Perpendicular rocks on each side and Sweetwater stream passed through awful rocky looking bed of the stream.[18] A good many tried to go to the top of the Devil's Gate but it was so steep and dangerous that none got up from the side. Some few went around and got up—they looked awful small up there waving their hats and shouting. That afternoon we lay over on the Sweetwater River and celebrated the Fourth of July with anvils and guns, and rested our stock.

We met a large party of miners on their way back from California; they were riding horse or mule back, and had lots of pack animals to pack their provisions, blankets, clothes and Gold. They told us of some new discoveries in Northern California at Shasta and Yreka and Scott River and Trinity and Salmon Rivers, and they had done well. Some were from Missouri, Iowa, Illinois and Indiana. We were awful glad to see them, for they were the first and last we met on the route.

They cautioned us to look out for Indians, and be very carefull to keep close together, and keep good guard out always.

One night we got to a place called Plum Creek Mountain (because of wild plums that grew on the hill and creek) and we camped in a little valley, a nice looking place and plenty of grass for the stock,[19] of which we had over 160 head. Capt. Hill[s] told us he campt there in 1846 or '47 with a train on their way to Oregon, and at night they had some large fires built and the families were around the fires, some cooking, and some eating, others sitting talking of the Future and the past, when some Indians shot a shower of arrows into camp, one of which struck a young lady just married a short time in the breast and killed her right at her

---

[18] Reinhart has these famous Oregon Trail landmarks properly ordered. His reference to Frémont's Peak is in error: he means Independence Rock. Frémont's Peak is a part of the Wind River Range in west-central Wyoming, some fifty miles from Lander. Riddle, *Early Days*, p. 12, states the party celebrated the 4th at Independence Rock.

[19] George W. Riddle records that three days after passing through Black Rock Canyon, the party traversed High Rock Canyon, then into Surprise Valley. From there they crossed "a spur of the Siamavada [Sierra Nevada] Mountains, then called Plum Creek Mountain and camped near what is now known as Fandango Creek, said to have taken its name from some emigrants having a dance when they were fired on by Indians." *Ibid.*, pp. 20–21.

husband's side, and with her father and mother along too. Her death caused great grief in the whole train, and they buried her next day right there in the valley on a knoll close by.[20]

We were all talking about it, and we would not have a fire after dark. I was on guard that night, and we had Double Guard around the cattle and wagons, for it was considered a dangerous place for Indians. We all felt a little squeamish on account of the young wife getting shot there, so we all felt a sort of awestruck. We tied one of Hill[s]'s mules at the mouth of the Corrall; he was good for watching and kept guard like a man. He did not like to smell Indians and after dark (we had him tied with a rope to the first wagon) he kept all at once running into the corrall, then he would go out and feed awhile, and all at once come running in again. It was pitch dark and raining a little, and I kept my eyes on the mule, most of the time watching his ears, and I was glad when we were relieved at midnight by the next guard, and as soon as possible found my partner. We had to splice and borrow some bedding, for Lockridge and I had lost all ours, and John Welch had taken his mattress and put it in Riddle's spring wagon. He could not have anyone sleep with him on account of his arm, so Lockridge and I slept in a tent on the outside of some other wagon (not my own).

After an hour or two we heard an awful shouting and all of the cattle out of Corrall stampeded over the wagons and upset five of them, and run through wagon sheets and bows. The spring wagon John Welch was laying in was one of the five wagons upset and he got his arm hurt again bad. Our heavy wagon was upset and in straightening it up the grub box opened and some things fell out. My fifteen or sixteen half-dollars done up in a rag were missing (some one had stole them in straightening our things up). No one could account for the stampede but the cattle must have got scared at something and just run over anything in their way. They got off several miles before they were stopt and all the folks thought that the Indians had made a break on us. [They] were huddled around with their guns to see what was up. And it was nearly day before things became quiet again. I found I had not a cent left in the

---

[20] Hills came west in 1847. His party was led by Lester Hulin who had served in Frémont's 1845 expedition. Leaving St. Joseph, Missouri, April 23, the company took to the Oregon Trail. On September 29, Hulin recorded in his journal: "This night we were sadly visited by savages. They approached, and, finding they could get no cattle, vented their spite at a young lady who had been baking and was then by the fire. They shot 3 arrows at her: two of them hit her, one passed through the calf of her leg and the other through her arm into her side. We fear she is mortally wounded, but hope for the best. Her name is Ann Davis." Subsequently, Miss Davis recovered, and later married C. Hendricks, proprietor of Hendricks' ferry on the McKenzie River. The incident took place near Plum Creek (see *note 22, post*). Margaret Skavlan (ed.), "Over the Westward Trail," *The Oregonian*, May 3, 1931; *The Oregon Journal*, December 31, 1932. This report is more to the truth of the matter than Reinhart's "romantic" account given in the text.

world, but my old rifle that I had when the Indians attacked our wagon and robbed it.

Next day we got all fixt up again and started on our way. Welch suffered a great deal with his arm, and the weather quite hot. We passed through the Carson Desert and had to drive two nights and one day to cross without water.[21] (We saw thousands of dead bodies of cattle, horses, and mules, wagons and blacksmith's tools and every kind of machinery left and broke up or burned.) It was over eighty miles acrost and was awful hard on our stock, some of which was left. One day we passed through High Rock Canyon, about twenty or twenty-five miles long, and the sides from 100 to 700 feet high.[22] We were in dread of the Indians hemming us up in there or firing down on us, or rolling rocks down on us, for it was a dubious place to be hemmed in. We had to camp in it, as we had eight or ten miles to go to get through. Found good grass and water and a large flat to camp on.

After supper Mr. Briggs (Capt. Hill[s]'s father-in-law) was looking for a wagon tire for one of his was very thin and there were generally some left at every camping place. He saw a hole where someone had dug and thought maybe someone had buried a good wagon out of sight. He looked down and scraped away some dirt and saw the end of a barrell. He got a shovel and uncovered it and found 5 or 6 barrels and one keg. The top of one of the barrels had been cut into with a hatchet and he scraped away the dirt and put his hand in the barrel and found it contained High Wines but the water had got into it and made it weak and not fit to drink. The news soon spread of the find and soon everybody was helping to uncover the hole the barrells were in and when exposed found 5 whole large barrells and one half barrell. Before now every team was said to be loaded as heavy as ought to be, but after tasting the High Wines everybody emptied their syrup kegs and jugs and all the water kegs were put into use to hold some of the Liquors. Beckworth and Riddle found room for all the Brandy contained in the half barrel, it being buried had taken the fiery taste out of it. I do not know now if any of the barrells were hauled by any of our train, but every vessel was filled and by mixing $\frac{3}{4}$ water to $\frac{1}{4}$ high wines made good whiskey and plenty strong to drink. Capt. Hill[s] said he heard that a Government Train with supplys had lost most of their teams and were overloaded had buried the Liquors and had left some clothing and blankets too somewhere, but

---

[21] The party passed through Black Rock Desert. From subsequent descriptions, the Reinhart party did not venture far enough south to hit the Carson Sink. Riddle, *Early Days*, p. 17.

[22] After passing through Black Rock and then High Rock Canyon, the party entered Surprise Valley. From there they crossed a spur of the Sierra Nevada at Plum Creek Mountain, then on to Fandango Creek. From there they traveled to Goose Lake, thence to Tule (or Rhett) Lake. It was here, at Bloody Point, as Reinhart later notes, that the party met a band of friendly Indians. *Ibid.*, pp. 20–24.

we hunted all around that night but could not find anything more. Some of the Party that had buried the barrels had on their return lay down and struck the one barrel and cut the hole in it and taken out what they wanted but had neglected covering the hole again (it had been all left there in about 1847).[23]

We started on next morning, everyone in good spirits (high) and strange to say nobody got drunk or disagreeable by taking too much of the spirits at once, and nearly every wagon had its flasks filled and handed around every little while, and some had filled their Coffee pots and carried it in them.

After crossing the desert we got to some hot springs and some of us had picked up some ox horns, good shaped for Powder horns. We stuck them on the end of a long pole and run them down into the hot springs and left them over night and in the morning they were nice and soft so that we could cut and scrape them down thin so that they were transparent and put in bottoms and made nice Powder Horns. We camped all night at the springs; they were boiling hot and clear as crystal and so deep you could look down and see no bottom. It must have been where Virginia City, Nevada, now stands; it then was called Black Rock Springs[24] and we found a lot of trunks and boxes containing clothes and shirts but they had been exposed to the weather and were moldy and rotten. Lots of iron of wagons lay scattered around and there was good grass and a good camping place and dry willows for fuel. I think we lay over a day or two at the place to rest our stock after crowding them the last 2 nights and one or two days.

We next got to some place that the ground just glistened like gold and some of us thought that it must be gold, but found it Mica or Isingglass. We got to several alkali lakes where you could shovel it up by the shovel full. It looks and tasted like saleratus and the air was thick with the fumes and the sun evaporated the water. Some of us took some along to bake with. We had to keep our cattle from drinking the water and in spite of our care they would find some ponds and drink and they would scour and get weak and some would have to be doctored with soap or grease and fat pork. Some had to be left.

We got to some lakes on the Lawson road to Feather River that were

[23] This discovery was not far beyond Black Rock. Three barrels of whiskey and one of brandy were uncovered. "From the fact that a heavy government wagon had been left standing over the cache, the conclusion was reached that it had been abandoned by a government train several years before." *Ibid.*, p. 20.

Alonzo Delano, under date of August 23, 1849, records meeting "a party of men and wagons going east. . . . We found it to be a relief party from Oregon, going to meet troops on the Humboldt with supplies. . . ." Joel Palmer was the guide. This could possibly be the train referred to here by Hills. The meeting place was in the vicinity indicated by Riddle. Alonzo Delano, *Across the Plains and Among the Diggings*, p. 86.

[24] This is an erroneous identification. Black Rock Springs is north of Mud Lake, which is a considerable distance north of where Virginia City, Nevada was situated.

just alive with wild geese or Brants, a specie of Wild Goose, and lots of wild ducks and our boys had lots of shooting them. The lake would be covered with large lots so you could see nothing but wild geese (Brants), the sky dark with them in places called Lawson's Meadows. There we got into the Applegate Route to Oregon.[25]

We did not see Indians for a short distance till we got near Pitt River (or Lost River).[26] When we were on a high long ridge we could look down on Pitt River Valley, and we saw Indians, some on foot and some on horses making for the road ahead of us. We stopt and all caught up to each other for fear of Trouble. When we got off the Ridge into the Valley we found the Indians had mown some gras[s] and had it cockt up to dry. Our loose cattle could hardly be kept away from the hay, they wanted to tease and horn it around. Some Indians came up and called out, "How do! How do!" but our Capt. Hill[s] kept motioning them back. But the old chief threw his gun and bow and arrows on the ground to show us he was unarmed and wanted to come up to the train to talk to us, but Hill[s] motioned them back all the time. The old chief got mad because of the stock hooking at the hay and wanted us to pay for it, and talkt and flew around like a crazy man, and came up close to talk and threaten, but Hill[s] took him by the neck and collar and pushed him away. We lookt for trouble with them; they were bold and saucy and we could see them for quite a distance trying to cut in ahead of us on the road. These Indians had a bad reputation, and had cut off a train of Emigrants from Oregon to California and had killed a lot of women and children of a train at the place we came down into the valley (and the point of the ridge we came down was called Bloody Point).[27] Next year Capt. Ben Wright from Yreka, California,[28] was there with two companies of miners and fought

[25] From the description, Reinhart has confused the trail. Lassen's Meadow is situated at the point where Lassen's Road leaves the Humboldt River; there is no lake near at hand. Since the party was passing through Black Rock Springs, this would place them north of Mud Lake on the Applegate Road, considerably north of Feather River.

[26] Reinhart confuses these two different rivers. The Pitt (originally spelled with only one "t") rises in northeast California and feeds the Sacramento. Lieutenant R. S. Williamson (1855) gave the river its name because of the Indian pits dug along the banks to snare game. Located in Klamath County, Oregon, Lost River was discovered by Frémont in 1846 who named it McCrady River after a friend, but to no avail. During its course through Langell Valley, the river goes underground for a few miles, hence the name.

[27] In September, 1852, the Modoc Indians massacred a party on this so-named promontory which juts out into Tule (or Rhett) Lake near the Oregon border. Only one man survived. Sources vary on the number slain, ranging from twenty-three to sixty-four.

[28] Yreka, derived from *Wy-e-kah*, the Indian name for Mount Shasta, began in 1851 as Thompson's Dry Diggings; later that year it was called Shasta Butte City, and in 1852 was given its present name by the California legislature when it was made the seat of Siskiyou County.

Benjamin Wright, Indian agent at Gold Beach, located at the mouth of the Rogue River, was murdered by an eastern renegade Indian, Enos, February 22, 1856. He had helped raise the Yreka volunteers who set forth to avenge the 1852 Bloody Point

them, and one Dave Jackson, who mined with me the winter and spring of 1852, was killed, and some others I knew in the same company were killed.

So we were afraid if the Indians could all get together, they might bother us before we got unhitched and in confusion, so we kept going so as to get up as near as we could to where we had to cross ... Lost River by a natural bridge or stone ford, and we drew up in a good safe place to camp, kept close watch of our stock, and in the morning early started up the river towards the Natural Bridge.[29] The Indians did not see us in time day before or no doubt they would have got together in force quick enough to have made an attack on us, for these Indians were the Pitt Rivers, and afterwards became notorious as the Modocs and Klam[aths] and fought the United States troops in 1854–1855,[30] and afterwards the Modocs fought and killed [Canby] in the Lava Beds in 1875 or '76. And at several different periods have caused our Government a great deal of trouble, and killed Indian Agent Lockhart and his brother, and afterward Indian Agent Meacham.[31] But we saw no more of them that night.

When we had got two or three miles from camp we saw a large party of horsemen coming towards us on a fast gallop, and we could see their guns glisten in the sun. At first we thought they might be miners, until they got closer and by their fast riding we could make out that they were Indians. How we all felt at that moment is undescribable! We saw that there were at least 60 or 75 of them, all well mounted and well armed with Rifles and Revolvers, coming full tilt on the run.

The forward teams halted and the rear ones came up. We all got our guns ready, expecting to sell our lives Dearly, but with little hopes as to numbers and their being so well armed. Our Capt. Hill[s] called them to halt at a distance, but they paid no attention to his motioning them off and still came thundering on towards us. But they commenced to call "Klahiram! Close Klahiram!" and Capt. Hill[s] understood the Chinook Indian language for "How do do! Good How do do!" and recognized them as friendly Kijus and Klickatat Indians.[32] When they got up to us they

---

massacre. Frances F. Victor, "A Knight of the Frontier," *The Californian*, IV (1881), 152–162.

[29] This famous natural stone bridge was used by the Applegate party to cross Lost River, July 6, 1848. It is located near Merrill, Oregon.

[30] Ethnologists consider the Pit (Pitt) River, Modoc, and Klamath Indians as separate and distinct tribes. Both regular and volunteer troops fought them intermittently from 1854 to 1855, in what was dubbed the Rogue River War.

[31] On April 11, 1873, General E. R. S. Canby and the Reverend Eleazer Thomas were killed by the Modocs while holding a peace parley. Indian Agent Alfred B. Meacham was wounded, but recovered. Keith A. Murray, *The Modocs and Their War*, treats many of the Indian events mentioned here by Reinhart.

[32] Reinhart is referring to the Kuitsh and Klickitat Indians. The Kuitsh lived on the upper Rogue River eastward about Table Rock and Bear Creek in the vicinity of Jacksonville. The Klickitat inhabited the valley of the Umpqua.

jumped off their horses and shook hands with us and said they were General Palmer's scouts and Body Guards and hunters, of the superintendents of the Indian departments of Northern California and Oregon, and that they were hunting and that Genl. Palmer was at Yreka, California, about 60 miles off by trail, but about 250 miles by the way we went by wagon road.[33]

They wished to trade us game—deer or antelope meat—for ammunition or tobacco or sugar, but we did not dare to trade powder or lead to Indians, no matter if they were friendly, but we were so glad that they were not Enemys or Modocs that some let them have sugar and guncaps for meat. They said they would like to come acrost some of them wild thieves of Indians and they would clean them out for us. And I should not wonder if the Modocs knew that the Scouts were close by, and it kept them from bothering us the night before.

So the Friendly Indians escorted us a few miles acrost the Stone Crossing or Naturell Bridge of Lost River, then bid us "Highip close Klyhiram," which means "a very Good By." They took off south east down the valley, and we went northwest on our way, feeling better than when we first saw them coming towards us (not knowing but they might be wild Modocs and Pitts and Klamaths).

In a few days we got to Klamath Lake, and saw where some white men had campt [who had] been prospecting in some gulches for gold. We all commenced now to talk of what to do. Some were for going right to California as soon as we struck the California road to Oregon, which we expected to get to in a few days. About four days from the Klamath Lake (we had crost the Klamath River near the lake), we struck the wagon road over the Siskiyou Mountains. At the Yreka and Rogue River crossing there was a place or farm and Ranch taken up at the foot of the mountains, and we found good water and grass and campt over night. Some of the young men took one wagon and a set of blacksmith tools, and some six or seven of us went with the wagon and two yoke of cattle and took our little provision and about four gallons of the high wines in a large jug with us. Clark Richardson and James Douglas, James Downer, John Lockridge and myself and two others went with the team. We bid the balance of the train goodbye next morning. They said they would go to Oregon and winter, and if favorable news came, would come out to California in the spring when they heard from us.

Charley, my brother, kept with the train. He wanted to take up a land claim close to old Wm. Riddle, who had a couple of Girls along.

[33] Joel Palmer was appointed superintendent of Indian affairs by President Franklin Pierce in 1853. He was made a major general in the Oregon militia, 1865. At this time, 1851, he was commissary general for the volunteers who had been mustered in the face of increasing Indian unrest. Born in 1810, he emigrated to Oregon in 1845 for the first time. He died June 9, 1881. Harvey W. Scott, *History of the Oregon Country*, V, 208–210.

So next morning we all pulled out, us to California and the train to Oregon. We took to the Siskiyou Mountain, and when we got to the top we saw Mt. Shasta, a large Snowy Mountain, 17,000 feet high.[34] We had been seeing it before, but now we were going right towards it. The Siskiyou Mountain is the dividing line between Oregon and California. At the summit a large rock is supposed to be the exact line.

[34] Mount Shasta (14,162 feet), a part of the Cascade Range, lies about sixty miles east of Redding, California.

# 2

## MINING APPRENTICESHIP: SISKIYOU COUNTY

(1851–1852)

The second day we camped on Cottonwood Creek, and at noon we nooned with some men just on their way to Oregon for cattle. One name was Dr. Harding (he was killed by the Indians three or four years after in Rogue River Valley).[1] They showed us some nice sacks full of gold dust, and were very careless with it and let us handle it and look at it as much as we wanted, and gave us a great deal of advice and some good information in regard to Mining Localitys. We parted in good spirits. We crost the Klamath River on a ferry close to the mouth of Cottonwood, and the fourth day got to a place called Cottonwood Ranch,[2] twelve miles from Yreka. We campt and it being Saturday, the boys concluded to stay over till Monday and give their cattle a rest. But I took my rifle, all I had, and a shirt or two, and struck out Sunday morning for Yreka.

I crost Little Shasta [River] clost to town[3] and got in in the evening and found a lively noisy town of 5000 to 8000 inhabitants. Sunday was

[1] Reinhart's reference should possibly be to Dr. John R. Harding, who was ambushed and killed near Willow Springs, August 10, 1853. *Oregon Spectator*, September 16, 1853. The spelling of the last name is open to question, but should not be confused with Benjamin F. Harding who rose to legal and political prominence in Oregon in the later decade.

Another contemporary source notes the fact that "L. L. D. Harding" was mortally wounded by Indians on August 10, and died on August 14. Oscar O. Winther and Rose D. Galey (eds.), "Mrs. Butler's 1853 Diary of Rogue River Valley," *Oregon Historical Quarterly*, XLI (1940), 352–353.

[2] Rechristened Henly in 1861 after a prominent citizen, today it is called Henley. [Harry L.] Wells, *History of Siskiyou County, California* . . ., p. 210. The ferry operated several miles to the east of the settlement. Reinhart's memory is at fault here —the places indicated and the distances involved would hardly require four days travel, even with the incumberance of the cattle being driven by some of his party.

[3] Reinhart must have crossed the Shasta. The Little Shasta is located south of Yreka.

one of their best days. I found a boarding house kept by Dr. Adams and Wilson, and got my supper and went out to see the sights. I found Musick in a large round tent full of Gambling Tables all around and the tables full of Gamblers and Miners. One of the Violin players was Alvin Hill[s], a brother of our captain, [C. J. Hills], and I soon got introduced to him and his partner, a banjo player named Billy Pitt. Hill[s] was glad to see me and I gave him news of his three brothers. I had a good time looking on the sights and gambling. The miners were very liberell in betting their gold dust, and the dealers in the banking games treated their bettors and bystanders often to Liquor and Cigars.

After looking around all I wanted to, I went to the Hotell and found a man named Charley Johnson. He and a party of miners were prospecting some creek claims on Humbug Creek, a new discovery, in July, and it had prospected unfavorable then and the discoverers called it "Humbug Creek." But in the fall others had found plenty of good rich placers, Bar, Hill and creek claims, and quite a rush had taken place since the last of July, 1851.[4] Charley Johnson offered me $4 a day and board to work in his claim on the North Fork of Humbug Creek, about one and a half miles above Minersville, where there was a large trading post and miners' outfitting Establishment kept by the Dejarl[a]is Brothers.[5] So I went with Charley Johnson on Monday morning to Humbug Creek. He had a horse to ride and I walkt, and he let me ride a short time at a time and we got to Humbug City, twelve miles south of Yreka.[6] It was quite a little town, full of saloons, round tents and trading posts and butcher shops. We had still over four miles to go to where Charley Johnson's claims were located. We passed through another little town called Freetown, after Judge Free who had a trading store and butcher shop there. It was about half as big as Humbug City.[7] We found the whole distance all staked off into claims and men at work. We next got to Minersville, just started, but lively. De[j]ar[lai]s Brothers had a large pack train of 150 mules loaded with goods and liquor from Sacramento, California. They had all Mexicans to work for them. They are great packers.

[4] "Humbug" found considerable use by miners as a place name. The gold strike mentioned here was made at the forks of Humbug Creek, not far from Yreka.

[5] There were three Dejarlais brothers: O. Dejarlais ran a store at Scott Bar; N. Dejarlais operated one in Yreka, and A. Dejarlais at Minersville. The correct spelling of the name appears on an 1856 lithograph of Scott and French Bar (copy in California Historical Society).

[6] Humbug City was situated at the creek's forks, ten and a half miles *northeast* of Yreka. In the fall of 1851 it had a population of around 600 miners and was incorporated on June 7, 1852. Wells, *Siskiyou County*, p. 71.

[7] Freetown was two miles above on the north fork of Humbug Creek. In 1851 there were two stores and several saloons. The town vanished in 1854 and the site was mined away. J. N. Free, after whom the town was named, ran for Siskiyou County Judge and Coroner in 1852, but was defeated for both posts. The records fail to disclose any subsequent source for the title, "Judge." *Ibid.*, pp. 72, 209.

We got to Johnson's camp in the evening of Monday, September 16, 1851. He had for partners Kirkpatrick, a school teacher, Fred Coffin, a miner, and Tom Hart, all from Indiana, but had mined one year at Scott River.[8]

I worked a while with Johnson and Company until they thought their claim would not pay. I had not yet learned much about mining to go to work for myself, so I got to work for another company just above Freetown. The Company was called Captain Jack's, and Mr. Weston, Mr. Conners and Steven Creamy were four partners. They had four or five hands besides myself and paid us four dollars per day and board. They done well and took out a large amount of gold out of the claim and then sold out to Captain McDerm[i]t and Judge Smith.[9] They offered to keep us hands at $5 per day all winter, but we were excited and wanted to take up claims and work for our selfs. (But we had better kept on a while at that!) So we all quit and took up claims at different places.

I went in with a young man named William Latham, from Plymouth, Massachusetts. He came around the Horn on a vessel in 1849 from Boston and went by the name of Boston Bill. We took up forty feet front each of a bank claim running back without limit. We made from $4 to $6 per day.[10] Above us a man named Oliver Cantlen and Harvey Green and Charley Pope were partners, and one more, Dave Jackson, afterward killed by the Modoc Indians at Bloody Point on Pitt River in 1852, was in the same company. Their claim paid big and we expected to strike the same lead of pay dirt. They made about $800 one day, the gold coarse and black-looking.

In February I put up a log house for a bakery, but I never started in the business at Minersville. There came in an awfull rush of miners from

---

[8] This river, whose name perpetuates the occasion of John W. Scott's 1850 gold strike, is located in north-central Siskiyou County, and empties into the Klamath just east of present-day Hamburg.

[9] Charles McDermit was born in Cambria County, Pennsylvania, May 7, 1820. During the Mexican War he served as first lieutenant, Company D, Second Pennsylvania Infantry. In 1849 he came to California via Vera Cruz and Mexico to supervise "the building of the United States barracks at Benecia, built and operated a saw-mill at Bodega, and went to Trinidad in the spring of 1850." He was among the first to move to Yreka, building the town's first log cabin. May 3, 1852, he was elected Siskiyou's first sheriff. Subsequently, he served two terms in the state assembly, 1859–1861; then volunteered for Union service during the Civil War. He served his time in Nevada and died there in an Indian battle in Green River Valley, near Owyhee, Nevada, August 8, 1865. Wells, *Siskiyou County*, p. 171; J. Roy Jones, *Saddle Bags in Siskiyou*, pp. 22, 24.

Dr. A. M. C. Smith, born in Ohio, arrived in San Francisco in late 1849. He appeared early in the Yreka gold rush and opened a medical practice with Dr. J. Lyle Cummings (or Cummins). Later he joined Gus Meamber in running a pack train from Sacramento to Yreka. He died in Lodi, California, April 6, 1905. Yreka *Journal*, September 25, 1899; April 19, 1905.

[10] An early report from Yreka stated that the miners were averaging $12.00, about one ounce of gold. *The Oregonian*, April 17, 1852.

the Southern mines of California, large trains of pack mules and loaded with provisions and miners supplies and about 1,000 miners with them. They commenced to try and cut down the size of our claims and jumped some. They had lots of Mexicans and gamblers with them and there was fighting and miners' arbitrators meetings every day. My partner and I went in with some 12 others and took up a large lot of creek claims down on the lower main Humbug Creek 4 miles below Humbug City. We took up 1,400 yards of 50 yards apiece and some of us went right to work to dig a canal and a smaller ditch through a bar called Rocky Bar. It took us 3 months to dig the canal and dikes and 2 dams and a tail race. We got tools and put in pumps and water wheels to do the pumping by May 8th, 1852, and got the lower end opened up to drain and got sluice boxes and a good prospect for Gold.

Our Company was composed of ole miners who had mined 2 years and understood the business, and I took for granted all they done; some of the names I will mention here for I will come acrost them again in this Book: first, James Saunders from Boonville, Missouri, he used to be Sheriff there in same county; Jobe Frazer, too from Boonville and he was Government teamster and had deserted in Rogue River Valley; John Smith; C. Westfall, a Swede New York clerk; Boston Bill; Dave Jackson; A. Tirrell of Texas; Joe Nitsell, a Prussian from Danzig, next year killed at Yreka by a Spaniard; Bill Woods from Missouri, and myself. . . .

We got all ready to commence on Saturday night so as to get in a good week's prospecting and work to commence on Monday morning. Sunday morning some of the boys went to Yreka to see a horse race and the balance of us went up to Humbug City to get provisions for the week and to see what was to be seen. Sunday was the holiday of the miners, and everyone went to town to see what was going on, gambling, drinking, fighting, and to see old friends. I saw John Lockridge, James Downer and Clark Richardson. They had come on to Yreka, then on to Humbug Creek and had taken up 4 or 5 creek claims on the left hand fork of Humbug below the forks. They had built them a log cabin and had commenced to open out their claims. They were glad to see me.

Humbug City was full of miners drinking and gambling and carousing. There was a man named Nobles keeping a drinking saloon and he played on the banjo and gambled, and he had a Fancy Woman, very good looking, and she attended bar and sold drinks at 50 cents per drink. She was a stout robust splendid looking German girl from Strausburg, France, and a many a miner paid 50 cents for a drink to get to see her. Some of the men said they had not seen a woman for 5 years or more.

About noon it commenced raining very hard and some dams some distance above on the creek gave way, then several before it got to ours which was the strongest and largest on the creek. The Head dam gave away with the amount of water dammed from above, and when our head dam gave away, all the back water carried away the banks on each side and all made a clean sweep of our lower dam and carried off our pumps, wheels, tools and sluice boxes and covered up our train and tail race level as before. The bank at the head dam was all washed away so that we could not put in another near it without fluming quite a ways at a big expense.

So when we all got home to our claim that Sunday evening we were badly discouraged and most of us were willing to abandon the claim for we were not willing to risk it again. It would cost more than before to rebuild the dams, so we all abandoned our claims and scattered out to other places. If we had only known, that Rocky Bar we had ditched through afterwards turned out rich! But we did not think of the bar. We were after the bed of the stream and just missed it by not giving the Rocky Bar a good prospecting.

While I was working at my bank claim above Minersville with William Latham (Boston Bill), there was a circumstance took place at Freetown that showed how miners can be influenced by excitement and self interest. Close to Freetown there were 4 men mined on a bar and creek claim, and they had their log cabin a short distance off. The men's names were Hurst, Fay and his son, a man named Howe, and (I am not certain) one name Thom Hart was with them. Hurst and Fay was down at the claim to work and Howe had just left them to go up to the cabin to see if everything was safe. He had seen some friendly Indians the day before at Freetown begging food and old clothes. He saw them that morning at the different camps still begging grub and he saw them go toward his cabin. So he hurried up and saw them at his cabin [as they] were just leaving it. He saw they were carrying things with them. He stept into the cabin to see if he could miss anything that the 3 or 4 Indians might have taken and found they had taken some fresh beef, flour and sugar, some buckskins belonging to Howe, and some powder, lead and a box of caps out of his bullet pouch and some shirts belonging to Hurst and the Fays. So he run down to the claim and told Hurst and Fay and they came up to the cabin to see what the Indians had stolen. Howe and Hurst said they were going after the Indians to get their things back, the buckskins and shirts anyway. So when they were about to start, Old Mr. Fay asked them if they had any arms along. They both said they had not so Mr. Fay told them they had better take along a gun and pistol for fear the Indians would resist and not give up the things they had stolen. So Hurst took a Navy revolver and Howe took his rifle and they found the Indians at Freetown where a man named McDonald kept a bowling alley. The Indians were just going into the building when Hurst called to them and asked about the shirts and buckskins they had stolen. He had some hot words with them and followed the Indians into the ball alley. The door was open and Howe saw his partner quarreling with the Indians. Howe was outside and wanted to go in too but McDonald would not let him go in to quarrel with the Indians in his building. Howe could see Hurst and the Indians scuffling and tried to go in, but McDonald held him back. Howe was a small man and McDonald large and he held onto Howe by the shoulder and rifle to keep him out. When Howe heard some gun or pistol shots fired in the building and heard shouting, Howe thought the Indians were killing Hurst, so he jerked away his gun from McDonald just as the door was burst open and 4 or 5 Indians run out of the ball alley. Howe just jerked his gun away from McDonald, leveled the gun and shot one of the Indians through the body as he run. The Indian run about 60 yards and fell dead. Hurst had shot one of the Indians in the building with his pistol.

So the miners all gathered around and arrested Howe and Hurst for the shooting and afterward went to arrest Fay for aiding and abetting the shooting of the Indians. There were a lot of Indians, some 30 or 40, with their squaws and children all around begging and trading fish to the miners. The party that done the stealing belonged to the same lot. They were Klamath Indians and the two shot had their squaws along. They soon sen[t] to the Klamath River 12 or 15 miles below and had the Chief and a lot of warriors to come up. They threatened that if the other one shot by Hurst would die they would kill all the white men on the creek and burn up their houses. They scared the whites pretty bad for there was several prospecting parties wanting to prospect on the Klamath River and if the Indians were not pacified, [they] would be apt to keep the whites from prospecting. The white men reported that the Klamath Indians were very powerful and could gather over 1,000 Indians and exterminate the whites.

The Indians built a big fire just below Freetown and they burnt up the body of the Indian killed by Howe, and his widow took his ashes and put them on her head and covered the ashes and hair and top of her head with tar, a practice they have of mourning, and all squaws widows mourn that way so they call them Tar head Indians. Some of the female relations do the same with their heads.

So the excitement was kept up by the white miners that wanted to go prospecting. [They] were afraid ... [unless] the Indians were satisfied and a Treaty of Peace made with them. The Indian Agent, Judge Snelling, from Scott's Valley, had been sent for by Indian runners.[11] He brought along his interpreter, named Swill, a big Klamath Indian over six feet high; he was the best bow-and-arrow shot I ever saw; he would shoot by elevating his arrow and strike a stump-center a hundred to a hundred and fifty yards away, and the arrow could scarcely be pulled out again, it was shot with such great force. The would-be prospectors

---

[11] R. B. Snelling was elected "county judge" at Scott Bar in 1851. Wells, *Siskiyou County*, p. 91.

(miners) went all over the Creek to notify the miners of a meeting to be held next day at the Justice of the Peace to try the men, Hurst, Howe, and Fay, for the killing of the Indian and shooting of another for stealing an old buckskin. They belittled the amount stolen and magnified the shooting into a great inexcusable crime. Some of their advocates were in favor of delivering the white men, and even Fay, to the Indians, and let them punish them as they saw fit. But most men got quite mad at that, and said they should have a fair and honest trial, and if found guilty of any unnecessary cruelty, they should satisfy the Indians in either pay or provisions, and if the white men could not appease them with a fair, sufficient amount, a collection could be made among the miners who were willing to see the fair thing done to the three prisoners, and most all approved of the plan.

So next morning several thousand miners got together. Hurst, Fay and Howe had attorneys engaged to defend them, and the Indian-scared sympathizers prosecuted the case. Some very foolish propositions were made to the prisoners, and the Indians got quite bold and were encouraged by their "Lo! Poor Indian" sympathizers, that most all the miners got disgusted with the trial.

The Indians kept threatening as to what they would do if justice was not given to them, and kept saying they would break the Pipe of Peace and prepare to redress their own grievances by killing every white man on the Creek. When the white men saw how saucy and overbearing the Indians were getting, most of them went and armed themselves, and prepared to give them a good reception in case they did break out or got too obstreporous and saucy. When the Indians saw the feeling was going against them, they quieted down and let their Indian agent, Judge Snellling, try and effect a peaceable settlement with the whites and their chiefs; and after all day wrangling, the Justice found all of the men guilty, and they were to be bound over for trial at the District Court to be held at Yreka (formerly Shasta [Butte] City) in the spring. Fay was to pay costs and some $300 or $400 to the Indian widow and the mourners. He made arrangements to settle with the Indians, and Hurst and Fay were put under guard that night. Capt. Ben Wright was to take charge of them and take them to Yreka next day (He was captain of the Government for Judge Snelling).

But in the night someway Howe made his escape down the creek. Ben Wright heard he was in a miner's drift or tunnel down the creek. But he got away next day and was seen beyond Yreka, and Ben Wright, after following him three or four days, lost track of him away up near Klamath Lake. It was in the middle of winter and Howe had only a blanket, gun, and hatchet with him, and what little provision he could carry conveniently afoot, to keep from being overhauled by the Indians with Ben

Wright. But I will speak of Howe again further on, and of Capt. Ben Wright, too.

During the winter the miners would all go to the saloons and Trading Post of nights and spend their money at cards and to look on and see them gamble.[12] Drinks were fifty cents per drink all over and a great many miners played poker and rounce, 7 Up, Yuka [Euchre], Monta, and Faro for money or whiskey, or something to eat, such as sardines, Oysters, or potatoes, onions, apples, or groceries or clothing, and when a man lost a whiskey game or the sardines, it come to from two to five dollars to pay, but men were restless and did not care for money.

After our abandonment of our creek claim, Boston Bill and I took up a couple of Bar Claims some four or five miles further down the creek. There was a very rich bar discovered a half mile above ours, called Ross's Bar, [where] a man named Brown kept a store, and he had kept at Scott Bar before, and he was well acquainted with the miners. His clerk, John Down, I will speak of hereafter. They kept books to loan at fifty cents per week each, by leaving the price of the first book as security with them, and return in good order. We kept ourselfs in reading books that way to pass away the nights and when we did not work.

We carried our dirt in hand barrows some fifty to seventy-five yards to the creek where we had two sluice boxes and a long Tom to wash the gold.[13] One dug down the dirt and helped carry the barrow, the other helped carry the barrow and washed the dirt. We made from $4 to $6 or $7 per day, each of us, and if we had known how to wash with sluices and had brought water in, could have made $15 to $30 per day easier than with barrows. We did not stay long for our pay dirt did not all prospect as well.

When we were at Minersville a man kept a store, named George Rogers, and some packers from Oregon came there and stayed a few days till they sold out the loading of their train; they had potatoes, onions and flour and bacon. Flour was worth $60 to $75 per hundred, Potatoes and onions $65 to $70 a hundred lbs., beef thirty to forty cents per pound, bacon seventy-five to eighty cents a pound, apples a dollar and a quarter per pound, or six for $1.25. The Packers' names were [John] Pool and

---

[12] The interested reader will find "Trumps," [*pseud.* for William B. Dick], *The American Hoyle,* useful in providing descriptions for the card games mentioned in Reinhart's narrative.

[13] The long tom was an improvement over the crude rocker. The latter is a device made of a wooden hopper set over a canvas apron which emptied into an open trough. By agitation, the gold could be separated. The long tom lengthened the hopper into a trough and was fitted with "a heavy sheet-iron sieve emptying into another long trough fitted with the usual riffle bars." The major improvement was that a steady stream of water replaced the bailing process and made the mining chore less manual. John W. Caughey, *Gold Is the Cornerstone,* pp. 161–165.

James [C]lugage, the last from Chicago, Illinois.[14] I got well acquainted with them both.

They told that there had been some new diggings discovered in Rogue River Valley just close to Jacksonville. A gulch called Rich Gulch had been discovered by their partner named Skinner.[15] He was the son of the then Indian Sup. Agent of Rogue River Indians,[16] and they, Pool and [C]lugage, advised me to go there as there had been two or three rich creeks struck at Jackson's Hill or Jackson's Creek on both forks,[17] and there had been some miners with three or four pack animals come to Yreka for provision and [one] had paid in very nice coarse gold, some different from the Yreka gold, and when the merchants tried to find out where he got his gold, would not tell. But he got drunk and spent his money so freely and kept saying he could get plenty more where that came from, that some men concluded to watch him, and one night he struck out on the sly with his animals loaded with provisions and new tools. The men that had been watching him followed him with pack animals, and they had along tools and provisions. . . . The same night there were fifty or sixty men all followed right after each other and followed the man that they were after right into his camp and found that two or three others were mining on the sly. [They] went to work prospecting and found a rich creek and in a short time a thousand miners from California had overrun the whole country. (How[e] was one of the discoverers.)

After Bill Latham and I got tired of our bar claim, we sold out to a New Yorker named Pat Ford; he put in a ditch of water and sluices

---

[14] The two men mentioned should read John R. Pool (sometimes spelled Poole) and James Clugage (infrequently spelled Cluggage). Reinhart consistently spells the latter name "Glugage." Pool in association with Henry Kippel laid out the town of Jacksonville, summer, 1852. *The Oregonian*, December 7, 1930.

[15] Most sources credit Pool and Clugage with the gold discovery in the vicinity of Jacksonville, January, 1851. The name Sykes is sometimes included. James Skinner, a nephew of Judge A. A. Skinner, was an associate as well. By February, 1851, the rush was on. A[lbert] G. Walling, *History of Southern Oregon* . . ., pp. 337–338, 359–360. Hubert H. Bancroft, *History of Oregon*, II, 186, credits Sykes, an employee of A. A. Skinner, with the initial discovery, with Clugage and Pool as second, while Rossiter W. Raymond, *Mines and Mining of the Rocky Mountains* . . ., p. 215, credits Clugage, and dates the discovery in the autumn of 1851.

[16] Alonzo A. Skinner, born in Ohio, 1814, emigrated to Oregon in 1845. He served as a judge under the provisional government and in the second judicial district, 1869–1870. He was appointed an U. S. Indian commissioner in 1851 to assist in the negotiations leading to the extinguishing of Indian land titles in Oregon, his commission dated October 25, 1850. His colleagues were J. P. Gaines and Beverly S. Allen. He then served as Indian agent, at Rogue River, 1851–1852. He died, April 30, 1877, at Santa Barbara, California. *Oregon Spectator*, September 23, 1851; *The Oregonian*, May 21; October 5, 1877.

[17] Subsequent textual references to these place names used by Reinhart indicate they were located in Jackson County, Oregon, in the vicinity of Jacksonville. They should not be confused with identical place names in current use. Raymond, *Mines and Mining*, pp. 214–232, gives a good summary of Oregon gold mining.

and made good pay. Bill, he stayed, but a man came and told me there was a big rush at the Jackson Creek mines and that my brother Charley, with another of our boys that had crossed the plains with us, named Eli[sha] Hammer, were keeping a Bakery and Boarding House at Jacksonville and doing well, and that he wanted me to come out there, for he had left a Land Claim in the Cow Creek Valley, South [Umpq]ua River,[18] and he wanted to go to see to it, and wanted me to come to Jacksonville.

We used to buy milk for our coffee; it was $12 per gallon; it had to be brought on pack mules from Yreka, about 14 miles. We would buy a quart and keep it in pint bottles in the cold creek water, and it would last us four days. Shovels were $12, picks $12, gum boots $32, hats from $5 to $8, socks $2, blankets $8 to $16 per pair; sardines half-boxes $3; whiskey fifty cents per drink.

I had worked till the last of May when I heard of the Jackson Hill excitement and concluded to go there, and if I took a cut-off, I would not go by Yreka and save some forty miles in a little over one day. So one morning I took right down Humbug Creek till I got to the mouth where it emptied into the Klamath River, and then went up the Klamath River. I was all alone with no arms but a butcher knife, (and a sling shot that I found hanging up in my bank claim) and I had a roll of blankets to carry. So I just rolled my blankets up shot-pouch fashion, and struck for the Klamath Ferry (where the road from Yreka came in and crossed the Siskiyou Mountains)[19] and [where] I expected to stay overnight the first night. The Indians on the Klamath were considered dangerous but by good luck I got along without being seen by them.

I had two very narrow escapes that day; the first, I had to climb around a very high point of rock at the edge of the river. I had to hold on with both hands and [had] my blankets in a roll shot-pouch fashion (or over the one shoulder and under the opposite side arm at the hip). In making a turn on the summit (the rock was flat) I some way stubbed my toe or foot, and fell with my shoulder just at the edge of the precipice. My roll of blankets striking first saved me from going over into the river and rough rocks eighty or a hundred feet below. I lay so frightened that I did not get up for a few seconds. If I had not had the blankets as I did ... my shoulder would have struck first, which would have thrown me a foot farther, or over the cliff anyway.

My next escape was when I got to Shasta River. It was very high and how was I to get acrost? I could not swim a stroke. The water from four to seven feet deep, very swift and rocky and large boulders close to the

---

[18] Cow Creek, located in southern Douglas County, is a tributary to the South Umpqua River.
[19] The ferry operated a few miles north of the confluence where the Shasta River joins the Klamath.

Klamath River so that if I could not make the land on the other side I would be swept into the Klamath River where nothing could save me. I walked up and down to see how to get acrost. I had no axe or hatchet or rope, nothing to do me any good; and to go back I would lose 45 or 50 miles. So I saw no help but to go as high up the Shasta River as was favorable for me. I took a stick to steady me, tied up my pants and boots in my blankets over my shoulder around my neck to keep them dry (the water was ice cold) and I put my stick from rock to rock. The water would wash me down every rock I would hold on, and several times I thought nothing would save me from being washed into the Klamath. I would hold on one rock, jump and get washed down, and catch on another rock, and so on until I just did make the other side of the Shasta at the very edge of the Klamath River. My hands were fearfully scratched and my feet, too. I just lay down awhile to recover my breath from the fright.

Then I began to think I still had a long way to go and it was getting late and I had nothing along to eat and did not know if I could make the Klamath Ferry. My mouth became parched; I could not swallow. I drunk water but still my throat and tongue and jaw muscles pained me. I was still afraid of running into a hostile Indian camp or village. And I alone with some gold but no arms would leave me completely at their mercy, so I kept on keeping a good lookout ahead until about sundown I saw the mouth of Cottonwood Creek on the other side of the Klamath River, and some persons quite a way off, but I could not make out if white or Indians. So I kept out of sigh[t] and in about one hour made the Klamath Ferry. I was awful glad to get through in one day. Going around, I should have been at least three days of hard walking, besides my expense for meals and lodging.

I asked the ferryman if I could have some supper. He made me some tea and he had some fresh pork boiled, and I thought I would make a good supper, for I felt hungry and thirsty and sore and tired. But I took a mouthful of meat and could not swallow a morsel of anything, only a little tea. The man said I must have a touch of the scurvy.

I slept well that night and next morning started on my road to Jacksonville. When I got near the summit of the Siskiyou Mountain I had a splendid view of the Mount Shasta. I met a party coming from Jacksonville. (The one man I knew was De[j]arl[a]is, who kept a store at Minersville on Humbug Creek, and at Yreka. There were three brothers of them, French-Italian-Jews; the oldest was afterwards, in 1856 or 7 burnt to death in a store at Scott Ba[r] and the younger one was killed at Yreka in 1853 or 4).[20] The one I met told me of a murder at Jacksonville

[20] O. Dejarlais was a victim of a fire which leveled the town. He tried to save his goods which were housed in the only brick building in town. The building survived, but

and of the hanging of the murderer by the Vigilantes. I knew the man that was hanged; he had a claim next to mine on Humbug, above Minersville; his name was John Brown from Pike County, Illinois.

He had come out in 1849 or 1850, and his father and brother had gone home, back to Illinois, from the Southern mines, and he . . . came north and worked on Humbug Creek in 1851 and 1852. . . . Brown was a gambler, and I heard that he used to go with the Indians a great deal before he went to Jacksonville. He was a good foot racer and rassler, very stout build, about 24 or 25 years old.

De[j]arl[a]is saw Brown hanged and he says if he had a cool, fair and impartial trial he might have been cleared, but he was tried by excited miners who worked up a prejudice against the gambler, and Brown was called of that class, then very obnoxious to the miners, who had lost money with them, and were mad at them for beating them out of their money. But if they had won it would be all right. But Brown was in a manner justified, for he had great provocation. I will here state the case as I heard it afterward in Jacksonville.

John Brown was considered a gambler and he run with gamblers, but he was sick and could not work. The man he killed was a large, robust man, over six feet tall and weight over 200 lbs., [who] was drinking and blustering about in Jacksonville that he could out-run, throw down, or whip or out-jump anyone in town. Some men in fun called to Brown to come and run a foot-race against the big Missourian. (Brown was quite fast on foot.) But Brown said he was not well enough to run. But the Big Missourian thought that Brown was afraid of him, and he could bluff him because he was sick. So he commenced to dare him to run, and as Brown would not, and turned away from him, he said Brown was nothing but a horse-thiefing son of a b—— and he could whip him or outrun or throw him down, and abused Brown to what he could lay his tongue to.

Brown left the crowd and went and got a navy revolver and put it in the bosom of his shirt and down under his belt, and he come back where the Big Missourian was still blustering around cursing and swearing at everybody. Brown walked up to him and asked him to repeat his words that he called Brown. The Big Missourian came toward Brown and as he came he unbuckled his belt containing his pistol and knife and handed the belt to some friend of his, expecting that Brown had come back to fight him. He came to Brown and said, "You are a lying, gambling, horse-thieving son of a b——," and doubled up his fist to strike Brown when Brown had his hand in his bosom on his pistol, and drew his pistol and shot him through the heart, and he fell back and soon died.

---

Dejarlais suffocated to death. Shasta *Courier*, October 13, 1855. Wells, *Siskiyou County*, p. 218, states this was A. Dejarlais.

Map 3. Western Oregon and Northern California

Brown gave himself right up to the City Marshal and it caused great excitement. The miners were for lynching Brown right off; others formed a Vigilance Committee to try him next day. Some thought Brown had done right; some thought not. And what made it worse was that the wife and three children of the Missourian had come out and got there a short time after he was shot, and the miners all sympathized with the widow and her three children, and that her husband had been cruelly murdered by a gambler was adding to the crime against Brown. But he, Brown, still claimed he had done right, that if he had been well he would have fought him, but being weak and sick, he could not look over the names that the Missourian had called him and his mother, and if he had it to do again he would do the same every time. . . . But the friends of the killed man said he was drunk and should have been excusable for what he said.

So next day the miners all met; they had a Vigilance Committee and a judge and jury elected by the miners to try Brown, and after a little while the jury found Brown guilty of murder and he was to be hung next day close to town. A guard of fifty men (Vigilantes) guarded the jail, and next day he, Brown, was hauled in a wagon to the place of execution. His hands and feet were bound and several hundred horsemen armed with rifles and shotguns ranged along the wagon until they got to the gallows, hastily constructed, not far from Jacksonville. They were careful not to take his fetters off his feet and hands until they got on the gallows, for it was Brown's only hope that they would untie his feet and hands and he would have made a rush for the hills, and in the confusion he might have got away, for no doubt he could have outrun any man afoot on the ground, and they knew it, and kept him tied until they put the rope around his neck. They askt him if he had anything to say; he said only a few words—that he thought he had done right, and if it was to be done over would do the same; bid all friends goodbye and died game. The authorities at Jacksonville apprehended a rescue by the California friends of Brown, but they did not get the news until he was about hung, and it was too far to go to avenge his death by what few friends he had around Yreka or Humbug Creek.

From the summit of the Siskiyou Mountain I was nearly three days to Jacksonville; the last half-day I got quite unwell and I got overheated walking, and awful tired. So when I got to Jacksonville I was nearly down sick. My brother had gone down to Cow Creek Valley on the South Umpqua river to look out for his land claim in Douglas County; he heard some one had jumpt it. Elisha Hammer, Charley's partner, askt me to stay till Charley should get back. Their baker had gone on a spree and left Hammer because he did not like him. So I took his place in the bakery, baking pies, until Charley came back, and then when they came to settle up their business, Charley found that Hammer had been swindling and collecting money and not accounting for his collections.

He had collected some four or five hundred dollars from the stores retailing bread and pies, and he had a purse of $600 or $700 in gold dust. But he claimed that it belonged to a young man, that he was keeping it for him. So Charley bought out Hammer and he went to mining. Charley got a chance to sell out his bakery, boarding house and saloon to a man from Oregon named Pitney. In digging a well on their lot where their bakery was located, they found a good prospect and many claims were taken up, and workt drifting and tunnelling—too deep to strip.

For a few days after Charley['s] selling out, we took walks up the two forks of Jackson Creek and saw some rich claims of Shively, Amos Blue and Newt Bra[m]son, who had a rich creek claim on the left-hand fork.[21] Blue and Bra[m]son took out over $2500 in one day. One piece weighed $1800 and more, the largest ever taken out of the Southern Oregon mines. We were somewhat acquainted with Amos Blue and Newt Bra[m]son in Oregon.

I liked the mines around Jacksonville and wanted to go to mining there, but Charley wanted me to go down and locate a land claim near his, if he could get his back. I went up and saw the rich gulch near Jacksonville where Skinner, [C]lugage and Pool first made the discovery of coarse gold. It was very shallow and they used knives and spoons in the crevices and took out from $500 to $700 per day, nice bright coarse gold like drops of moulded lead.[22]

They got to work with rockers until they got water to work sluices and long toms. They took up great long claims and smuggled some between them, under fictitious names, until the miners got to cutting them down and jumping some of them. But it being so shallow, [mining] was not very extensive and they soon skimmed off the best of the rich gulch.

There were very rich discoveries at Sterling, Gold Hill, Applegate, afterwards in that vicinity, and I would like to have stayed, but I allowed myself to be persuaded by Charley to go and take up 160 acres donation claim. He said that on the south side of his, one was vacant and just suited as it was most all prairie, and his part timber and right along the river or creek. The valley was a beautiful valley some five or six miles from the mouth of Cow Creek up to a canyon. There were some ten or fifteen claims taken at that time. Timber [stood] along Cow Creek, and the mountains on each side were covered with pines, firs and cedar.

Charley bought four head of horses and saddles and we packed one with our blankets, clothes, and some provisions, and Charley took along

[21] "Old man Shively" reputedly amassed $50,000 and there upon returned east. Walling, *Southern Oregon*, p. 338. Newtown Bramson and his brother, George, worked their way to Oregon as teamsters for William H. Riddle.

[22] Reinhart is given to exaggeration here. The Skinner, Clugage, and Pool diggings were shallow, but they yielded only around $100.00 a day. Skinner's mine was called "Hard Scrabble" or "Rich Gulch." *The Oregonian*, June 5, 1852.

the baker that had gone on a spree while Charley was gone to Oregon while in with Hammer; he was a good ship carpenter in Europe and Charley wanted him to build a house on his claim. He had been on a continued spree for some time and we thought by taking him away from town we could break him of his fearful drinking. Charley let him have about three or four drinks a day to sober him off gradually, and it came near killing him, but in the end got all right again.

When Charley had gone down to Oregon a month or so before, the Indians along the road were quite bad and troublesome, and on Grave Creek the chief's son, named Warty, had robbed some travelers and stole some horses, and scared some men out of their provisions. When Charley got to Grave Creek, the same chief's son, Warty, came out to the road and spoke to Charley, and sold Charley a double-barrelled pistol. He bought it merely to humor the Indian; he gave him a blanket and some two or three dollars in money. It was loaded and Charley never fired it off on the trip—he had a shotgun along.

When we three left Jacksonville we traveled slow, for George Williams, the carpenter, could not ride fast, and he had been near the Delirious Tremors for some time. We stopt every eight or ten miles and let George have a drink of whiskey, where they had it for sale. We crossed Rogue River at Perkins' Ferry, some 16 miles from Jacksonville. There was some mining, a big bar eight or nine miles from Jacksonville on Rogue River, and there were a great many men mining on the bar: Capt. Jane, a Fancy, and her niece Tony [who] had got married at Yreka to a man named Burgess; and Dutch Louis and Nobles had gone from Humbug City to Yreka, and from there to Jacksonville, keeping a saloon, and Nobles and Louis had boarded a while at Charley's Bakery, and then took up some farming land on Rogue River, near Big Bar, where they had some mining claims too. We found miners prospecting in gulches all around Jacksonville and all the farming land had been taken up for homesteads in the last four or five months and were already held at high prices. At Ev[a]n's Ferry we found a hotel and saloon and quite a little village.[23] We stopt first night at a creek called Jump-off-Joe; the second day at Grave or Woodpile Creek; third day at noon at Hardy Elliff at the east side of the big canyon.[24] We had dinner and fed the horses, and George was quite tired. So we started after noon to go through the canyon; we were to cross the creek 102 times in 12 or 14 miles, and it must have been a fearful drive to go through with oxen and heavy wagons.[25] We found it hard enough with saddle horses, and George kept

[23] Davis Evans operated a ferry three miles above the creek named for him. He was one of the first Douglas County commissioners in 1852.
[24] Lewis A. McArthur, *Oregon Geographic Names*, pp. 273, 331–332, describes Jump-off Joe and Grave Creeks in Josephine County. Hardy Elliff lived in Cow Creek Valley. Walling, *Southern Oregon*, p. 424.
[25] If Reinhart traveled northwestward via Grave Creek, this would be Canyon Creek

falling back and did not keep up well. . . . We pushed on to get to Mr. Knott's Hotell and Sawmill at the north of the canyon, at the west side of the mountains, at a little town called Canyonville.[26] We got to Knott's just before sundown and got our horses taken care of; then ordered supper for the three of us. When it was ready George had not come yet. We concluded to eat and by that time he might come up, but when through, still no George, and it was getting dark. So I thought I would fire off a shot or two out of a pistol, and see if he would not hear and answer, or that he would know we were not far off. Charley gave me the double-barrelled pistol he had bought from Warty, the Grave Creek Indian Chief's son. I stept out from under the Hotell porch to fire it off (there were several men sitting around under the porch) and I held up my hand to fire. It went off with a loud report, but it rebounded with quite a jar with a heavy load. I happened to look at the pistol and saw one of the barrels had burst and gone, so I held up the other and fired with the same results as before. I lookt and the other barrel was gone. The pieces had flew in all directions so as to miss everybody and had not even hurt my hand. We all came to the conclusion that Warty had loaded them very heavy and then had drove the top of the balls with some iron, so to wedge in the bullets, that it might kill the person who fired it off. We were very lucky not to get hurt.

George still had not come, so two or three of us went back a ways on the road, and at last met his horse with his saddle and bridle on all right, coming towards us. George had rode on a double blanket, and after shouting and getting no answer, we came to the conclusion that he had let the horse go, and had taken the blanket and layed down and was asleep, too tired to come through the canyon to the Hotell. So we went back to the Hotell and went to bed.

In the morning early George had not come up yet. I got on a horse and took his horse for him to ride and started back, and got about one or two miles when I met him. His face and hands were all bloody and scratched up. He said some three or four miles back he took a sort of fit and he let

---

Canyon, a defile which opens into the valley of the South Umpqua. Early pioneers called it Umpqua Canyon. The 1846 emigrant party did suffer terribly in trying to take wagons through.

[26] Joseph Knott settled Canyonville in the summer of 1851, taking out the area's second land claim. In 1852 the town site "was marked simply by a log house [Knott's Hotel?] and a smith shop." Walling, *Southern Oregon*, p. 425. An 1853 traveler observed only a mill and a few settlers there. Mary M. Dunn, *Undaunted Pioneers*, p. 28.

George W. Riddle, *History of Early Days in Oregon*, p. 31, recalled: "Knott was a man of intelligence and energy but of domineering disposition. He sold out the Canyon location in 1852, settling upon a donation claim near Sutherlin afterwards moving to Portland where he and his sons operated the first steam ferry on the Willamette river." He also served in the 1858–1859 legislature.

Since many of the 1851 overland party had settled in Douglas County, Reinhart probably renewed their acquaintance.

the horse go and took the blanket and lay on it. He took so bad he was afraid he would die; he kicked and scratched around among the rock and gravel and became insensible. When he came to he was covered with blood and so sore he could scarcely move. When daylight came he washed the blood off and took his blanket and came on. He was awful weak; he had not eat much at his breakfast or dinner day before, and had had no supper or breakfast, so I got him on his horse and came on to Canyonville and had some breakfast for him at old man Knott's. After he got rested up a little we got some few things at the store and went over to Cow Creek Ferry, about six miles off, kept by Thomas Smith[27] and Clement Glasgow. Both [were] batching it and keeping a sort of store and Liquors and the Ferry.

Cow Creek was a large stream ten months in the year. The rains would raise it in one night so that it was quite a river, but the banks were high, and good farming land lay on both sides of the creek or river. We stopt that night at the store to find out what was to be found out about claims around where we wanted to go, three or four miles above. Smith and Glasgow had each 160 acres, making 320 together. [It was] fine land, and they made money out of their goods, ferry and Liquor.

Next day we went up Cow Creek Valley to where my brother had taken up his last claim. The other had been jumped by a man named Nichols. He had married Wm. Riddle's youngest daughter, Isabell, and then jumpt Charley's claim, saying he heard Charley had abandoned it and was staying at his bakery and saloon at Jacksonville.[28] And Charley had not made any improvement, nor even laid the foundation or frame of a building; he had merely staked it off and asked old Bill Riddle to hold it for a short time, until he would come back and build a house on it. Riddle owed Charley for one yoke of cattle and some work, but after Nichols married Isabell, Riddle denied [he had] cheated Charley out of what he owed him. Charley was discouraged for he was expecting to marry Isabell himself. So he did not try to get back the Nichols claim but just took up one not quite so good, adjoining Riddle on the south; most of it an oak grove right along the banks of the creek.

In the oak grove was a good building place for a house, and I took up my claim adjoining Charley on the south. More prairie. We put up our tent and made camp in the grove, and Charley and I went back next day to Canyonville to order a bill of lumber, some of which we had to wait

[27] Thomas Smith, born in Campbell County, Kentucky, September 14, 1809, came west from Texas to California via the southwestern route in October, 1849. After a variety of experience as a gold miner, he settled in Cow Creek Valley in November, 1851. He became one of the first Douglas County commissioners in 1852. Walling, *Southern Oregon*, pp. 529-540.

[28] I. B. Nichols came from Iowa to California in 1847 and engaged in pack-training supplies from Oregon to California gold mines. He settled in Cow Creek Valley in 1851, marrying Isabella Riddle in June, 1852. *Ibid.*, p. 424; Riddle, *Early Days*, pp. 28, 39.

some time for before they could saw it. Charley traded for a wagon and harness and we took some lumber back with us to build a shed to work, sleep and cook under....

Right along the road from Cow Creek Valley to Canyonville we passed through a dry gulch where some one had prospected and found a good prospect for paying gold diggings, but no water—only in winter when it rained; then a person could work a rocker. I took out a few pans of dirt and put it in a sack and we hauled it to a creek about two or two and half miles away. I panned it out and was surprised at the nice gold. The amount I got would pay well with sluices, and with plenty of water a man could make from $7 to $10 per day. By hauling the dirt to the creek where I panned it out, a person could make three or four dollars per day in wagons.

We found after we got to camp that we could not get our timbers and lumber for one or two months, so many orders were in ahead of ours, so we concluded to go prospecting up Cow Creek above the falls and canyon. We took our mining tools, guns and provisions and blankets on pack animals, for you could not go through the Cow Creek Canyon with wagons, and we were to look for some cedar trees to make shingles and clapboards.

We pitched camp some eight miles from the ferry, or four above Wm. Riddle's at the head of the Cow Creek Valley. We found splendid prospects of good mining along the bars and creek. Right in the main Cow Creek, in the rock, on ripples, we found pot holes in the bed rock that would hold from one to four pans of sand which contained fine light gold, from twenty to a hundred dollars to the pan, but hard to save, for the sand was nearly as heavy as the gold and it took great care and still you could see the gold wash out of the pan and lodge on the rock down the stream. By panning the same sand over and over a man could make from $3 to $7 per day with the pan only. [We] could not save the gold at all with the rocker, and we knew nothing in those days of quicksilver or we might have done very well in mining.

The water was still too high to mine in the creek, but some miners had taken up some creek claims, just ahead of Canyon John Kitchen and Co., but the water was too high, and we took up some bars and creek claims to work after the streams would get lower.

We found the finest cedar trees I ever saw, yellow and red, and the logs were thirty to fifty feet without a limb and could be slit with just an axe and wooden glut. You could strike in an axe and slit twelve feet as straight as a ribbon. A good many men that had taken up claims below in the valley cut cedar logs and floated them down the creek on high water to their donation claims. A single man could only take up a quarter section, or 160 acres, but a married man was entitled to 320 acres, or a half section, by living on it and improving the place for four years.

We concluded to go to the gulch between Cow Creek Valley and Canyonville, and take up three claims of fifty yards each, for my brother Charley, George our carpenter, and one for myself. One of us to dig the dirt out of the gulch; one to haul the dirt in the wagon, and one to wash it on the creek with some sluices and long tom where we had our camp. We worked some two weeks, but could not make much for we could not get dirt fast enough to make wages for three of us and the team, and provisions were very high.

Some other miners from Oregon on their way to California stopt to prospect and quite a lot of miners, some from Jacksonville and Rogue River, were running around prospecting. Some of the Oregon men, one named Pete Fulkerso[n] and his partner Van Sickle, went with us to prospect up Cow Creek where we had taken up our creek claim, and a young man named Bill Clark, or Dancing Bill, was stopping (on the creek where we had been washing our dirt hauled in the wagon) with a family named Yokum who had taken up a half section.

Before I go any further I will tell of an adventure I had while I was washing the dirt for gold at our camp. One day a man, a blacksmith and saddle-tree maker from Canyonville, was at our camp. Charley was driving the team and George was working in the gulch. I had got through dinner, and the man, Ashcraft by name, was alone with me, when two squaws and a little girl came to camp. One was an old woman; one about twenty, blind of one eye, and the little girl about seven or eight years old. They were begging for bread, flour, or sugar. Ashcraft for fun asked them some questions, and the old woman said the young woman would for some bread and a handkerchief or some sugar. They sat by the fire and eat some bread and meat we give them, and Ashcraft went off with the one-eyed one. [He] had not been gone long when he came back alone. And the one-eyed squaw came in from acrost the creek.

I noticed an Indian man coming towards camp. The old one and the little girl had seen him first, and called to the one-eyed one, and she ran around, crossed the creek, and came into camp from the opposite way she had gone. The Indian man was the husband of the one-eyed one and he suspicioned her of mischief and accused her and Ashcraft. But she denied [it] and so did the old one and the girl. At one time the Indian drew his bow and arrow on me, but I grabbed my pistol and he dropt his bow. But he then caught his wife by the hair of her head and she done the same to him. But the old woman and girl at last made him believe all was right, and they left together. I found out that she had been unfaithful to him before and she had been punished by the loss of an eye, which was customary punishment for adultery. I saw several afterwards that were marked that way.

I met that same Indian after that several times and he always asked me about that time and said he still believed her guilty. That same afternoon an old man Indian came and told me he was chief and she was his daughter. He said Mr. Knott at Canyonville never molested their squaws and he wished all of us to let them alone.

When we went up Cow Creek again we had along four or five others: Peter Fulkerson, Van Sickle, Bill Clark, John Wagner, and us three. (This was July 1, 1852.) We all took up claims and commenced to cut a large ditch, which by cutting a few hundred yards ran into an old creek channel. . . . We commenced to dig the canal or ditch and found we had to dig it wider and deeper and through hard rocky and sandy places. When we thought we had it low enough, we cut down trees and split slabs to build a large head dam to turn the main channel of water through the canal or ditch. But we found the bed of the stream deeper than we had anticipated, and it backed up the water and raised in the dam, so we could not get it tight. We worked in water up to our arm pits, putting in brush of cedar and firs, and wheeling in dirt in wheelbarrows and hand barrows, we could not get it as tight as it ought to be. We dug the canal deeper and wider, but the water was still too high and so cold that we could hardly work in it (in July!): it was melted snow from the mountains above.

So we concluded to wait a while longer and make some sluices and prospect a bar in the mouth of a little creek. The gold was very fine and the sand heavy and so rich with gold that you could see the gold all through the sand. When we panned out the water out of the hole, the gold settled on the bottom and looked very rich. We run the sluice and Tom, but after a half day's work washing found we did not save the gold. We panned over our tailings and found them the same as the dirt not washed at all. We had to give the bar up for the present.

So [we] went down to our claims in the valley and commenced on Charley's house. We built it on the line of both our claims so that one building would answer to hold both our homesteads.

I used to go and prospect with different parties; once we took a trip with some fifteen or twenty men in the big bend of Cow Creek up at the two Forks. We had to go to the summit of the mountain range and then take a divide. We were tracking some parties that had gone ahead, three men Eph Kitchen, Re[nnick] Cole, and Dad How[e]. We found their tracks on a divide going to the forks. We did not get down until the third day and found it awful work to get our horses down to the Forks. We found where they had prospected and worked some, and [then] had crossed and left Cow Creek for the head waters of the Co[qui]ll[e] River, some sixty or seventy miles off. But we had only a few days provision left and could not follow any more, so we prospected the bars and creek and went back to Cow Creek. Two of the men with us from Jacksonville, one a butcher named George (he was English) and a young man named Bill Hill, a brother of Flem Hill of Winchester, where he kept a store with old man Knott that kept the Canyonville Hotell and sawmill. [He] killed a man in the store with a four-pound weight and the Vigilantes came near hanging him, but old man Knott had plenty of money and saved him. But he had to leave and so did Knott.[29]

When we got back to Cow Creek Valley most of the party went in different directions for their homes. We found a great excitement among the settlers for they were expecting an Indian attack from the Grave Creek Indians.[30] I spoke before of the chief's son Warty (so named for a large wart on his cheek and a large scar on his neck where a year before a white man caught him stealing his horse and he cut Warty with a big knife, and he was expected to die for a long time. The old chief, his father, said that if he died he would kill every man, woman and child in Oregon. But by good or bad luck Warty did not die then).... It seems that Warty and another Indian stopt at a man's home named Adams to get a drink of water. He found that there was no one to home but a daughter of Adams, about eighteen years old. Her father and brother had gone to some other neighbors, and when Warty got the water he said she must get him something to eat. She gave them both some bread. Warty made some insulting proposal to the girl and threatened her with his knife. The other Indian took the knife from Warty, and Warty grabbed a rifle. The other Indian took hold of Warty, to keep him back by taking the gun away; while he was taking the gun away the girl run out of the door and run towards the next house, kept by Billy Weaver.[31] Warty ran after the girl, and the other Indian after Warty to keep him out of mischief. The girl got to the house first and there were several men with Mr. Weaver.

The girl told them the Indian was going to kill her, so they grabbed their guns and went out and surrounded Warty.... He had been drinking whiskey and was supposed to be drunk.

The men took him prisoner and sent off to the neighbors for more

[29] Reinhart's memory is in error here. Knott killed James Hill in his store on election day in June, 1853. He was brought to trial before the second term (the first term for the newly appointed judge, Matthew P. Deady) of the U. S. District Court at Winchester. This was the first homicide case in the district—the case of *The Territory* vs. *Joseph Knott*. The tragedy produced great excitement, including talk of lynching the accused, "but this was promptly suppressed by the better of citizens and the efforts of F. R. Hill, who was a brother of the victim." The prosecution was headed by Sim, Harding, and Sheil; the defense by O. C. Pratt, R. E. Stratton, A. C. Gibbs, and S. F. Chadwick. The trial resulted in a verdict of acquittal for the defendant. Shasta *Courier*, August 20, 1853; Elwood Evans, *History of the Pacific Northwest: Oregon and Washington*, I, 409; II, 372–373.
[30] Members of the Umpquas, their chief was called Milwaleta.
[31] William Weaver and his wife, Anna; his two sons, John W. and James B., and his daughter, Harriet, lived three miles south of Myrtle Creek. Near at hand resided the Adams family: John, his two bachelor sons, Henry and John, Jr., and a daughter. Together they founded Missouri Bottom in 1851. The Weaver brothers subsequently opened a store in Myrtle Creek. Riddle, *Early Days*, pp. 31, 61. A longer account of the Weavers is to be found in *Portrait and Biographical Record of Western Oregon*, pp. 977–978, in a sketch on John W. Weaver.

men, and Mr. Adams and sons. They all come and organized a committee of Vigilantes and tried him, Warty, by a jury, and found him guilty of intending to ravish, and sentenced him to be taken out and hung. Warty seemed to take it all in sport and said they would not dare to kill or hang him for his father, the chief, would kill every person living in the valley.

But there had more settlers come in, and they took Warty and put him on a horse, and his hands were tied behind and a rope around his neck. [They] took him on a hill where there were lots of trees, back of Weaver's house, and throwed the rope over a limb and told Warty he must die. But he only laughed at them; he thought they were only intending to scare him. So they just run the horse from under him and left him hanging until he was dead.[32]

Some of the Indians started off two runners to let the old Grave Creek chief know that his son was dead, and the white men sent all through the valley to prepare to meet the Indians when he came, for they all had heard what he threatened to do if Warty had died the time his throat was so badly cut. We heard that the Indians were coming, so quite a lot of whites gathered at Weaver's place to protect them. Some of the prospectors from Jacksonville that had been out with us were there to help guard Weaver's.

Warty had been cut down when dead and laid on some boards near the house. Some Umpqua Indians were there, too. The miners had a large fire built nearby and stood around it, when someone made some remark of burning Indians . . . the English butcher George from Jacksonville grabbed aholt of Warty's body and just threw it on the fire, and they let it burn up. We were mighty uneasy all around that part of the country for fear that the chief would retaliate for the death of his son.

But they did not until two years after when the regular Rogue River war broke out and old Gen. Joseph Lane fought them at Table Rock.[33] The Indians got rather the best of the fights with the whites, and many were massacred and property destroyed, but some of their chiefs were taken prisoners and taken to San Francisco. One, old Joe, got loose from his fetters on the steamship and they had quite a time to get him in irons again. The Indians in that vicinity were then put on a Reserve ten miles from Jacksonville and treated very stringently.

[32] Warty was the son of Warta-hoo, head of a band of some twenty-five Indians in the vicinity of Cow Creek, east of present Glendale. Riddle, *Early Days*, p. 56, writes:

In 1852 a young Indian, a son of Chief Wartahoo, was hung at the William Weaver place. It was claimed that he had insulted a young white woman by an indecent gesture. Within four hours he was hung. This might have been considered justifiable from the white man's point of view, but to the Indians, the boy's fault would not compare with the treatment their women had received from drunken white men.

[33] The battle of Table Rock, located in the Rogue River Valley near Medford, was fought on August 24, 1855. Subsequently, Joel Palmer, superintendent of Indian affairs, dictated the terms of peace which were signed on September 10.

# 3

## GOLD RUSH DAYS IN SOUTHERN OREGON

(1852–1853)

After we stopt mining in July on Cow Creek on account of high water, we stopt on our ranches and completed Charley's house. Peter Fulkerson had turned out a fine gray horse in a herd running between Cow Creek Valley and the other little creeks emptying into Cow Creek, while he was helping us cut our canal. His horse had been quite run down and poor by his trip home from the Willamette Valley. . . . Six weeks or two months grazing on good grass made him fat and hard to catch, so he traded him to Ashcraft, the saddler, for another gentler horse, with some boot. Ashcraft thought he could catch the gray, but after fooling around about a whole day, he could hardly get up in sight of the gray—he seemed to have run wild.

So one day Ashcraft told us he would have some fun and if a lot of us would take our horses and run the gray so that we could catch him, he would make each of us a nice saddle tree, and they are worth from five to eight dollars apiece. So some six or eight of us concluded to try and have some fun running him. Two or three of us commenced to run him with some other horses he was running with, until we drove him from his companions, and two of us would drop off and two men with fresh horses would keep the gray going, so as not to let him rest or catch his wind. But he was a powerful horse, and having no one on him, had the advantage of our horses, by our weight riding. He seemed to keep fresh when our horses would commence fagging.

I was riding a splendid gray Spanish saddle horse; he could outrun the loose horses and turn him easy (he was a stock or lasso horse and you would not have to turn him by his bridle, all you would have to do was to watch his quick turns after the other horse, or you would get thrown off). John Wagner had a fine cream or dun; he was with me a great deal, but he stepped into a gopher hole and went over end over end. John Wagner was thrown on his shoulder and hurt quite bad, [but recovered].

Green Heron and Glasgow and Kitchen all tried, but we could not catch him the first day. The next day we tried again; some of us had changed our horses and kept changing, and some had lassoo ropes, but still could not get near enough to throw the rope until dusk. He [then] commenced weakening and some 5 or 6 of us got close to a high bank of the river on one side and we surrounded him from every other side and at last got a rope on him and led him to the store or ferry of Glasgow and Smith. It was dark, so Ashcraft staked him out that night and next morning some one should ride him.

So next morning bright and early we got up. The horse ... snorted like a porpoise, and no one would venture to ride him; all were afraid he would jump stiff-legged and buck them off. Ashcraft at last offered me five dollars to ride him an hour or two, so I took him up. There were plenty of better riders than I there, but they would not risk him. So I got a good double-cinched or girt saddle, a Spanish bit, and coiled up a lariat and held it in my hand, got up to the horse and put on a halter and bridled him and put my handkerchief on for a blind. Someone held the horse and I got ready to mount. I had on moccasins and a tight belt around my waist, and a handkerchief around my head, for I did not wish to bother with a hat. Every one, some fifteen or twenty men, looked on expecting to see a little of the hardest kind of bucking and running.

I put my foot in the stirrup and quickly threw my leg over and gathered my bridle and rope. The man let go. I raised the blind and struck both spurs into the horse to start him on the run so that he should not stop to buck. But [to] the surprise of all of us, he would not budge one step, but laid back his ears and sulked. I kept spurring him to no purpose. I struck him with my rope, but to no purpose. So we all had a big laugh. Some took brush and hats and hooted to make him go, which he did at last on a walk. I could hardly spur him enough to make him trot or gallop. He seemed to be completely disheartened and never afterwards had any ambition in him, and was not worth fifty dollars. At first Ashcraft refused $200 for him. ...

Our party came to the conclusion that we could not save the gold in our claims, and we all abandoned them for the present. In 1854 or 1855, while I was at Gold Beach at the mouth of Rogue River, I saw how they worked quicksilver. ... I saw that if we had known how to work it on Cow Creek in 1852, we might have made a good thing for us all. For on Cow Creek we could have worked all day and left our sluices and claims overnight, and next day could have commenced again where we left off.

*Gold Rush Days*

But at Gold Beach, we would have to set sluices strip and work inside of every tide of six hours out and in.

Some time during August [1852], Clement Glasgow, [and] Sam Hawkins, an old miner who used to work on Scott River, California, and run a butcher shop there, [and who] located a homestead claim in Cow Creek Valley not far from us, were going down to the Willamette Valley to buy seed wheat, flour, butter, potatoes, and onion seed, and asked me to go along and see something of Oregon and its valleys. Glasgow took some 8 or 9 mules and horses; Hawkins 3, and I took 3 or 4 of Charley's. We were to pack them with pack saddles. Most all of them were fresh, and quite wild, and some of them had never had a pack saddle on them. . . .

We did not get started out of the corrall at the Ferry until about 3 or 4 o'clock, and as soon as they got out of the corrall we had all we could do to keep them from getting away from us and running back to their range and the hills. But by hard riding and whooping and whipping we got started and crossed the South Umpqua River by fording the stream, and drove some 3 or 4 miles west of Myrtle Creek and it got about dark. We concluded to camp.

We found a good place and took our cooking kitt off our le[a]d horse. We had two large sacks made out of raw hide dried, made into square pockets or sacks to hang one on each side of the pack saddle on the horse, and we had our bedding or blankets on another with our provision in the pockets, or called Parfleshes. But when we came to catch our horses and mules we had drove, to take their saddles off, we could hardly catch them, they were still so wild, and we had to get on our saddle horses to chase them, and we had to let some 6 or 7 go with their saddles on all night, for we could not catch them, it being too dark. So we turned out our riding horses with picket ropes on, and staked some of them to keep them from running back home, made our fire and got supper and went to sleep.

Next morning as soon as daylight came we got up and two of us went out to find our horses. We found them at last quite a ways off [. . . and started on our way.] We passed through Roseburg and had to cross the North Umpqua River at Winchester, the county seat of Douglas County.[1] We had to drive on to a large Ferry boat, for the river at this place is quite a stream. While we had all the horses and mules on the boat, which had a high railing around it, we got a chance to get up to some of our wild horses and readjust their saddles and change some that we wished to ride so as to gentle them while on the road. We put up our packs and got

---

[1] Winchester was laid out in 1850 by Addison R. Flint. It served as county seat until 1854 when Roseburg was chosen for that distinction. Roseburg was founded September 23, 1851, by Aaron Rose who emigrated from Michigan.

out of the boat and went on a few miles west of Winchester and passed the famous place of Genl. Joseph Lane, of Indiana and Mexican War fame. He had been Governor and Senator of Oregon.[2] He had a large farm here, but his land in wet weather was the stickiest land (Blue mud) I ever saw, and when the mud came off of a horse's legs the hair came with it, and it was nearly impossible to go through his lane in wet weather. [It was all] the old gent could [do to] keep people from throwing down his fence and going through his field, and [this] kept him in hot water about it.

We only went about 5 miles beyond Winchester to where we concluded to camp. We got down and commenced to unpack our kitchen horse, that had on the square sacks containing our bacon, coffee, sugar, bread, and cooking utensils, when the horse became frightened, . . . got loose and stampeded with the rope dragging and flying after him. . . . Two of us got our horses to follow him, but after a mile or two we lost sight of him going up the North Umpqua River, so we came back and got our horses all unsaddled and built a camp fire, but we had nothing to eat for our provision was all on the horse. A couple of us went out and tried to track him, but it got dark and we could not see, so came back to the camp, made up our beds and laid down hungry to sleep, blessing our runaway horse. . . .

So we done the best we could, and before daylight we all went out to hunt our runaway and at last found his track and followed on and found one of the rawhide sacks, but everything had been scattered to the four winds of the earth. We then came back and got up all our horses and one of us went on horseback to follow the track. We had . . . went back to some house and got some bread, butter, meat, and eat a breakfast, and after we got ready to go Glasgow came back with the runaway. He had found him four or five miles west on our road, in a man's corrall or lot— he had taken him up early that morning. We got to the house and got the other sack and saddle and horse and got some breakfast for Glasgow. (The man's name was Eleza Bun[ton]; he was a hard case and I will speak of him and his family in after years at different places. He had a nice young daughter at that time about 16 years old—her name was Minerv[a]—and he had some three sons; one named Bill was afterwards in 1863 hung by the Vigilantes of Montana.[3] After old Eleza's death his

---

[2] Joseph Lane, native of North Carolina, was born December 14, 1801, in Buncombe County. Later he moved to Indiana. His political career was launched in 1822 when he was elected to that state's legislature. He served with the second regiment of Indiana Volunteers during the Mexican War. President Polk appointed him governor of Oregon in 1848. On March 2, 1849, he arrived in Oregon City and took office. The following June he resigned, but subsequently served four terms as Oregon's territorial delegate to Congress, and as U. S. Senator, 1859–1861. He died in 1881.

[3] Bill Bunton was hung, January 25, 1864, at Hell Gate, by a party of Vigilantes from Nevada, Montana, for cattle theft and robbery. Thomas J. Dimsdale, *The Vigilantes of Montana*, pp. 180–182, 271.

*Gold Rush Days*

wife and daughters lived neighbor to me, from 1862 to 1865, on the Walla Walla River when I lived on Dry Creek, Umatilla County, Oregon, 12 miles from City and Fort Walla Walla, Washington Territory.)

After starting on that morning we soon got to Oakland, where there was a fine flour mill and a small town close to Yonc[a]lla Valley.⁴ We camped on the head of the creek (Yonc[a]lla) for the night. . . . After supper I took a pan and scraped up some dirt out of the creek we camped on, and I panned it out and got a good color of gold in the surface of the creek. And some year after there were good prospects found high in the mountains of the North Umpqua River.

We had crossed the road or trail that went to Scottsburg, a harbor at the mouth of the Umpqua River, afterwards called Gard[i]ner City.⁵ The merchants in Southern Oregon and Northern California got their goods shipped from San Francisco, California, to Scottsburg and from there on they packed on pack animals to their destination. It is the lowest seaport or harbor on the Oregon Coast, next to Crescent City, Del Nort[e] County, California.⁶

That day we found some fine farms at the foot of the Cal[a]pooya Mountain and we crossed the mountain at Cartwright's Crossing or Pass.⁷

Next day we got to the Willamette Valley. At noon we stopped at a farm to see if could get some fresh beef, but they had no fresh, but some dried jerked beef, but did not like to spare any but would let us have a few pounds for accommodation at 30 cts per pounds. We took some 5 or 6 pounds. Hawkins did not happen to be with us when we got the beef; he had stopt at a house before to look at some cows—he wanted to buy some good milk cows—but when we stopt at night, we cooked some of our jerked beef, and after eating we remarked it was awful tough beef and poor at that. Hawkins asked if we got it at that certain house and Glasgow said yes. Hawkins said he had passed that house a week or so ago and saw a sick steer close to the house and he should not wonder if

⁴ Oakland lies sixteen miles north of Roseburg. Dr. Dorsey S. Baker settled there in 1851 and built a store and grist mill. A[lbert] G. Walling, *History of Southern Oregon . . .*, p. 428.

⁵ Scottsburg, as it is still called today, was founded in 1850 by Levi Scott, an 1844 emigrant. Gardiner is situated on the north bank of the Umpqua near its mouth and was founded in 1850. Reinhart has confused the two. Scottsburg is inland near the river and was the key transfer point for goods brought in at Gardiner. *The Oregonian*, June 23, 1855; *The Oregon Journal*, November 24, 1937.

⁶ Crescent City was established in 1852. Soon after, because of its favorable location, it displaced Gardiner, Scottsburg's harbor, and became the center of transhipment of goods to southern Oregon.

⁷ Named for B. D. Cartwright who came to Oregon from Syracuse, New York in 1853. He settled on the upper reaches of the Siuslaw River in Lane County on the east side of the territorial road leading into Willamette Valley. He opened a hotel and stage station which he called Mountain House, and which still stands today. Born February 7, 1814, he died July 29, 1875. A[lbert] G. Walling, *Illustrated History of Lane County, Oregon*, p. 482.

the steer had died and they had cut up part of the meat for dog meat (For they kept a lot of hungry looking dogs at that house).... Hawkins asked if we saw the hide of the steer, and Glasgow said it hung in a granary and gave a description of the color, and Hawkins said it was the same he saw nearly dead. So we felt like throwing up our supper, and we threw away the balance of our dried jerked meat. Glasgow swore he would get even for it on some of them farmers for selling us dog meat that had died of sickness "for accommodation" at 30 cts per pound.

We come in the valley a little below where Eugene City (Oregon) now is, and next day passed through Marysville, Benton County (now called Corvallis), quite a nice large town.[8] At low water it was the head of navigation of the Willamette River. At high water Eugene City was the head.

We went over the Willamette River range of mountains to a fine little valley called King's Valley,[9] between the coast range and the range we came over, and found that King's Valley was one of the richest valleys in Oregon.

Old rich settlers had taken up large homesteads in an early day, when a head of the family could take up a section of land, and a single man a half section. Land was very rich and well watered and plenty of grass for stock in the hills, and timber—the world has no better than Oregon, California and Washington Territory. These farmers had plenty of fine stock and raised as fine wheat and oats as I ever saw, and onions that beat any I ever saw. We stayed a few days and got some seed wheat, butter, and onion seed, and priced cows in case we should wish to come back after some.

We were detained one day by some of our horses straying off, but found them all right and left King's Valley by another road that brought us to the Rickr[e]all [Creek] where Senator Nesmith had one of the best Grist and Flour Mills on the Pacific coast at that time.[10] We loaded several of our animals with flour and then started to come home. We crossed several fine streams emptying into the Willamette, such as the Big and Little Luck[ia]mute, and we passed several little towns along the river....

All the streams in Oregon and Washington Territory, Idaho and Northern California are just swarming with the best salmon in the world, and so plenty that they can be bought from the Indians for a piece

---

[8] The city was settled in 1846 by Joseph C. Avery; the name was officially changed in 1853 by the legislature.

[9] Named for Nathan King, an 1845 settler.

[10] James W. Nesmith, a Canadian by birth (1820), made his way to Oregon in the 1843 "Great Emigration." One of Oregon's most influential political leaders, he served as Indian superintendent, 1857–1858. He died in 1885. Harvey W. Scott, *History of the Oregon Country*, V, 172–183, presents a concise biography.

## Gold Rush Days

of bread or a little flour or sugar. [Some are] as large as 30 or 40 lbs apiece. At high water they run up from the sea and they keep going up as far as the water will allow them to go over dams and falls. They can get up steep falls that a person not knowing would not believe it. They have a hooked bill or nose to hold by, and they jump and twist over high rocky falls, and after they get up stream and can go no further, spawn and die. Their meat is death to dogs for they cannot digest the bones of the fish, but will eat and live on them during the season. We boil them or fry them and they are the best fish when fresh there is, but when they go high up rocky streams they become bruised and spotted and not good to eat.

Deer, elk and bear are very plenty in the hills on the Willamette or the Umpquas or all the different valleys in Oregon, and in the morning, grizzly bear tracks are as common as hog tracks; so are cinnamon and black bear, and they like to go to farmers' hogpens and kill hogs, quite often.

Coming back, we . . . [went through Oakland and Winchester]. One day we were out of meat and we had stopped at several houses to buy some, but no one would spare any, so we camped at noon on a small creek about 100 or 150 yards from a house. A sow came along with 6 or 7 nice shoats that would weigh about 30 lbs apiece. Glasgow had a hatchet—he had just been staking one of our horses—and he drove the sow and the pigs right into our camp. He called to Hawkins and I to head them around, which we did, and he threw the hatchet and struck one of the pigs and knocked him down, but the old sow came toward him with her mouth wide open and the pigs squealed awful loud, but we all rushed in and got the pig Glasgow had struck with the hatchet and he killed it right in sight of the house. The old sow and pigs made an awful noise while we were killing the pig. I was afraid they would hear or see us from the house, but no one molested us, and we skinned the pig and burned up the hair, skin and offall, and cooked some for dinner, and while we were eating, some of the men and boys of the house came down and saw us eating dinner of their pig but they did not know it. I thought it was quite a cheeky trick but we claimed among ourselfs that we were justified, for no one would sell us any meat, and the only one that had, sold us a piece of steer that had died of some sickness "for accommodation" at 30 cts. per lb, and we thought we were getting satisfaction on some of the farmers that had played us the trick. Our pig lasted us a few days till we got acrost the Cal[a]pooy[a] mountains and while at camp one evening a large fat hog came along, and Hawkins run around and throwed his hatchet but missed him, so we had to wait until we got to Oakland, where we got some bacon to last us to Winchester on the North Umpqua River and we got some meat to last us home.

We stopped at Rosebur[g] a short time; the old Col. Aaron Rose was keeping a store there.[11] He had some fine-looking daughters. We went on to Round Prairie where old Col. Burnett kept a hotel.[12] (Myself and brother and his wife afterwards rented the Burnett House, and I was appointed Post Master of Round Prairie Postoffice in 1859.) Col. Burnett had a fine daughter, Molly, too.

Next day we passed old Lazarus Wright's on Myrtle Creek,[13] and we come to some fine land called Missouri Bottom,[14] between Myrtle Creek and the South Umpqua, where Adams and Weaver lived, where Warty, the Grave Creek chief's son was hung, and we got home all right and sound, although I forgot that we camped out so much in the old camping places that we all got lousy, and in coming home on the road in the warm weather, Glasgow and Hawkins would frequently while riding along, pull off their shirts and catch lice and throw them away. One morning in camp on the Cal[a]pooy[a] Creek, Glasgow caught one off of him and Hawkins caught one of his, and they took a hot frying pan, turned it bottom side up, made a mark and ran one louse against the other for the drinks. Glasgow's was black, or the Indian, and Hawkins was white, but the Indian could outrun the white.

We were all bachelors; so was Tommy Smith, the partner of Glasgow's. In fact, there were but few white females in that part of Oregon in 1852. But there used to be a great many Indians on Cow Creek and on the Umpqua River, and they brought fish, deer and elk and bear meat to sell to the whites, and a good many loose squaws would come around the ferry to beg and trade, and they liked whiskey whenever they could get it. Glasgow and Smith sold considerable to the settlers and miners, and made lots of money....

The Indians of Oregon, Northern California, Washington Territory and Idaho and Montana have learned a language or jargon from the Hudson Bay Company Employees. It is composed of Chinook Indian, French and some Spanish, and partly made up of English and most all Indians had learned part of it, to trade with the Hudson Bay traders, who are French, Scotch and English, and many of the men, mostly French, had married squaws, and many Scotch had done the same, and some few English. All the trappers and hunters of the Hudson Bay company had a certain interest in the company and drew rations of provisions and a

---

[11] Rose founded the city in the fall of 1851. He came west from Michigan in that year. Born June 20, 1812, he died March 11, 1899. The town was first called Deer Creek, being at the junction of Deer Creek and Umpqua River.

[12] James D. Burnett was born on March 12, 1822, in Blunt County, Tennessee. He reached Oregon in 1850 and settled in Round Prairie two years later.

[13] Wright bought the property in 1852 from James B. Weaver who first settled there the previous year.

[14] A sort of valley situated a half mile from Myrtle Creek, it derives its name from the home state of its first 1851 settlers, the Weaver and Adams families.

certain cut of dividend at their trading posts in Oregon, Washington Territory and British Columbia. The company had a lease for 100 years of part of Oregon and Washington Territory and the lease expired the summer of 1858 or 1859, and the United States Government would not extend their lease, for it would interfere with the American Fur Company which was chartered in the United States and which was composed of American citizens.[15] The settlers of Oregon and all the Pacific States had many of them learned this jargon or Chinook talk or tongue and all the young folks tried to speak it as a secret way of sly speaking because so few could speak it, and us boys in the Oregon mines used to speak it and it caused a great deal of fun, for if you did not sound certain letters right it would give it a vulgar meaning in English, and many would do so a-purpose. I became quite a proficient at it and we talked it to all the Indians around us who had learned it. (Captain Hill[s] spoke it to the Modocs at the Pitt River Valley when we crossed the plains a year before.)

After my trip to the Willamette I found a gulch up a small creek at the mouth of which lived a family named Flannery.[16] They had come from the Willamette Valley and taken up a homestead claim. They had some 2 or 3 nice girls and a lot of cattle, and they used to kill and sell fat meat of their steers.

We built a cabin and mined and made 2 or 3 or 4 dollars a day in nice coarse gold. We bought all our beef at Flannery's. About in October some men had crossed the plains in 1852, the northern route, and were on their way to the mines at Althouse[17] and Sailor Diggin[g]s[18] discovered [in] the fall of 185[2]. The winter coming on, they concluded to prospect and mine a little on the creek where I was. I showed them some good ground and they took up some mining claims and built them a cabin. Their names were Alex Fuller, Jefferson Howell, two brothers Haskins, and a man named Simons or Simmons [Simonds], all from Iowa, and a young man named George Fetterman from V[e]nango County, Pen[nsy]lvania. He went to work with me on my claim and lived in our cabin where my brother would sometimes stay with us. We all mined until the last of November.

[15] Hudson's Bay Company's Crown charter expired in 1858. Its rights in Oregon and Washington were safeguarded by the 1846 Treaty between the United States and Great Britain. The Company failed to get a renewal of its charter. In 1869, through negotiations, what property it had left in Oregon and Washington was purchased by the U. S. government. Mary A. Gray, "Settlement of the Claims in Washington of the Hudson's Bay Company . . . ," *Washington Historical Quarterly*, XXI (1930), 95–102.

[16] More likely this was H. B. Flournoy who settled in Cole's Valley in late 1852.

[17] Named for Phillip Althouse who discovered gold in the vicinity in 1852. Situated in Josephine County, the town is several miles north of the California line. Reinhart consistently uses "Aulthouse." Perhaps this was the original spelling; "Althouse" being a result of usage. Shasta *Courier*, May 14, 1853. In the June 4 issue, the paper reported: "Both forks of our stream have now bared their beds to the gold seekers and they are reaping a rich harvest."

[18] Subsequently renamed Waldo, it became the county seat in 1856.

Charley took a couple of horses and helped a man namel Abel George to move his family and stock of cattle and horses to Rogue River Valley, 3 or 4 miles from Jacksonville. He came [down] from the Willamette Valley and bought the place he was going to in 1851, and he had improved the place and next spring he was offered $6000 for the half-section (320 acres) but he would not sell. My brother got so much per day to help with his two horses.

It was a hard winter at Jacksonville and all the southern Oregon mines and many miners had to lay over their claims and go down to the Willamette Valley to winter, for the snow was too deep for work, and no provision to be bought at a living price. At Jacksonville flour got so scarce that it sold, the winter of 1852, for one dollar or a dollar and a half per pound. Potatoes 80 cents to a dollar a pound. Salt, none to be had. Apler and McKenny, a merchant, had a keg of very salty butter, too strong and old to eat. He rendered or melted the butter and the salt settled at the bottom; he sold [it] for three dollars per ounce to the miners who had no flour and had to eat poor deer meat (caught in the deep snow) without bread, pepper or salt.

The Hotell at Jacksonville sold meals for $1.50 to $2 per meal and kept a doorkeeper at the dining room door to collect in advance as they went in to eat, and those who had no money were clubbed away from the door. Lots of men only had a square meal once or twice a week, and lived on lean deer meat straight the balance of the time. We in Cow Creek Valley had to live on beef alone for several weeks until flour, meal, and potatoes could be had. We eat about 4 to 5 lbs per day to each man. We would boil meat for "bread" and fry the fattest to eat with the lean (or "bread"). We paid from 10 to 12½ cts per lb. in the mines at Jacksonville, Althouse and Sailor Diggings. Good beef was from thirty to fifty cents per pound, but we had salt with our meat cheap. Charley, when he came back from Abel George's trip to Rogue River Valley said he thought if they did not get provision, there would be great suffering during the winter. . . .

We used to have lots of fun with the Iowa boys, and as the winter advanced there were a great many miners passed through our valley to the Willamette Valley to stay till spring and the waters got lower so that they could work some claims they had taken up on Althouse Creek and on the head of the Illinois River where there had been some new and rich digging discovered by the Althouse brothers; and the Iowa boys stopping with us on our gulch thought they might take up some vacant claims, or buy cheaper than they could in the spring. So after a while they got ready and started, promising to write to us if there was anything favorable and we could come out there too. There were five of them, two Haskin brothers, Simonds, Alex Fuller, and Jeff Howell. George Fetterman continued work with me, and as the weather got too cold to work, we would go and stay with Charley on our homestead claims.

One day while stopping with Charley, old William Riddle and his son-in-law Nic[h]ols and some others staying with Riddle came down to our house and said they had heard that some one had been traducing his two daughters [and told] as to how they had misbehaved on their trip crossing the plains and said they thought my brother Charley was the one that had traduced them. And if they should find further evidence, they were going to punish him some way. I asked my brother about it and he knew nothing about it. So I thought I knew what they were trying to do. Old Riddle was owing Charley some money for his yoke of cattle and work, and they got Nic[h]ols to jump his claim while he was keeping the bakery at Jacksonville the spring before, and now they wanted to scare him so he could not claim what Riddle owed him, and maybe get him to leave his claim again, so that the man who was wanting to marry Riddle's oldest daughter, the widow Chapman, could get Charley's claim. So I just got up and told the whole crowd to get out of the house, that they could not come around trying to bully me, and I made a break for my shotgun, and they concluded to get out. I was just mad enough to give the crowd a few shots for their cowardly way of trying to beat Charley out of his money and land claims.

We found that Ed Northcut and his two brothers had been joking Nic[h]ols about Riddle's daughters, and in fun told him he had heard Charley speak of the girls, and when I saw the Northcut brothers at their ranch in Democrat Gulch, near Althouse Creek,[19] and told them, they said they only were joking Nic[h]ols, for they were all packing together, and happened to camp close by Nic[h]ols.

After working in the gulch until we got tired I come to the conclusion to go to Althouse Creek and take George Fetterman along and see if we could not prospect and take up some mining claims, and leave Charley my brother to hold our two homesteads. So we were to take one horse, saddled, and tie on our blankets and some little jerked meat—we had no flour.

We found a fat calf near our cabin. It had strayed from the herd. I threw a rope around his neck, George knocked it down with an ax and Charley cut its throat. We skinned it and took its hide, head and offal and threw them into a deep prospect hole and threw some dirt on it. We cut it all up in strips that night and jerked or smoked it and dried it so as it would keep and we took a lot of it along with us. We took a pick,

---

[19] E. J. Northcut, A. G. Walling, and a man named Bell settled the region in 1852. They formed a joint partnership, Walling and Company, supplying the miners. Reinhart confuses Northcut's associates, calling them two brothers. Walling, *Southern Oregon*, p. 454.

shovel and pan, and a little ax, and would ride in turn, and we could make splendid headway. The streams were still high and it took us some 5 or 6 days to get to the Northcut ranch and store on Democrat Creek, 3 or 4 miles from Brown[town][20] on Althouse Creek. Charley and I were acquainted with the boys in Jacksonville and Umpqua Valley.

Ed said I had better try to bring my horse back from Althouse Creek and he would keep him for $3 per month. We stayed all night at Northcut's; next morning we went to Althouse Creek and looked around until afternoon. I took my horse to the ranch and bought 7 lbs of flour at 80 cts per lb. and went back to Althouse by way of the creek, and noticed as I went along for favorable places to prospect. I found two places I liked the looks of, and when I got to where I had left George Fetterman and our camping outfit, he told me he had some bank claim in view to try and he could get it if it suited us at $65. So we concluded to try it with a rocker next morning.

That night we got leave to stop in a miner's cabin with Dr. Pitts and Bob Dudrick, a gambler who was half Cherokee Indian. In Brown[town] at night George and I went to a large store, and a man named Barnes and a partner in whipsawing lumber wanted George and I to play them a four-hand game of Euka or Yuka [Euchre], for a pound of coffee, or $1.50 worth, whatever we wished to get in goods or groceries per game.

I had but three or four dollars left, and George not a cent, but I was satisfied that Barnes and his partner played by signs, and I could post George to beat them, by the signs I could learn him. So I took George and spoke to him a while how to play and we went in and played and we beat them four games in succession at their own game. They became confused and did not understand each other, which made it worse for them than if they had played on the square. So we took our goods and never lost a game.

Next morning we went to prospect the bank claim George wished to buy. We took a rocker and rocked out over sixty buckets of what we thought the best dirt, but it would not pay. I told George he might buy if he wished to, but I would not. I would go and prospect the two favorite places I saw down the creek as I came up, the evening before. So George said he would go with me. We got to the first place I had picked, on the head of a low and part high bar. We crossed over to it and found two or three old prospect holes there, but I liked the low bar best, and . . . I commenced [to dig]. After getting down two or three feet, I washed out a pan of dirt and got a good prospect of coarse gold. The creek was

---

[20] Reinhart mistakenly calls the town Brownsville in his manuscript. The latter was a city established in 1853 in Linn County. Browntown no longer exists. It was named after Henry H. "Webfoot" Brown. *Ibid.*, p. 455. Brown later moved to Yreka, California, and became, with J. Tyson, publisher of the Yreka *Union* in 1858. J. Roy Jones, *Saddle Bags In Siskiyou*, p. 73.

quite high, it being about the last of February or the first of March (1853), the highest time for high water of the year. We kept bailing out the water and digging down to about three feet and a half. It was so good a prospect that I concluded to stake off our claims.

Just as I had passed out our last pan of dirt, a Chilian named Emmanuel and a partner of Barnes, the whipsawyer, came along and saw my prospect. I used to know him on Humbug Creek, California. He commenced to sink a prospect hole a short distance down the bar below ours. George took his claim fifty yards from our prospect hole up the bar, and I wanted to take most of his claim down the bar and mine just below his, but Emmanuel had commenced his not over sixty yards below ours, and George was satisfied to take his above and mine fifty yards from our hole down the bar, fronting the creek, and Emmanuel took three claims of 50 yds apiece from my claim down for himself, Barnes and Braziel their partner. I then dug down some lower in our hole and washed out the dirt. George dug and I panned out some four or five dollars in a short time and quit work for the evening. It was not over five hours from the time of starting to dig our hole till we had located our claims and quit for the evening.

So the next day I borrowed a rocker and started a larger hole near the other, and we made something over $12, and we were well satisfied with our claims. The second day we took a walk down the bar to see what it looked like at the lower end, and to my great surprise, found two log cabins and some five men working the lower part of the bar and doing quite well. They had taken up five bar claims and five creek claims extending up the creek fronting on our bar claims. They had located their claims the fall or summer of 1852 and were making money. The men's names were Dusenberg, Beerup, and Nat Childs, and two others whose names I do not now remember. The bar was named the Dusenberg Bar after the discoverer Moses Dusenberg, who I found in after years used to live close to us in Lake County, Illinois. He had left his family on his farm and went to California in 1850, and rented his farm to a German family named Schertz and one of the boys, John Schertz, went to California the same year, 1850. (I got acquainted with him in 1856 on Indian Creek, California, and in 1857 and 1858 worked with him as my partner on Sucker Creek, Josephine County, Oregon.)

The next partner of Dusenberg was Mr. [Daniel] Beerup. He was one of the men we saw at Lytton's in Iowa the winter of 1850, when he advised us to come to California. He had a son about ten or eleven years old, called Charley (a spoilt boy). Dusenberg was off in California and I got a chance to stop and sleep nights in Beerup's cabin, while we were building a cabin on our own claim on the hillside back of the bar. We had a couple of boys helping us build our cabin; they had just come from the Willamette Valley. One's name was James Summers, the other William

Fo[u]ntain (or Dancing Bill). They were prospecting for claims and helped us with the large logs in our cabin, and when completed, stopped with us quite a while.

About the fourth day of our work on the claim, the weather being very cold, and the water was from the snow in the mountains and as cold as ice, we had only commenced cutting logs for our cabin, and our shoes were in pieces and we could not work long in the ice-cold water, for we had to bail out every time we commenced work and keep bailing while we were at work. On that day a pack train got into Brown[town] with boots and picks, shovels and axes and groceries. The day was cold and blustery but we concluded about noon to go to work and see if we could not make enough to buy us a pair of boots each, at $16 a pair. . . .

So we went to work with our almost naked feet and bailed out the cold water, for the hole would fill up to the level of the creek when we quit work or quit bailing the water out. We worked until it got so cold to our feet that we could not stand it longer, and about two or three o'clock washed up and took our gold over to Barnes' cabin acrost the creek to weigh it, and found we had made $88 in beautiful heavy coarse gold, good work for two of us in two or three hours.

So we went up to Brown[town], about a half mile above, and bought us each a pair of boots and warm new woolen socks, a sack of flour, a shovel, pick and ax, and some groceries, all for a few hours work. We were in the best of spirits and we got Jim Summers and Bill Fo[u]ntain to help us with our cabin and let the work on the claim wait till the high water run down and we could work to better advantage.

One day we again went to work, it being about our 6th day's work in the claim. We worked the biggest part of the day in a crevasse and at night weighed our day's work and found we had made $154 nice coarse bright gold. We kept our gold in a pan. I done the rocking and carrying the dirt in buckets; George Fetterman worked in the hole digging the dirt, and shoveled it into my two buckets. Sometimes I would have to help bail out the hole, it came in so very fast in the crevasse, which was very rich, and the water bothered us a great deal. We got from $8 to $20 to the bucket. One piece in the pan was as round as a bullet and must have weighed some $7 or $8. Emmanuel came and looked in the pan, and I think the round bullet was still there when George Fetterman had the pan and panned out some of the sand that was in the gold pan where we kept our day's work in—and that was the last I ever saw of the round, bullet-shaped piece of gold. George tried to lay the loss of it to Emmanuel, but he was a good honest boy, and I afterward became satisfied that George Fetterman, my own partner, stole it. (He was already half-owner, and only stole the other half from me.)

After a few weeks my brother Charley came from the Umpqua Valley (Cow Creek) and he had been stopping at Mr. Yo[k]um's Ferry on the

South Umpqua, just below where Mr. [S. B.] Briggs and Sons had put in a bridge acrost the South Umpqua River, all the latter part of the winter and the commencement of spring.[21] He said some stranger had come into the Cow Creek Valley, and married Mr. William Riddle's oldest daughter, the widow Art[inec]ia Chapman, and had jumped my claim (my homestead). My brother could not hold it for me, for I was gone, and they claimed it was abandoned by me. So he sold his own house and left his claim too. He said he did not care for his after mine was jumped, for mine was mostly prairie and his mostly timber. Together they would make a good half-section (320 acres) of land, a good farm, but one from the other separated was not so good. So he came back to Althouse Creek where I was.

My old neighbors and settlers in Cow Creek Valley wanted me to come back and they would run the man off that had jumped my claim. (Many of them would like to have had the excuse for doing so, as they were down on him for getting the Widow Chapman from right under their nose, for several had been hanging around her too long for them to ask her and this stranger just come in and jumped my claim and the widow too. It made some of them quite wrathy and they would like me to come back and claim my claim and they would have protected me in my rights and run him to hell, if necessary, to get even with him. But the settlers did not know that I was not of age (21) and could not have held it by law if I would wish to do so. But I would not then have left my mining claim on the Dusenberg Bar on Althouse Creek for the whole of the farms in the Umpqua Valley. At that time I thought I could make all the money I would want and go back to our old home in Lake County, Illinois.

So we let our land claims go, for I would not be of age until Dec. 31, 1853, which was about nine months more. So Charley looked around town (Brown[town]) and came to the conclusion that he wanted to start a bakery and saloon in Brown[town]. So he and I bought two bank claims, or mining claims, which made the two lots we wanted to build our bakery on for $80, and we hired some help and put up a large house and bakery, built a bake oven and two chimneys in the house, and had them nearly completed and the house ready to open up the next week, when Webfoot Brown, the storekeeper right acrost the street, came and claimed our lots as building lots and produced a contract he had made with a carpenter to build him a store on the lots. We contended we bought the lots and mining claims of the miners owning the ground, so the owners we bough[t] of had to see about the title of the grounds, and they agreed to leave it to an arbitration of three men.

Now Brown had a man of family living and running his store and doing business for him, named Grimes, from Iowa. And knowing as soon

[21] Samuel and "Sandy" Yokum were members of Reinhart's emigrant train as were Samuel B. Briggs and his family.

as we started our bakery and saloon, he would be damaged in his trade, . . . he said we would have to move out [of] our buildings or pay him $250 for the lots our buildings occupied. Now our buildings and oven and chimneys and bakery could not be rebuilt for that amount, and we would be so much longer until we could open up, for now we could open in a few days more if we got our difficulties with Brown settled.

The miners got a man to arbitrate for them, and Brown got a man named Chapman, who was keeping a restaurant at the upper part of town, and his wife baked pies, and our starting up interfered with him, and another man kept grocery store in the Round Tent, name of [George E.] Briggs; he was a personal friend of Brown's, and we had nothing to do about it. We left it to the miners to defend our title for us, but we could not recover any damage from them, and it was decided in Brown's favor. It was the grossest kind of injustice to us, for we bought and done our building in good faith and thought our lots all right, but the prejudice of the pie baker and the restaurant keeper, they decided against us to keep from injuring their business. So Brown give us so many days to move our buildings or pay him $250 for the ground, which in the long run we found was the quickest and cheapest way to open up our business. The miners we could do nothing with, for they had nothing, but we owned the mining claims the buildings stood on and we had to do some work on them every three days to hold them from being jumped as mining claims. We went to a great deal of expense to build our houses, for we had to get our timbers on the hill and carry them on our backs and shoulders to the place. We had to buy our clapboards and siding and our lumber cost us 25 cts per foot, and it took more than 2000 feet for flooring, counters, shelves, doors, window-frames and paneling around the sides and in our bakery.

Bill Fo[u]ntain and Jim Summers and Abraham Niecely, myself and Charley were all busy when I could get off from the mining claim. When it was too wet to work we could work in the saloon, sewing lining and tacking up ceiling. Our lining was calico or curtain chintz, only 16 inches wide, and cost us 60 cts per yard.[22] Our six decanters cost us $10 apiece, common Rib such as now can be got for two or three dollars per dozen. Our tumblers were $1 apiece. Whiskey and brandy and wine from $6 to $10 per gallon. But then we had no license to pay and sold our liquors at 25 cts per drink.

So we had to pay Henry (or Webfoot Brown as they called him) $250, and we worked the harder to open our Bakery and Saloon as soon as possible. One day it looked like it would rain. We were all living in our cabin on the Dusenberg claim. I told George if we did not go to work in the claim I would go uptown and help Charley as much as I could, so as

[22] "Our saloon lining was the bed curtain chintz with Peacocks, roses and bright flowers and monkeys." *Reinhart Manuscript.*

to open our saloon the sooner. (Charley had been wanting to take in George Fetterman into our bakery and saloon, but I opposed it for I did not think he was honest.) So Charley and I, with two others helping Charley, went uptown and went to work sewing the lining for the ceiling, some time after or about noon. I did not know but George was somewhere about town, when someone came up from the claim and told me George had gone to work. He had not said a word about it, but slipped down and went to work.

We had a place already stript off. I went down and asked him why he did not let me know that he was a-going to work. He said he thought I was busy and did not wish to disturb me. He kept what he made that day, which was not very honest in him, for I had helped strip the ground and that was the hardest part of the work. One Saturday afternoon we stripped off a small place about 4 by five feet, and on Sunday morning George proposed to go to work and clean up that small place. I did not want to give him another excuse to work by himself, so we went to work and cleaned it up in three or four hours, and we got $98 for breaking the Sabbath! The claim was very rich. If we had known more about mining, as we did in a few years after, we could have worked the same ground to a better advantage and made from $300 to $400 per day, but we would have sooner worked it out....

My brother Charley on his way from the Umpqua River Valley stopped overnight at T[w]ogood's and [Harkness] on Gr[a]ve Creek, keeping a hotel,[23] heard someone speak of two young fellows striking it quite rich the second or third day after coming on the creek, and he said it just struck him that it was us two, only he heard that they were both last year's emigrants (George only being one)....

We opened our Bar and Bakery the 10th day of April, 1853, on Charley's birthday. We had two violinists to play for us. We paid them $4 a day apiece and board. One name was Dick, or Richard, Jacobs, and the other James Cranston.

We kept open day and night and Sundays and took in from $80 to $200 per day of 24 hours. We sold drinks at 25 cts, cigars 25 cts, pies $1, or $8 per dozen to wholesalers to retail at $1. Cider 25 cts, cards $1.50 per deck. The day we opened our Bakery, Brown's occupation as a bakery and saloon was gone. We did not take in George [Fetterman as a partner] until the 17th of April, and he could help me tend bar nights and Sundays, when he would not be at the claim to work. We hired two men to help George mine; their names were Bill Fountain and Ab Niecely. Every Saturday night George was to turn in to me the gold taken out

---

[23] James Twogood and his partner, McDonough Harkness, called their place (originally built by a man named Bates) "Leland or Grave Creek House." The former was the first official name for the creek, but never became public vogue, the latter being preferred. Walling, *Southern Oregon*, p. 462.

during the week and settle up our hired help and the expense of the claims. In a few weeks George failed to make his customary returns or pay in the proceeds of the claim, but I thought maybe they had not taken out much and he would carry it over to the next week.

One day a sleight-of-hand performer, or juggler, came to Brown[town] and give some entertainments. One day he came to our saloon and weighed some nice gold dust. He said he had bought it. It weighed about $8 and was very nice and smooth. I asked him where he got it, that it looked like the Dusenberg Bar gold.

Charley Beerup, son of the old man in the claim below Barnes happened to be there, and he said yes, that it come out of our claim, that Belknap the performer was acquainted with George Fetterman and he came along by the claim. George had the gold in the pan, and Belknap begged him to sell him the gold and gave George a $5 gold piece for it. I did not say anything to George about it until the next Saturday when he gave in his amount dug for the two last weeks. He said he did not get much the week before, but said nothing about the gold he gave to Belknap until I saw he was inclined to steal it. I then asked him if he had not sold some gold to Belknap. He colored up and said yes, he had give John Belknap a few dollars in some specimens, and Belknap presented him with a $5 gold piece, but he thought that there was not over $2 worth, and that he need not give account of the small amount for he had got more than double the worth of it. I told him I had weighed the gold and there was over $8 of it. He was so surprised at the amount, but still did not propose to pay in the $5, so we let it go and concluded to watch him. But I was satisfied he would steal, after his mean actions already mentioned.

We took in a great deal of money and we credited considerable to miners for pies, drinks and cigars, but our expenses were high for flour was $60 to $75 a hundred, sugar 50 to 60 cts per pound, butter 80 to 90 cts per pound, cigars $80 to 120 per thousand, syrup $1 to $1.50 per qt, lemons $3 to $6 per dozen, eggs $1.50 to $2 per dozen, salt, dried apples, and all in proportion. Beef forty to fifty cents per pound, lard fifty to eighty cents per pound. Labor $4 per day and board; our music $8 per night; cards $4 to$16 per dozen; or whiskey and brandy and gin from $6 to $8 per gallon. French cordials and champagne, also port and claret and fancy drink were awful high, and had to be packed on mule trains from Portland, Oregon, and Scottsburg at the mouth of the Umpqua, and packing was worth from 12 to 30 per pound for freight alone.

In May there was some mining discoveries on the creeks running into the Illinois River. Many miners went to prospect there. James Downer, who crost the plains with us and lost his sack of clothes by the Indians on the Humboldt River, was at Brown[town]. He and his two partners, Robt. Shaw and John Whitt put up a Ball Alley and saloon near us in Brown-[town] and done a good business. Alex Fuller and the Haskin brothers with Jeff Howell all located mining claims above Brown[town] on the main Althouse Creek. They used to come lots of times to our saloon.

When a new saloon was opened, drinks and cigars were free at first night. We had some five or six rooms private for gambling, and we had several tables around the room for banking games, and in another end of one hall, card tables for miners to play cards for drinks, cigars, or pies and cakes. A quarter-pie was 25 cts, a whole one a dollar. They played seven-up or all 40 Yuka [Euchre], polka [poker], rounce, freeze-out, Pedro, and all kinds of games in cards Spanish or American, and there would be lots of gold won and lost. We hardly ever had any trouble with drunkards; when they became noisy we got rid of them and refused them more drinks.

Like many others I felt like wanting to take a trip and prospect on Briggs Creek on the Illinois River. I had been up so much nights I wanted a change. I had been in the House steady since we had opened, and I left George and Charley to run the saloon and bakery. I was gone about seven or eight days and did not find anything to pay over common wages, so when I got home I fixed up our books and found some of the merchants had not paid up for the pies they had sold from us.

One, named Dunlap & Co., we thought owed us some $65. I took their bill and found they said they had a bill of goods we had charged to us of over a hundred dollars. I was surprised and asked for the items and they made it out—a lot of fancy bowie knives and silk sashes and silk handkerchiefs, some pocket-knives and miners' belts. I knew of no such things and took the bill down to Charley and he did not know anything about them, so asked George Fetterman and he said he got them. I asked what he had done with them and he said they had been played off at freeze-out, polka [poker], and most had been lost and the balance was credited out on the books. I just went and settled the $95 we owed them instead of collecting the $65 due us as I thought.

[I then] went to all our customers and found we had run large bills while I was gone and Charley was too busy baking to be in the saloon much and had no idea how George was running the saloon. I had a settlement with our two men working in the claim on Dusenberg Bar, and they told me that they knew George was stealing from us when we first started our saloon and bakery. They had come up from the claim behind him and saw him pick out pieces of gold and put them into his pocket instead of bringing all of it to be weighed every night and put into the company purse of the mining claim. We had some words with George and accused him of it to his face and he did not dare deny it, so Charley and I come to the conclusion that he must buy or sell to us, for we did not mean to stand it longer. If he would not agree to a fair thing I would have him arrested and with the proof I could bring by Bill Fountain and Abraham Niecely he would stand a chance of being lynched by the miners. So he came to the conclusion to sell out his interest to us, and we paid him some $250 in gold and divided up what cash there was on hand, and give him some $250 in book accounts to collect for himself and got rid of him.

He afterward changed some $600 worth of gold dust for gold coin and went up the creek and bought into another mining claim. He was a notorious thief and I will show later how he was caught at it and how he suffered for it.

Some boys some time in June wanted me to go with them on a little prospecting expedition and Charley got someone to take his place baking and he tended the saloon. I was to be back in 8 or 10 days and our party was composed of John Denny, Al Lee, Robert Tripp, myself and one or two others. We first went up the left-hand fork of Althouse Creek and prospected, not finding anything to pay. We went over the divide north and struck down on the right hand fork of Sucker Creek; there was no one on this fork, but some 4 to 5 miles below on the main Sucker Creek there [were] some few men at work prospecting the banks.

We went down to the mouth of the right hand fork where it came into the main creek and found a place where we could turn four or five hundred yards of the creek by cutting a shallow ditch through a low bar and letting the water from the ditch come out into the main Sucker Creek. So we worked three or four days hard to turn the creek and then we prospected the bed of the stream and found it very deep, and large boulders, and we did not get much of a prospect. So we concluded to go right opposite of the mouth of the right hand fork, where there was a large bar on the main Sucker Creek, and a low bar favorable if there was any gold in the fork we had turned, and . . . sinking a hole there would be the same as prospecting both creeks, but we found it a difficult job.

We started our hole not expecting to go over 8 or 10 feet deep, but found so many large rocks we could not get out, we had to make it smaller as we went down, and we only got down in one corner about 18 feet and got some dirt and kept prospecting but we got nothing but fine gold and not enough dirt and too much rock. And the creeks were still too high to work to any advantage, and at last we concluded to lay our claims over until the water was lower. We had each taken up double claims, or one for discovery each, and we had been gone nearly two weeks from Althouse and I wanted to go home and if my partners in the claims wished to work on, I would furnish my share of expenses and labor. But we all went back to Althouse awhile. Sucker Creek was a great deal larger stream than Althouse Creek, with only a high mountain between, but it was about four miles from Althouse Creek to the top, and awful steep mountain at that. It took a man's breath to go up, and then down to Sucker Creek it was short but steep. It emptied into the Illinois below

*Gold Rush Days* 69

Althouse Creek a few miles with some good ranches and farms after it got into the valley.

When I got back to Althouse I concluded not to prospect any more just then. I had got tired of camping out and hard work. I liked keeping bar best. One of our violin players used to get on some pretty big highs, and we could not depend on him, so we hired a new one named [Ch]ester Eastman, a Californian. He was a splendid player and him and Jim Cranston played well together. Dick Jacobs, the one that left, was a son-in-law of old General Palmer, and he was Indian Superintendent on the Pacific coast.[24] Dick got on a spree and left his wife to go to the mines. I will speak of him again that fall and again in 1861-65 . . . at Walla Walla, Washington Territory, where he had a large grocery store with his brother Cyrus.

In June, 1853, old General Joseph Lane ran for Congress against Gen. [A. A.] Skinner, who used to be Indian sup. Agent in Rogue River Valley in 1852.[25] And they were both stumping Oregon. They both met at Althouse and had a red-hot time politically, and the miners got to drinking considerable. The speaking was in the Ball Alley saloon kept by Downer, Whi[tt] and Shaw. The miners got to arguing, as well as the two aspirants for office, and some got overbearing, one in particular, a newcomer named Judge Young, from Arkansa[s].

He and some seven or eight old California miners had come to Althouse a week or two before from Trinity River, Northern California, and they had some fine Trinity gold with them. It was higher-colored and finer than the Althouse Creek dust, [so] that unexperienced merchants or saloon-keepers did not like to take it, and change it for coin. Judge Young was a large robust man of six feet and two or three inches, and somewhat overbearing in his opinions. Another man named Ben Sykes, a Boston engineer,[26] got to arguing politics, and the lie passed. Sykes was as tall as Judge Young, but spare and muscular, and they two come to blows. The house was just as full of men as they could stand up, and when the fight commenced, the crowd from the street rushed in. They, the fighters, could not stand up straight for the crowd. And no doubt friends on both sides tripped up the men, and some say even outsiders kicked them while down, so they could not get up. It was an awful rough and tumble fight and after a long time it was stopped by one of the men. Judge Young became insensible, and Sykes was taken away from him. He was not much hurt, but Judge Young had one of his ears cut or bit off and his face fearfully bruised up. Some say he done it on the edge of the Ball Alley, others

[24] Joel Palmer was superintendent for Oregon, 1853-1856.

[25] In the territorial delegate election of 1853, Lane handily won over Skinner, 4516 to 2951. George H. Williams, "Political History of Oregon From 1852 to 1865," *Oregon Historical Quarterly*, II (1901), 4.

[26] Possibly the same Sykes credited with the joint gold discovery at Gold Hill, Jackson County. (Chapter II, *note* 15, *ante*.)

that Sykes bit it off. During the fight old Joe Lane tried to stop it but could not, so got outside and come down to our bakery in his shirt sleeves with his plug hat in his hands, wiping the sweat and dust off his face with his handkerchief. He said it was the hottest damn place he ever saw and the roughest fight too. Someone brought down a piece of what looked like raw dirty meat, and it was the whole rim of Judge Young's ear; it was laying around the street for quite a while.

Sykes had to skip, for someone was going to have him arrested for maiming Judge Young, and he kept away from Althouse Creek for a while until it blew over. We had no city marshal, only a justice and constable, and there was no law or order at that time.[27]

Althouse Creek proved to be a rich stream, and I knew Dr. Savage at Humbug Creek in California; he and his company took up some creek claims and was already working the bar claims, and they used to take out $700 to $1000 a day, and did not work to good advantage, it being deep diggings and hard to work, and the water kept them from work a great deal.[28] There were several large nuggets taken out near his and on his claim. His bar was called Grass Flat, only half a mile above Brown[town], and it was the place where there was a little town called Grass Flat. Justice of the Peace Charles Walker, a lame man, kept a little store there. He was a tailor by trade, from Wisconsin.

Our town, Brown[town], had a population of about 800 to 1000, and some seven or eight saloons, three Dry Goods and Clothing, two bakeries, four or five restaurants and hotels, seven or eight groceries, one bowling alley, two or three butcher shops, and three or four blacksmith shops. Two Dance or Fancy houses, music in ours and the dance houses, sometimes in the bowling alley.

About the 20th of June Charley thought he had better take some men and go over to Sucker Creek, where some more discoveries had been made, some three miles above the lower crossing, and a town had been laid off and was being built up quite fast. Our business on Althouse Creek had slacked up a little, and there was a prospect for a good town on Sucker Creek, so he took Jim Cranston and Chester Eastman, Bill Fountain, and some other, and they got out the timber and shingles and clapboards and built a Saloon and Bakery and got it opened by the Fourth of July, 1853. They done a good business and James Cranston tended bar. When about completed Charley sold the half of it to a packer named Dupuy; he . . . run a large Pack-train of mules and took goods into Alt-

---

[27] Josephine County was not created until January, 1856. The first court was held in the fall with Judge Matthew P. Deady presiding. It was rather common practice for each community to provide some degree of legal protection for the common peace until established judicial procedures were instituted.

[28] Possibly James Savage. Born in 1833, he arrived in California in 1853 and soon after moved to Oregon. He died at Salem, March 19, 1917. *Transactions of the Oregon Pioneer Association, 1917*, p. 389.

house and Sucker Creek. He stocked up the saloon in liquors, cigars, can fruit, oysters, sugar, butter, flour, lard and everything; and some clothing. After running it as a saloon and bakery, Charley and Dupuy concluded to put up a bowling alley in connection with the saloon, so they got Bill Fountain and someone to furnish them in lumber for a double alley 98 ft. long building. The alley 72 ft. long, at 25 cts per foot or $250 per thousand. They whipsawed it out of sugar pine logs, and then had to kiln dry it so that it would not shrink after being laid. Mr. Lee, a carpenter who had been of our party when we had prospected the right hand fork, in May, done the work laying the alleys. I and Charley owned one half together and Dupuy owned the other half. They commenced the bowling alley in August, but the lumber all had to be whipsawed out by hand and then kiln dried, then dressed, and the building got ready to lay the alleys in. With but one man to dress all the alleys alone made slow work, then we had to send our order to San Francisco for the balls and ten-pins, all took up time.

In the meantime I have some experience of mine to relate: There was a man named Dr. Black, an old Pen[nsy]lvanian, he had bought some creek claims just at the upper part of the town bar, but he could not work them, for the company just below him had the oldest claims and they put in a canvas flume to work their claims, and they put in a dam at the head of their claim and it backed up the water most all over the creek claim, so he would have to wait until they worked out their claim before he could commence.

He used to be around my saloon a great deal. He was a stubborn conceited old codger, and tricky and dishonest, I afterward found. He used to play cards for drinks and sometimes if he found anyone he thought he could beat, would play for money, but not for high stakes, say only fifty cents to a dollar per game. There was a little dark half-Mexican, half-Indian who used to play cards more tricky and sharper than Dr. Black, and he was considerable of a card sharp, and ran and associated more with the gamblers than miners. He was cooking for some men on Grass Flat; one of the men he worked for was named Dutcher, the constable. Jimmy, as we called the little half-breed, came to my saloon and played, but he and Dr. Black in playing always wrangled so much over their cards that I told them to quit playing together if they could not play quietly.

So one evening they sat down to play together (I had not noticed them —I was busy around the saloon) when my attention was drawn to them by their commencing to wrangle over their cards again, and a lot of outsiders not in the game were attracted by their quarrel. They both drew pistols but were afraid to shoot. The crowd kept aggravating them by shouting to them to shoot, and making all kinds of sport of them, until I got tired of it and went to put them out of the house. I got Jim the half-breed to the door, and Dr. Black put his pistol to Jim's breast, but when he had raised his pistol the cap fell off and he did not know it until his pistol snapt. He turned to go back to get a cap; as he turned he ran against me, I was shoving him out. Jim was outside and fired past me, I being between both, and shot Dr. Black in the arm, breaking it above the elbow. Dr. Black fell, I caught him and laid him in a corner on some blankets until a doctor came and examined his arm and dressed it. He could not be moved to another house so I had to give him a room and bed for several months until he got able to get around. Jim gave himself up to Squire Walker, the justice of the peace; that night, he had an examination and was cleared. Both parties were armed and had their pistols drawn. It learned me a lesson, not to go between men about to shoot. For I was in the most danger, Jim shooting in from outside, and Dr. Black inside, me between trying to stop them, both afraid the other would shoot first and both excited—it was a wonder I did not get shot instead of Dr. Black.

That evening Dutcher the constable and I examined the pistol Dr. Black dropped and I took out three large bullets, three charges of powder, and thinking that all the powder was about out, put on a cap and snapped it—and it went off, and more than a double load in, and it kicked awful! If the cap had not fell off when Black raised it, no doubt the pistol would have exploded (and all of us wished it had) and killed both, for they were mean hounds. . . .

Now I will have to go back a few months to tell of some exciting times we had on Althouse Creek and Brown[town] in the last of April or fore part of May. One day a miner came in from Trinity River, Klamath County, California. He was acquainted with the party that came with Judge Young (who had the fight with Sykes at the bowling alley when Joe Lane and Gen. Skinner were stumping together). This Trinity miner got drunk and went to all the saloons and gambled and lost some money and paid it out of his purse. He came to my saloon in the afternoon and played Polka [poker], and Charley Miller, a miner, won some from him. He wanted me to weigh out four ounces of his dust and give him the coin and silver for it, but I was not posted on the Trinity gold dust, and it was so much finer and darker than the Althouse gold, I refused to take his gold and mix with mine and it was no benefit to me to take their dust. So the men left my saloon and went up to the bowling alley of James Downer, John Whi[tt] and Bob Shaw. They had some Spanish Monta Banking game going on in their saloon, and the Trinity miner commenced bucking at their game and he lost some, and weighed the gold at the bar, for they knew the kind of dust. When he weighed out some $50 of his gold at my place for Charley Miller (for I would not buy it) I judged that he had some $300 or $400 in his purse.

While playing at Monta a dealer named McCloud was dealing the game. [Also at the game was] a boy named McCoy, lately from San

Francisco, where his father was night-watchman or police, and the boy was on the nigh[t] watch with him. The boy stole something and he had to leave and he came here. He was a hard case, but he was a splendid ten-pin-roller. He saw the drunken miner's sack of gold and he and McCloud made up a plan to beat him or steal it from him, and McCoy and McCloud would both go down to Scottsburg, at the mouth of the Umpqua River on the coast, where there was a town just laid off, and they would build a ball alley and saloon with the money. That night the miner got very drunk and laid out all night in the street. In the morning he waked up in the street and found his purse of gold gone. He came to some men and told them that somebody had robbed him of $400 or $500, and the last he could remember he was in Downer, Whi[tt] and Shaw's bowling alley and saloon playing at Monta.

The miners thought that McCoy was around him a good deal and he might know something about it, so commenced looking for him, and happened to see him just going up the mountain, just on the road to Sucker Creek, four miles north. They took after him and brought him to town and accused him of stealing the man's purse of gold. The miners, Judge Young and his partner, scared the boy into telling all kinds of stories and proposed to hang him or make him tell. The miners organized a Vigilance committee and all was excitement. The boy, about 18 years old, was afraid they would hang him, but said he knew nothing of it, and he had left the drunken man at the McCloud Monta bank playing Monta, and saw no more of him until then, and he commenced to tell he thought McCloud must have got the purse, that they had planned to take it, and if it was gone, he thought McCloud must have got it. So the Vigilantes arrested McCloud. He said he knew nothing of it, that the man had left before he closed his bank, and that he had his purse with him the last he saw of him, and if it was gone, young McCoy must have taken it. They were both searched but nothing found.

All was red-hot with excitement. Jim Bean was constable for the Vigilantes; Dupuy, our packer partner, Jack Driscoll, Judge Young, Constable Dutcher, and Samuel Cowen and about thirty men well armed with rifles shotguns and pistols were rushing around making speeches against the gamblers, and took McCloud and McCoy and tied them to trees and whipped them. Each thought the other had the purse and both denied knowing anything about it, so they put the prisoners into a log cabin (both had been whipt until they were bloody and scarred all over their backs in deep cuts from the cat-o-nine-tails and switches in the hands of the strongest men until they were tired whipping, and McCloud fainted with pain, begging to be shot) and they put a strong guard, to see what the committee would do.

They got the boy to write to his mother in San Francisco stating that he was to be hung for stealing a purse of gold, and he bid his mother goodby, that he was innocent and was sorry he had been a bad boy in San Francisco, but he would die innocent of the stealing. The guard got him to write to his mother that he would be hung right away, then took him to a tree close by and put a rope around his neck and over the limb of the tree and raised him up and let him down again and asked him to tell, if that was all he had to write to his mother and father, that would be his last chance. So he finished the letter. They raised him up four or five times; the last time he was nearly strangled; when they let him down he fainted away. We got some water and brought him to. He said he knew nothing of the money, so the guard took him back to jail. The Vigilantes then read the letter to his mother; a good many believed, some did not.

While they were arguing what to do Jim Bean let McCloud go, not knowing what to do with him. McCloud was making his home at Cocoran's Ranch out at Democrat Gulch (the old Northcut ranch)[29]— he started to go out home and had gone some ten or fifteen minutes when the Vigilantes asked for McCloud. Jim Bean told them he had gone home, that he had let him go, not knowing whether they would want him any more. Jack Driscoll ran to catch up with McCloud. Just beyond my bakery and saloon he met Captain Bob Williams coming in from Cocoran's Ranch where he lived. Driscoll asked if he had met McCloud. Williams said yes, that he had told McCloud to go out home. Driscoll said he was going to bring him back. Williams said he should not. Driscoll started, and Williams brought his rifle down on him and told him to come back. Driscoll ran back into Brown's butcher shop (adjoining my bakery) and grabbed up an ax and went for Bob Williams with it. Brown caught the ax and Williams struck at Driscoll with his bowie knife. Somebody had caught Williams' rifle but the crowd rushed in and parted them, and the Vigilantes sent out Dutcher the Constable after McCloud the next morning and they tied him up again to a tree and whipped him some more, and McCoy too, until they did not know what to do.

Bob Williams came along the trail from the ranch and McCloud called to Williams for God's sake to shoot him and not let them whip him any more. The Vigilantes then thought a man by the name of Elmer James Roades had been around with McCoy, McCloud and the Monta banks when the drunken miner was playing, but Roades was present and told the crowd that he did not know anything of the lost gold, and if the crowd undertook to whip him he would die first, or if they did, he would kill every one, if it took him ten years to do it. So that with the opposition to the Vigilantes, and Roades' threats, kept him from being whipped. So McCloud and McCoy were told to leave the vicinity and Althouse Creek in 24 hours. Some few men that sympathized with McCoy gave him some money, and he left for parts unknown. McCloud went over to

---

[29] More frequently called "Walling's" Ranch. The sale was transacted in 1853.

*Gold Rush Days*

Sucker Creek, about 4 or 5 miles over the mountains, and staid awhile at our bakery and saloon, until he got well enough to travel (and he was innocent of the stealing).

When we started to build our bowling alley on Sucker Creek there were several hundred men at work close by, and by the time we got it completed there was quite a big excitement of the discoveries on Gold Beach at the mouth of Rogue River, and many left Althouse and Sucker Creek to go down on the Coast Beach to prospect there awhile. Jim Cranston, the fiddler, tended bar in Dupuy's place; he owned a half interest. Lee got the alley layed about the first of September; the balls and tenpins had been ordered at San Francisco a long time, but other orders were in ahead, so ours did not come to Althouse by a pack train until we had completed [building] and were waiting for the 30 balls and 20 pins. When they come in . . . it took 3 of us all day to carry them on our backs to Sucker Creek up that awful 4-mile uphill the whole four miles, and awful steep. We had to make a many a rest before we got to the top, and when we got over they commenced playing free until midnight, and we took in $80 between twelve at night and daylight. We got awful tired spelling each other setting up the tenpins. Several got tight, but the day we opened our alley there were not over 150 men left on Sucker Creek, and when we commenced building it there were three or four hundred and still rushing in from the surrounding country, with a good prospect for thousands of miners, but mining was mighty uncertain anyway.

Billy Clark, who used to be with us mining on Cow Creek, and lived with Yo[k]um at the ferry at the South Umpqua River, came to Althouse the spring of 1853 and had been working for Charley and myself at both our places, stayed and helped to build the bowling alley, and after set up pins all fall and winter.

Our bakery, saloon and ball alley cost us at least $1800 and our bakery on Althouse Creek with saloon and mining claims, creek and banks, about $3000 to build and buy, and in the fall we would have taken $1500 for all in cash.

James Cranston in the fall concluded to go to California and Chester Eastman went with him, which left Charley without music, but times were too dull to pay for it, and our partner in the ball alley and saloon on Sucker Creek got a man named John Whitt to take charge of his half interest in place of James Cranston. Dupuy was one of the leading Vigilantes in the McCoy-McCloud affair on Althouse, and he ran his pack train all fall and part of the winter to the different mines in Southern Oregon and Northern California, and when the winter set in rough, went into winter quarters down the Willamette Valley, where feed and grain was cheaper for his mule train.

(When I left off about George Fetterman, our old partner in mining and saloon and bakery, I had lost part of some notes of this book and did not get it in as I had it before, and then I found the lost part and now will

[add] some forgotten before.). . . . During June or July, a young man came to my saloon and asked me if I knew George Fetterman. I told him yes. Abraham Niecely who had worked for George and us was there. When he asked I had some one to attend to in the saloon, and left the young man with Niecely. He told him that George Fetterman was a partner of his in a bank claim a couple of miles above Brown[town] on the Althouse Creek, and that before George Fetterman had bought in with him . . . when he was alone he could take out coarse gold, but since George was working with him they did not get any more, and he mistrusted George, as he got to handle all the gold when they washed up, while the young man worked digging and shoveling in the dirt, and cooked while George cleaned up the sluices and panned out the gold. The man had a part-crippled hand. So when I came out to where they were talking together, they told me about it and wanted to fix up some way of catching George at his stealing. So Ab Niecely proposed to take a piece of coarse gold that we all could identify, and weigh and put a private mark on it with a knife, and we put the weight down in our books, and marked up against the outside of the bakery where we should know the exact weight. Ab Niecely told him how Fetterman had stolen from us in the claim and saloon. In a few days the young man came to Ab and me and told us that one day he took the marked piece of gold and put some clay on it and put it in the riffle of the sluice box to be sure it would not be lost, and when George washed up, sure enough the piece was gone. He waited a while before saying anything to George and when he saw that George would not show it, asked George what he had done with a piece of gold that he had himself put into the riffle of the sluices that day. George at first denied it, but when his partner proposed to search him, put his hand in his pocket and took it out and said he only done it for a joke. But the partner told him he did not relish his jokes and told him plainly he did not wish to work any more with a thief or man he could not trust. That he must either give or take, buy or sell out of the claim, and George agreed to sell for $200 his half, and he bought George out. George come down the creek and stayed till fall and winter at Democrat Gulch, and through some meanness had to leave there in the next spring, and I heard he went up the coast to Gold Beach and from there to Empire City north of the Coquille River, Oregon.

The young man in less than a week after buying out George struck it rich, and took out one piece of pure gold weighing over $600. He told me he was lucky that George was not in with him or he never would have seen the $600 piece, for George would have stole it before he could have got to see it. No doubt he was right, but I can imagine how George Fetterman felt when he heard of it, how he had missed it by his dishonesty, and it was a good punishment to his conscience if he had any. I doubt if he had.

# A PACIFIC RANCHER ON THE OREGON COAST

1853-1855

In the Fall times became very dull with me on Althouse, and the same with Charley on Sucker Creek, there were so many miners went to the new discoveries on the Pacific Coast Beach at the mouth of Rogue River and at the mouth of Sixes River and below Point Blanco, six miles north of Port Orford, and at the mouth of the Coquille River or Empire City.[1] The creek miners laid over their claims by recording them until they were workable next spring, when the water would permit of them being worked (We laid over those we bought of Doctor Black by recording them the same way). During the Fall I played cards a great deal at our saloon for drinks and money. I would have to stay up all night with the Polka [Poker] games to wait on them with drinks, cigars and cards and refreshments such as cove oysters, sardines, lobsters, pies, and cakes, and lunches, and I would have to make up the game when some would leave or all would have to quit the game and it would be to my interest to play to make money out of my goods. Some men would come in with their partners whom they had posted in signs to beat me and others, and I had a miner who could play with me, and we held well together and understood each other's play on the square as well as by signs if need be. His name was Duvall, and we won many games for drinks or cigars as well as money. Charley done the same on Sucker Creek; he opened up the alleys on the 16th of September and for about six weeks done very well, then it became an old thing, as all ball alleys become so in a small town where so few miners were left, and Charley got the Gold Beach fever and talked me into selling out one half of the ball alley (and we had bought out Dupuy a short time before). So we sold one half to John Whitt

---

[1] The various places mentioned are in Coos and Curry counties.

(That had tended to Dupuy's half before) and leased him the other half until May, 1854.

Charley and I mined a little and kept the bakery and saloon on Althouse until March, 1854; then we laid over our claims (creek) till June. Then Charley proposed to sell our house and lumber. We had paid $250 per thousand feet, and we sold at $60 per thousand, left the house and bakery to rent, and the tools and claims in care of Dr. Watkins, of the firm of Briggs and Watkins, Merchants at the Round Tent, Brown[town] (Althouse Creek, Brown[town], Josephine County, Oregon, 1854). They were to see to the claims and represent us in holding them until we came back when the season for mining commenced.

We bought three horses (they were high in those days), put on our bar fixtures, lining, and some stock in trade, and our blankets and provisions on pack saddles and struck out for Crescent City, and then to go to Gold Beach. We heard that the mountains were impassable but thought that by the time we would get there that the trail would be opened. We left over a thousand dollars worth of accounts on miners for pies, cigars and drinks to be collected with Dr. Watkins and Briggs. We got to Waldo (or Sailor Diggings) and stayed overnight. Next day we go to Col. Gates and Lafayette Gates ranch in Illinois Valley, close to the foot of the mountains.[2] Saw some men and they said the mountain was impassable to animals but men could go over on the crust of the snow. But there were six or eight pack trains going to open the trail next day. So we pushed on and went to a hotel on the mountains, as far as we could get for snow, and put our horses in the stable, at $1.50 per night each head, meals one dollar.

Instead of getting better it got worse, and we did not wish to turn back, expecting every day to get acrost, and more men came up from our side, and sometimes there would be twenty or thirty men of us with shovels shoveling snow from morning till night to dig a trail two or three feet wide and from three to fifteen feet deep. We were at Dr. Holton's Hotell over three weeks and paid out $125.[3] Got both our eyes sore and snow-blind. When at last a drove of beef cattle some 300 head, were drove through, which broke the trail, and some pack trains followed, and some men with shovels. It took us four days from Dr. Holton's to go to Smith's River. When we got to Crescent City (Del Nort[e] County, California) we were ragged and nearly blind with sore eyes from the sun and snow.

[2] Wallace and Lafayette Gates, in their early twenties, were living in Jackson County at this time. Reinhart apparently has confused them, since they later departed Oregon, with the Yamhill County Gates family, or with N. H. Gates, a prominent early figure living in The Dalles.

[3] Dr. David S. Holton, a pioneer physician, settled near Kerbyville in early 1853, on the banks of a creek which subsequently bore his name. He later was a member of the 1858 territorial legislature and served as state senator for Josephine County, 1860. A[lbert] G. Walling, *History of Southern Oregon* . . . , p. 508.

We put up with an old friend named Trask, who kept the No. 8 Hotell. We bought us some carpenter's tools and provisions and nails, and started for Gold Beach again in two or three days.

The trails up the coast and beach were very bad. Some places we had to go on the beach where the tide and breakers would wash our horses nearly away; sometimes would have to climb the hills out of reach of the breakers, and one of our horses fell off a high hill or cliff—we thought sure he was killed. The flour he was loaded with just rained in showers when he struck. He kept turning end over end six or eight times before he struck down on the beach. The last fall was all of twelve feet and he landed square on his back. He had on a pack saddle loaded with Flour, sugar, coffee, sheeting, nails and tools. When we got down he lay as still as if dead, and we thought he was. We took his saddle and ropes off—his loading had come off in his tumbling and we lost 100 pounds of flour. As soon as we took his saddle off he turned and jumped up and neighed to the other horses and commenced to nibble grass. He was all right—only his back was bruised from the saddle when he struck on the rock on the beach. He was blind of one eye; that was the cause of his fall.

At Whales' Head, a so-called hotell,[4] we heard very favorable news of the diggins and we hurried on till we got to the mouth of Rogue River and crossed on the ferry, and Prattsville, a small town, is but a half a mile away on Gold Beach.[5] We found a great many old friends there from Althouse and Sucker Creek and Jacksonville and different parts of Oregon and California. Ike Warwick and Peter O'Regan kept a store and a stock ranch and had lots of cattle, and many miners were doing well. Seavey and Collins were running a very rich mining claim (they had a store on Althouse at Brown[town], and so had Warwick and O'Regan). My brother and I were a sort of shy of mining, so concluded to go up the coast until we found a place and take up a ranch and keep a Travelers' Hotell and sell liquors and cigars and farm some and raise stock.

So after staying overnight at Prattsville, we started up the coast past Simmon[d]s' Bend, a rich place found by old Simmonds[6] (one of the Iowa boys or emigrants that came to me in the Gulch on Cow Creek, Douglas County, Oregon). With the Haskins, Alex Fuller and Jeff Howell, he came here in 1853 and discovered rich diggins in this end of the beach, made money and sold out for a big price, and had gone back to Iowa in the last winter. The Bend was named after him.

[4] Located several miles south of Gold Beach.
[5] Possibly named for F. H. Pratt who organized the first pack trains between Crescent City and Gold Beach. Prattsville probably was the first name for Gold Beach. Gold Beach was also early called Ellensburg (after the daughter of Captain William Tichenor, the settlement founder), but the former name is current today. *The Oregon Journal*, July 20, 21, 1927.
[6] Reinhart also spells the name Simons and Simmons. The correct spelling and identification are elusive.

Our next place was a little town called Elizabethtown just above Simmonds' Bend. From there we next passed Yreka Creek (from a tribe of the Yreka Indians who had a village there). When the tide is up or in the creek is very hard to ford. We next got to a house or hotell called the Three Sisters, so called after three large rocks in the sea that can be seen many miles off.[7] The Hotell was kept by John O'Brien, a Californian, and his partner, a prize fighter and boatman named Hough O'N[iel]l.

They got their goods from Port Orford in boats. A few weeks before we got there Hough O'N[iel]l and two others, one named King, got into the breakers with their boat, and King got drowned. O'N[iel]l was a good swimmer and got an oar and would dive when the waves would come towards him. The boat was broke up and the goods all got lost. It happened at a place a short distance above Simmonds' Bend, or Elizabethtown; we saw the pieces of boat as we came along, before we got to Yreka Creek.

We went 1½ miles further and found a creek called Mussel Creek, after a small tribe of Indians; then three miles above the Three Sisters, we got to a small creek and a nice bench of land right on the main trail from Gold Beach or Prattsville to Port Orford. The creek was heavy timbered, the further back the wider the timber land. It was yellow and red Cedar, White and Yellow Pine and Fir, some as fine as I ever saw; and the land lay in tables, some very rich spots. It was about a half mile down to the beach, very rocky and steep.

We built a house within twenty feet of the trail. We cut posts twelve feet long, five to eight inches through, put two feet in the ground and split out siding four feet long out of cedar or pine, nice, thin and straight; our roof of three feet clapboards to eight inches lap. (This was in May 1854.)

While getting out boards and building timbers we had a tent and kept our goods in and would be gone half a day at a time. While gone, Indians would come to the tent with fish, mussels, elk meat or bear or deer or duck's eggs or gulls' eggs to trade, and we never had a thing taken by them. I know as a general thing the whites would have taken things if they were placed in the same situation. The Indians were friendly and were well pleased for us to take up land and cultivate, and never offered to molest us. We did not have to pay any license to sell liquors or cigars and we had a large profit.

My nearest neighbors north was nearly three miles at the foot of the Humboldt Mountains. They were, one Frenchman named Francis Richards, and his five partners were half-breed Indians. They spoke French, English, and Jargon or Chinook. They had taken up a stock ranch, cultivated a small piece of land for a garden and kept Public House and

[7] These rocks are just offshore from present Frankfort.

## A Pacific Rancher

hunted and killed bear, elk and deer, otter, beaver and all kinds of small game, and sold the meat and furs at Port Orford, nine miles from them and twelve miles from me. The half-breeds' names were Allix and Frank Purier, John and Peter Grosluis, and Antoine Murain. Three of them, the former brothers, had some very rich gold claims near Empire City, Gold Beach, near the mouth of the Coquille River. They had made some twenty thousand dollars and went down to French Prairie in the Willamette Valley, Oregon, where their parents lived and where themselfs were raised.[8]

They left one brother each to work the claims, but the ones left kept getting drunk and gambled away all they took out of the claims and spreed on, and at last some white man named Dr. Bell swindled them out of the claims. So that when the three others came back from home and had spent and invested some at French Prairie, they found their claim being worked by others, and the two brothers they had left in charge of the claim nearly broke, having spent what little cash they got for the claims, and they had a note of $3000 from Dr. Bell, but he was not worth a cent, and he (the doctor) had sold it again and got killed. So the boys were discouraged and disgusted to be mistreated by the whites, for they would not credit them as they did when they had their rich claims making thousands of dollars a week out of their claim.

So they left the gold mines and come down here and took up the claims, and built them a large log house and hunted and caroused as long as their money lasted; then when they had no more whiskey and brandy themselfs, would come to our place (called the Pacific Ranch) and patronize me. They were very free-hearted, and five or six would come and treat each other one after another, not stop for two hours. Some would not take a tablespoonful at a time, but all would have to drink or give offense. We have taken in as high as $12 to $18 in one afternoon from them. They did not get drunk or quarrel, but when they got broke and had no money with them, they expected you to let them have it the same as if they had the money, and they would pay when they got the money.

We used to buy our groceries at Port Orford when the steamers landed from San Francisco and Portland, Oregon. We sold meals, cigars, liquors and lunches. In spring, summer and fall it was the best trade and business, but in winter it was very lonesome, and not much travel.

After we got our house finished and I could get along all right by myself Charley went to Prattsville, Gold Beach, and started a bakery and saloon—opened it in August or September.

I had forgot to say in my account at Althouse Creek, in about April I got my first letter from my sister, Mrs. Orson Breed, from Chicago. I had

[8] French Prairie, settled by former French-Canadian employees of Hudson's Bay Company, lies between the Willamette and Pudding rivers, north of Salem in Marion County.

wrote from Yreka and Humbug Creek, California, for there was no mails in those days to the northern mines, and we all gave our names to the Express Company to bring our letters to us. And I had left Yreka and Humbug and my letters had got there and no doubt were sent to the dead office or destroyed, being not called for. In Oregon at Cow Creek it was the same, and while at Althouse I give the express carrier an order to get them from Canyonville, so my first came in April and cost me $2.50, but I was so anxious to hear from home I would as soon give $20 for letters from home. . . . I did not get another letter, having left Althouse Creek and Sucker Creek and our letters were lost again until I wrote from Port Orford, and then I got my second letter after Charley was putting up the bakery and saloon at Prattsville (Gold Beach). . . . My first letter cost $2.50 and my second come by U. S. Mail and cost nothing, but about a year apart.

After Charley had gone to Gold Beach I got quite lonesome. I used to go to Port Orford; it was a lively town of about a thousand inhabitants. There was a garrison and fort and some troops stationed there, and many stores and hotells and saloons, and there was lots of business done by miners from Cape Blanco and the mouth of Sixes River and Floras Creek had some very rich gold claims all along the beach for ten or twelve miles.

While I would be gone the Indians nor anyone would molest my house or garden. The last of July a man named Johnson (they called him Coarse Gold Johnson) made some discovery of gold on the south fork of the Coquill[e] River.[9] He was a great bragger and while the excitement was up I went with a pack train of John Waddell's. It took us about five days from Port Orford, all mountains. We had a good look at [it], and camped close by Point Blanco. There was some very rich beach diggings just above and below Point Blanco, and a nice creek with some splendid ranches and farm land on it. The Point is the most western point of land on the United States coast, or continent, by the map.[10]

In going to Johnson Creek the mountain sides were covered with blackberries, raspberries, dill salad and serviceberries, and many kind I knew not. I never saw so many kinds of berries as there was along the coast and coast mountains.

We were disappointed in the diggings—everything had been exaggerated and but little gold, but I saw the biggest boulders and more of them and less dirt around them than any place I ever saw. We met lots of men coming back dissatisfied, but that is no sign, for the best diggings ever discovered were called Humbugs by many. So we had to see for ourselfs. Got in one day, looked around and left for home the next morning.

---

[9] The Coos Bay *News*, August 31, 1887, confirms Reinhart's statement. Johnson Mountain was named after "Coarse Gold" as well.

[10] The most westerly point in Oregon, but not in the continental United States (excluding Alaska): Cape Alava, Washington, holds that distinction.

There were many boatmen from Port Orford and San Francisco in the mines and a rougher crowd would be hard to find. I got back to Port Orford and I got acquainted with Billy Shepherd, a minstrel performer on the road. He was a bartender for Spanish Mary who kept a saloon in Port Orford and used to keep a fast saloon at Empire City. [The] Coquill[es], the half-breed boys, my neighbors, used to spend all their money at Spanish Mary's. She had killed a white woman in San Francisco and a mate of a steamer named Nollan got her away on the steamer secretly and she married him at Empire City. She always done a big business at Port Orford.

Then there used to be a merchant named Captain William H. Tic[he]nor; he had been captain of the vessel called the *Sea Gull*, lost as reported by his carelessness (he being drunk). He was elected to represent our county; he was very smart. He was a Methodist preacher in Iowa, and used to stop with me a great deal on his way back and forth from Prattsville, Gold Beach. He was a hard case to drink and spree and gamble, and had a nice family at Port Orford, where he had a large general grocery store. He owned and built a nice sloop which he called the *Nelly* [*Tichenor*] after his daughter, and carried freight to the small ports and rivers along the coast.[11]

When I was at Port Orford I stopped at the Pacific Hotel one night. A man from Dubuque, Iowa (where I had run a threshing machine) had a Keno game and bank (the first I ever saw). I played at the game and won $12 or $13, when the landlord and his wife got to quarreling (their names were Edson) and it broke up the game. Then there was a shotgun raffled and I won that and sold it for $10 and played Poker with some merchants and won about $20. One man was a butcher who used to be at Yreka, California and had an Indian boy living with him herding cattle. He was a moelack and could speak good English.[12] The butcher's name was Eddy of Eddy & Pratt.

Next day I bought about $15 worth of books and novels, and $8 worth of beads and handkerchiefs. I got all my fish, eggs, and meat and some

[11] Tichenor founded Port Orford in 1851, taking up a donation claim on July 26. Born in Newark, New Jersey, June 13, 1813, he early took to the sea. After service in the Mexican War, he was lured west by the gold rush. Twice elected to the territorial legislature, he served in the state senate in 1860, representing Coos, Curry, and Umpqua counties.
The *Sea Gull*, a steamer of 200 tons, sank in Humboldt Bay, February 26, 1852. Tichenor explains that the cause of the loss was "a defective engine, and a very heavy sea—south-west gale of wind. . . ." He was able to save his passengers by a feat of sound seamanship, in gratitude for which the passengers presented him with an inscribed gold watch. Capt. Wm. Tichenor, "Among the Oregon Indians." Manuscript, pp. 1–5. *Oregon Statesman*, August 19, 1851; Port Orford *News*, December 14, 1926; Binger Hermann, "Port Orford Homecoming," *Oregon Historical Quarterly*, XXV (1924), 327–328.
[12] The reference is probably to the Miluk Indians who lived at the mouth of the Coquille River.

money for the beads and handkerchiefs from the Indians. An Indian family lived about half a mile from me; he had got one of his legs crippled when he was a young boy; he had a son about 12 years old and he stayed with me a great deal. He said his father and some other Indians had at one time tried to take a steamer, and in the skirmish to board the steamer got some grape shot into his leg. The boy was quite smart and learned me a great deal of the Indian language, so that in seven or eight months I could talk quite well enough to be understood by them.

During the fall of 1854, Eddy & Pratt, the two butchers of Port Orford, gave me 40 head of cattle to ranch at a dollar a month per head. I used to take my gun and hunt and look to the cattle. I sometimes had Skeesy, my Indian boy, with me. One day I missed one of the steers and hunted all day and could not find him. Next day I took the boy again and hunted and found a trail through the timber and tracked a while, when all at once the boy ran on ahead and looked in a hole in the ground and then came back and told me he had found the ox. When I got to the hole, sure enough, he was in a hole about 8 by 4 feet and about 9 feet deep, and he stood in there and could not move.

I asked the boy how the hole got there, and he said the Squaws dug the hole, carried the dirt away off, and then lay four strong sticks each two making a fork or crotch at each end, so that if anything fell in they would fall between the forks and would be held up so that they could not stand on their feet so as to jump out again. Then they lay some small sticks acrost the top, then bushes and leaves on top so that you could not tell the difference in solid ground.

These holes are dug right in the trail going through the thick woods on timber land, and elk and deer fall in, and the forks keep up their bodies so they cannot jump out, and the squaws and Indians look to their holes every few days, and if elk, bear or deer are in them, kill them there and take them out.

The boy asked me if I was mad because the steer was in the hole. I told him No, that I would soon get him out. Then he said the Indians would be mad because I would destroy the trap. I told him I could not help that.

So I went home and got a shovel and pick and some rope, took the boy with me, and I dug a trench from the top sloping towards the steer down to his fore feet, and made it wide enough for him to walk out. But the steer was too weak and had not eat or drunk for two or three days and could not get out himself. So I had to get behind him and push him out with the help of the boy, and by hard work got him out. He was so weak he could hardly walk, and would reel like a drunken man. But the trap was spoiled.

Another time I was walking on a trail and the first I knew the ground gave away with me and I went down. The crotches were rotten and broke so that I did not get hurt. And once I found a young elk about a year old in the trap. He had been dead some time and was all decayed. The Indians had not found him in time to use it.

One day I had the chance to save an Indian's life. He was a Good looking young Indian, not over twenty years old, straight and over six feet tall. He came to my place on his way back from Port Orford and asked me if I would keep the Mussel Creek Indians from molesting him.

I asked him why. He said he was a Mackinoo Indian, and lived high up on Rogue River, and the Mussel Creek Indians were their enemies and would kill him.[13] Some three or four Mussel Indians saw him when he came to my place. I told him to stay until morning, and I took him past the Mussel Creek village on the trail. The Mussel Indians said I had no right to keep them from killing their enemy, but I told them that they should not, and I took him some three or four miles past Mussel Creek village. He had a rifle, and I had one, and a navy revolver. I told them I would protect him. So after a Pow Wow a long time, they let him by, and I kept watching so that they did not follow him for some time.

The same Indian showed his gratitude to me the next spring when I was up past the Mackinoo Village on Rogue River to see Francis Richards, the Frenchman (neighbor of mine) and a Canadian half-breed named Eneas, that had been a guide for General Fremont on his first trip acrost the Plains in 1847.[14] Richards was high up the river, hunting, but the young Mackinoo Indian knew me, made me go to his house and eat with them, and took me up the river to the chief and Eneas (who the same summer headed the Indian outbreak and killed so many settlers). He told his family and party how I had treated him, and they were very thankful to me, and invited me to eat acorn soup with them.

The Indians around Mussel Creek and Yreka Creek said they would like to work for me mining if I would give them a horse apiece every six months. I knew of several places on the Klamath River, California, where I could get plenty ground that would pay two to three dollars a day to the hand, but would not pay white men to work for that, and I thought it would be a good investment for from five to ten young Indians, stout and able to dig and work at sluices. I would have to board them, and a horse for six months work I could buy for $20 apiece. So I told them I would go in the spring and look out a location, take up claims or ground, and come back after them.

There used to be a good many pack trains to stop on our creek close to our house, and one owned by Jim Johnson and John and Gus Upton stopt after they had a store at Prattsville, Gold Beach. Sometimes they

---

[13] The Mussel Creek Indians belonged to the Takelma tribe. "Mackinoo" probably refers to the Mikono tunne who were members of the Tututni tribe.

[14] Enos was with the 1843 Frémont expedition, and later joined the Rogue River tribes.

would get one of us to go to Port Orford with them, and stop at our ranch as they came back and camp all night. In the morning we could help them load or pack up and they would drive clear to Prattsville without unloading and save some hands and labor. Several others done the same, and they would pay us well for it. Samuel Nicholls, whose partner I used to know on Althouse Creek (Steve Taylor), got me to go with him about twice a week.

When Charley got his Bakery and Saloon furnished at Prattsville, he took in a partner, a carpenter named S. M. Charles, and I got a pack train of 8 or 9 head mules and horses of John Waddell, who lived at Ellen[s]burg, below the mouth of Rogue River on the beach. He kept a boarding house and grocery, but he drinked awful. I knew him on Althouse Creek. I got his train to run for him on shares a few trips he had to stay at home, and Charley went with me to Port Orford and bought his flour, sugar, liquors, cigars, lining, and can fruit, sardines, oysters and everything for his bakery, and we loaded all the animals but one to ride in turn, for we had to cross a good many crossings of Brush Creek 5 to 8 miles from our place. We had a large silver water pitcher, and one of us had to carry it in our hands, so as not to bruise or dent it. It cost $16 and we put some paint stain dry in it to carry. We could drive the train load from Port Orford to our Pacific Ranch (12 miles) in four or five hours, but the packs would sometimes get loose and we would have to catch them and repack them.

We had a long stretch of beach or sand to go over, some two or three miles, and some points of rock to go around, and the tide would come up near the horses and as the breakers or waves ran back the horses or mules would get dizzy and follow the water out, and when the third big wave would come, catch the horses and throw some of them down, and their loading would get wet and too heavy for them to get up with it. We had to keep running from one to the other to keep them from getting washed out to sea. One horse got down by a breaker taking him in, and his loading, a large lot of bolts of sheeting on top of some other loading part spread out got so wet and heavy that we both had to run to help him up. Charley in a hurry set his big silver water pitcher down to help the horse up, and the next breaker was coming in and we just did get the horse up and out of the reach of the third breaker, when Charley saw his pitcher about to be washed away. He ran in and got hold of the pitcher just as the wave struck him and soused him all over; he had all he could do to run out towards the shore to keep from being washed out to sea. The contents of the pitcher got all wet and spoilt the paint and stain with the salt water.

It was a fearful time; the men would get dizzy and walk out after the water recedes, not knowing they are going out, and all horses and mules are the same. I made several trips with that train until Waddell got so that he could run it himself. I got good pay for helping the packers, and we lived higher and got in better flesh and health than ever I was or got after I left the coast. We had all the fresh fish we wanted; eggs from ducks (wild) or gulls, large and nice, and mussels, and with good ham, elk, deer, and bear meat, we could get up a good meal for anybody for one dollar....

All along the coast sea otter were very plenty and some old hunters made it a business to kill them and send their furs to San Francisco, or if they had many choice furs of fine prime otter skins, would send them to Paris, France, or St. Petersburg, Russia, where they got the highest price for them, from $50 to $500 apiece for extra ones, and the common price here was from $30 to $80 apiece. Some hunters made fortunes of it. They were very hard to kill, and after they were killed or wounded their companions or mates would carry them out to sea. If after they died they would wash in by the tide, someone else might pick them up, so the hunter would have to be on the lookout or lose them, for every morning by daylight the squaws would be out at low tide to get mussels off the rock, and if they found a dead otter or beaver or seal, would just put them in their basket and take them home, and the man that had killed or should have had them, would know nothing about them.

The half-breeds, my neighbors, used to trade with the Indians and make money out of their furs. I used to go down on the high rock in front of my ranch and see lots of otter, beaver, and seal, but could not get to shoot them. I had a nice view of the Pacific Ocean and steamers passing way out at sea, on their way and to and from San Francisco, Portland, and Victoria, Vancouver's Island (British Columbia) and on a clear day it looked beautiful.

I could see plenty of whales (the small humpback whale) in whole schools of them, blowing and spouting a few miles out from shore and lots of porpoises or sea-hogs tumbling over and over like an endless chain. Then sometimes the clouds out at sea would be dark and look like a high mountain and high hills, as if there was but a large river between, and sometimes when a storm was raging, and the breakers and waves strike the rock with the noise of a thousand cannons, and the water or spray would fly inland a quarter of a mile, and sometimes nearly to my house. But the house was at least 300 or 400 feet higher than the sea, and a half mile back from the beach.

In calm weather the Indians would go in their canoes to rocks out in the sea a mile or so, and have some of them keep the canoe off away from the rock (to keep the waves from throwing the canoe against the rock and breaking it or knocking a hole in it) and several would go on the rocks with sticks to keep the sea-gulls from striking them, and rob the nests of eggs. There would be wild ducks' and pelicans' eggs and they would nearly fill their baskets or canoe at a trip and I would trade with them for the fresh eggs, all I would want, for beads or handkerchiefs;

and they would catch canoes full of rock bass and other good fresh fish.

When I had nothing else to do in my garden, I would read a lot of novels I bought at Port Orford, but got awful lonesome sometimes and wished I had stayed in Chicago, or on our land in Lake County, Ill., and not spend the best part of the younger days of my life in that lonesome place, away from all female association. But it could not be helped. I had to make the best I could of it.

My Indian boy used to tell me that the Indians talked of killing all the whites sometime, for they had been mistreated by them. I laughed at him and told him the Indians would have to kill us in our sleep, as they were too cowardly to do so when we were awake. He said that was just the way they would do—when we were all asleep. I made all manner of fun of him about the Indian bravery, and told him that five or six well armed white men could just take any of their villages and do as they pleased, for all the Indians could do, and that was just my honest belief, and all miners believed the same. And we would not give them credit for what damage they might do under favorable circumstances—where we all made a fatal error. For in a short time they just done as the boy said, caught us all asleep and killed some 36 settlers in one night—but I must not anticipate my record.

Sometime in November, I took an awful toothache for two or three days. I could hardly stand it that long, so I concluded to go to Prattsville, 18 miles, and get my tooth pulled. I started about 2 P.M., and by taking a trail below the Three Sisters, would cutoff five or six miles by going over what was called the Devil's Back, a rock the point of which ran out into the sea, and on one side nearly perpendicular, 120 feet high, and only some crevasses to walk up in, but very dangerous. I was very feverish and nearly distracted with the toothache, and could hardly see. I mistook a place where some loose rock had slipped down for the trail, and after I got up about 50 or 60 feet, I found the rock overhanging, and could go no further.

I concluded to turn back (I had on a poncho or sort of blanket cloak)— I went to throw my poncho down so I could turn, when the loose rock under my feet slipped away, and to keep from turning [and] butting my brains out on the rock, I held out my feet and rested my hands on both sides to steady me to slide and keep from turning, and down I went over sixty feet to the bottom. I thought I was gone, and when I struck in the sand at the bottom, I was so badly scared that I could not stand up for some time. My hands were cut so that there were pieces cut out of the inside as large as quarter dollars and bled freely.

After I got over my fright I walked around and come to the conclusion to go back to the Three Sisters Hotell, and take the trail, over the mountains and ten miles around. And it would be dark before I would get over to Yreka Creek. But my toothache did not stop. I had walked back toward the Three Sisters a short distance and had become sort of cool and collected, and I knew that Indians went over the Devil's Back every day and carried heavy loads, and squaws with baskets full of mussels and children on top of the mussels went over every day, and that I could not have found the right crack or crevasse to go up. So I went back and examined close and found the right trail, but it was fearful. After I got to the top I lay down and held on to the rock and grass—I was so dizzy I could not look down. . . .

While I was laying down in fear of falling, and did not dare to stand up, I saw squaws come up the steepest place with a heavy basket on their back with only a strap of deer-hide acrost their foreheads, the basket full of fish or mussels and a baby laying acrost the basket with nothing to hold it on. Then the squaw would get to the top, stand and look around, and rest on her stick and look about on the very brink of the frightful precipice 120 to 140 feet high with all rock below, and seem as safe as if she stood on level ground, while it made me shiver and hold on to the grass and rock, not daring to look down, for dizziness.

I heard several men say after they had crossed the same Devil's Hump (or Back) that it was worth a man's life to go over it; some said they would not try it again for thousands of dollars.

But I got to Prattsville quite late, but as soon as I got there I went right in to Dr. Holton's office, and he sat me down on the floor in a pair of Saddle Machines, and yanked out that infernal tooth. I tell you I felt relieved, and then I went to Charley's and stayed all night. . . .

# 5

## RETURN TO SISKIYOU: INDIAN WARS AND MINING

(1855–1858)

About March 6th, 1855, Charley approved of my taking the trip to California to see about the claims to work the Indians on the Klamath River. So I took a good pony, my rifle and pistol and blankets, to prospect and take up or buy some good bar on the Klamath River. The snow was not off the Coast Range between Crescent City and Waldo, and when I got to Sailor Diggings I found the Siskiyou Mountains closed by snow, so that nothing could cross but on the snow crust. So I sold my horse to a Frenchman named Murain that lived and had mining claims at Elizabethtown, eight miles above Prattsville, Gold Beach. (He was packing here.) I stayed at Waldo with an old mining friend, a sailor named Pete Brown (he owed me a big whiskey and pie bill at Althouse Creek). I saw many old acquaintances from Althouse.

Now while I am on the subject, I forgot to state that our creek claims were jumped from Dr. Watkins, and he failed to let us know in time, and we thought our bank claims were not worth while to trouble about, so let the hide go with the tallow, as the saying is. But while I was at Waldo quite an excitement of discoveries of digging on Indian Creek had been made,[1] but the snow kept the men at Waldo until they could get over the mountains, and Waldo was quite lively.

A saloon and bowling alley was kept by a man I knew, and a prize fighter named Maguire from San Francisco, who owned part with the partner named Mellon. They got up an exhibition of the Manly Art of Self-Defense, or sparring, every week, and quite a number of shoulder-

---

[1] Indian Creek is southwest of Yreka and is a tributary to the Trinity River.

lifters and gamblers took part. I found our neighbor of the Three Sisters, Hough O'Niell near my Pacific Ranch, was giving lessons and training others, and I will here tell how he came to leave Port Orford and the Three Sisters.

At the time of the Johnson Diggings excitement he had gone too, and then come back to Port Orford where he had an interest in some boat in taking passengers and freight off the steamers, and he had some chums, very rough ones from San Francisco. One named Holmes had just got out of the Point St. Quinton Penitentiary. The steamer brought notice to the merchants here to look out, for he was a notorious burglar and safe robber. Only the merchants knew it. He got in with Hough O'Niell and they heard that some considerable money was deposited in a safe at the Pacific Hotell. So O'Niell went in the hotell and drew the attention of the proprietor and guests by giving some hints to the audience in boxing, with his back to the door in the office to keep any one out of the dining room where the safe was. Holmes got in a window from the side of the building and opened the safe, but Spanish Mary was quarrelling with Nollen outside on that side and saw Holmes go in the window, and Mary saw him and gave the alarm. The merchants had been watching and had removed the money already from the safe. The watchers rushed in and caught Holmes in the safe and put him in irons and whipped him and sent him back to San Francisco in irons. They tried Hough O'Niell as an accomplice, but did not quite prove it on him, and were afraid to whip him.

So they (the Vigilantes) notified him to leave Port Orford in 24 hours. So he went to the Three Sisters to his partner, John O'Brien, and made some arrangements with him and left for the mines and here he was at Waldo! He was glad to see me, but did not want me to mention the Port Orford affair when the snow got so that we could go over it.

Before I go further, I got acquainted with Tom Magin. He had a nephew—he called him Allick Williams, and he was in training to fight a little fellow named Shannon who was being trained by Hough O'Niell. It was only a sham match to make money out of the miners, and give exhibitions and make money for the bar, but it made me know the parties in future. A little English boxer named Howard was quite [sufficient] to[o] in boxing with Harry Shannon. He built a bakery at Indian Creek, and I will speak of it further on.

I started from Waldo (and I had not collected any of our old accounts I had with me, not even Peter B[rown]) and got to Indian Creek all right and stopped at Squire Charles Walker's (that used to live on Althouse Creek, and had Jim the half-breed who shot Dr. Black tried before him). He and Hugh Heaps, a packer, kept a large grocery store. I saw many old Californians and Oregon friends.

There was an awful excitement—there had been one man killed and his murderer was hanged by the Vigilantes and miners. I will relate the whole affair:

A man named John Pringle, or Squire John, or Yankee John, had a bowling alley and saloon. A man named Phillips, 26 years old, a miner, quite small, had a fuss with Pringle and sharpened his butcher knife on a grindstone and made threats he would kill Pringle or someone that day. In the evening after supper Phillips was at a store just opened by Pegleg Smith, second door from Pringle's Alley.

There were quite a lot of men in the store. Smith had a lot of new goods, just putting on shelves. A young Scotchman about six feet high and 23 or 24 years old and weight about 200 lbs, his name was McJanes —he was a fine peaceable man, sober and steady. He was joking with other men, and in fun took the pipe out of Phillips' mouth and put it in his own and commenced smoking it. Phillips asked him to give him back his pipe, but McJanes in a bantering way said he could not make him give it back, that he (Phillips) was too small. The first thing McJanes knew, Phillips drew his knife and stabbed him two or three times. The crowd did not know till McJanes said he was cut and fell. Phillips jumped out of the window and fell into a ditch of water and run off. Some men outside saw him fall into the ditch and laughed at him, not knowing he had cut any one.

As soon as the men in the store realized what had been done, they scattered out to catch Phillips. Some went up the creek and on to the trail (to Waldo over the mountains) and some went down the trail towards Happy Camp on the Klamath at the mouth of Indian Creek. But Phillips run up the creek to the Dutch Boy's Claim, they were Old John, Long John, Dutch John [Schertz] and Joe Grappy. He told them he had cut a man and begged for them not to give him up. They heard some one coming and hid Phillips. Some men came and inquired but they said they had not seen him. So they went on.

After they left, the boys told Phillips to strike for Happy Camp, down the creek or trail, but none of them knew how bad McJanes was cut, and Phillips did not himself. So he struck back down past town on the trail for Happy Camp, but only got one or two miles below town when in the dark he ran into a lot of men coming up the trail and Old Kentuck, a big old miner, took him prisoner. They carried torches and brought him back to town, and a guard took him in charge.

McJanes had died immediately after being stabbed, and next morning all the miners of the creek come together and they chose a judge and jury and appointed counsel to defend the prisoner and one to prosecute, and he was tried and convicted of murder in the first degree, and old Judge Taylor, a miner who had been a minister in Iowa, sentenced him to be hung next day and he was. Phillips said he had nothing against McJanes, only vext for aggravating him about the pipe and his smallness, but he did feel like killing John Pringle. McJanes was well liked, peaceable and harmless, but he did like to joke, and we can see that a light joke was the cause of the sudden death of the two young men.

I liked Indian Creek City, and I spoke before of Howard, the English baker, who had just completed a bakery and store and was ready to open in a short time. He offered to sell me a half interest in the property for $300 and to go in company with him, but I had not come to buy property on Indian Creek, for I had but $25 or $30 left after selling my horse at Waldo for $50, and I had barely enough to go to the Klamath and back to Prattsville....

The first night I stopped at Charles, or Squire Walker's. They had a poker game and three of the players were old acquaintances from Althouse Creek. They asked me to take a hand and make it up. There were John Myers, Lavollette Lindsey, a gambler named John L. Sands, a young butcher Jim ———, and myself, making it five-handed. We played awhile and I won about $75. Next day again and I was over $120 ahead.

That night I came to the conclusion if I could win the $300, I would buy the half of Howard's bakery. I had good luck, and had over $175, and I got an ace full. There were three of us in. The butcher had dealt the cards but dropped out after the first raise. John L. Sands bet me $100 better on the last raise. I hesitated, but thought if I should win the bet I would have $350, fifty dollars ahead more than would buy the half of the bakery, and if I lost, I was not over the $10 out that I started with, and I was playing altogether on their money, and I took more chances than if I had been playing on my own money. I studied how the deal was, and the dealer (the butcher) was not capable of putting up, or stacking, a hand of cards. So I just took $100 and called the bet. Sands had four nines, and my ace full was beat, and I did not get the bakery. The game was straight poker and not draw, and our hands were uncommon big ones. But the cards had run uncommon high and we had bet them very high. They all four made common cause against me, for I was an outsider or stranger on the creek, and they did not want an outside party beat them out of their money. They would keep trying to run me out by raising on every bet I made, so they could have the drop on me. And we played two aces, as good as a full hand in draw poker. For the hands run so much lower in straight poker. Sometimes I have thought that Sands, Myers, and Lindsey rung in a cold deck of cards and steamboated me on the hand, but in after years (2 years) John L. Sands stayed with me a great deal and run a Monte bank in our saloon at the same Indian Creek City, where Charley and I kept the Eldorado Saloon and Bakery, and he always assured me it was on the square, for it made no difference to tell me otherwise.

So next day after my game, I started for Happy Camp, at the mouth of Indian Creek on the Klamath River, some 12 miles. I got to the hotell kept by Henry Doolittle of Wisconsin. They done a good business and had a good stock of general goods and groceries in connection with their hotell and kept boarders too. I stayed overnight and inquired for the different bars on the river and heard W[i]ngate and Humphrey's Bar, below Clear Creek, prospected well, were large, and could be bought cheap, but that it would be troublesome to get water on them as they were quite high up from the river.

I took some tools and provision and got a boy I had known down Oregon (at Wm. Weaver's on the South Umpqua, and on Althouse Creek) to go prospecting with me on the bars in question. Found Wingate Bar would be best but would have to pay $300 for five claims of fifty yards each; it would last for years with ten or twelve Indians and could make $2 to $2.50 per day to the hand. And [we] could get sluices on the large bar right above Happy Camp. Some men got good prospects and were going to turn the Klamath River. I was there in time and took up a hundred yards running up the river off the bar and I got a good prospect.

One evening I went to the hotell and heard that the Indians above the mouth of Rogue River and Gold Beach—Joshuas, Toottootenays, Macanooes, Yewkas, and all the different tribes of Indians—had broke out, killed Captain Ben Wright (Sup. Indian Agent), cut his head off, and killed 32 ranch men and burnt up all their houses, drove off the stock and horses, and that Prattsville at Gold Beach was surrounded by the Indians and their supplies cut off; my ranch (The Pacific Ranch) and the Three Sisters, and all our neighbors and settlers were burnt up and the Canadian half-breed Eneas and young Capt. Jack of the Tetootenays led the Indians.[2]

A family named G[ei]s[e]lle, husband, wife and four children, lived at Elizabethtown (8 miles below my ranch) or ten from Prattsville. The Indians made the husband crawl into a burning log and killed and burned his three other boys; then took the wife and oldest daughter (about 15 or 16) prisoners with them.[3] One company of soldiers had been sent to the rescue of the people of Prattsville and one company from Crescent City and some more from Rogue River Valley under command of Genl. Persi-

---

[2] The Gold Beach massacre, February 22, 1856, resulting in the death of thirty-one settlers, was led by Enos and Chief John. Enos was hung at Port Orford, April 12, 1857, for his part in the affair. Chief John was shipped from Fort Vancouver to San Francisco on the steamer *Columbia* the same month. The Indians involved were the Joshua, Tututunne, Mikono tunne, Yukwitee, all members of the Tututni tribe.

[3] John Geiselle (or Geisel) and his four sons died; his wife and three daughters were made captive, so Hubert H. Bancroft, *History of Oregon*, II, 395, *note*, states. A more reliable report states that only three sons were slain; an infant child, a daughter (13 years old), and Mrs. Geiselle being made captives, but subsequently released. Binger Hermann, "Early History of Southern Oregon," *Oregon Historical Quarterly*, XIX (1918), 67.

fer [F.] Smith (afterwards in the great Rebellion).[4] He commanded both regulars and volunteers, and a regular Indian War had broke out in all Southern Oregon and Northern California.

So I thought I had saved my scalp by not being on my old Pacific Ranch. I sometimes think that Eneas would have warned me, for he did the half-breed boys, my neighbors, but being as I was white and they half-Indians might have made a difference, and I am glad I left and not took the chance to try them, or the Indians. For when they massacred the 32 whites, it was done simultaneously, three or four Indians to every house in the night, and not one escaped. My little Indian boy often told me that they would some time kill all the whites and I told him they would have to do it while we were asleep, and it was just the way they done it. But it was the white men's own faults, for we were too careless and unguarded with them and they got the advantage and took the opportunity.

In about a week after I got the news, my brother came to me at Happy Camp. He was at Prattsville when the Indians killed the 32 white settlers. Ben Wright [was] but two miles up Rogue River from Prattsville. They came and knocked at his door in the night, called him out to speak to him, when young Tootootnay Jack (young Chief) caught Ben Wright by the hair, others held him, and one cut his head off with an ax.[5]

Charley told me how it all commenced. Two white men (I knew one, named John Clevenger) were prospecting some distance up Rogue River, some 150 miles above. The Indians killed both of them and the whites heard of it and raised a company of volunteers and fought the Indians. Then Genl. Persifer Smith went down the river with a company of regulars (and one company raised at Crescent City) to go up the beach to the mouth of Rogue River and to camp there until other volunteers from above would form a junction with them.

A small company had been raised at Prattsville and the Crescent City company and the Prattsville company were camped three or four miles up Rogue River from Prattsville. One night there was a big ball or dance got up at Prattsville and all the volunteers except a guard of four men went to the ball. Some few of the Crescent City volunteers were up the river scouting. Nearly one o'clock at night a squaw, well known at Prattsville, told another squaw, the wife of Jim Hunt, a white man, that Indians were going to kill all the whites in the country while they were asleep

---

[4] Persifer Frazer Smith did not serve in the Civil War. He died May 17, 1858, at Fort Leavenworth. In February, 1849, he arrived in California to assume the duties of adjutant to Colonel Mason, and did not achieve the rank of brigadier general until December, 1856. At this time, 1856, he was commander of the Pacific Department. Colonel Robert C. Buchanan raised the Indian siege of Gold Beach (Prattsville). *Harper's New Monthly Magazine*, XVII (1858), 258, carried General Smith's obituary.

[5] Enos reputedly slew Wright. Chief John was chief of the Tututunne.

that night. The squaw was around the miners a great deal and made Prattsville her home most of the time, and she liked the whites, and the Indians did not think she would dare to betray them. So after dark she had left the Tetootnay village (about 4 miles above) to let the whites know, and told James Hunt's wife first. Jim Hunt took the squaw and brought her to a store kept by Gus and John Upton and James Johnson, and they called in other merchants and citizens and they questioned the squaw. She kept telling the same story, but the men just laughed at the idea that the Indians would dare to do so, and they actually proposed to whip the squaw for telling such a lying story. But at last let her go, and thought no more about it.

In the morning very early, about daybreak, one of the volunteer miners named Shaffer had been at the dance and started to go up to the camp, three or four miles up. When he got close to the camp, he saw everything so still around camp and no stir yet and no guards in sight. When he got to the tent, he saw all the four guards dead and scalped and lots of things gone. He just started and ran back to Prattsville with the news. He had to be careful so the Indians did not happen to see him. Everyone was astonished and confounded and raised men and arms and went right up and found Ben Wright had been killed. Then parties were sent out to settlers houses, and it was found all were murdered and the houses burned and stock drove off, and the Indians had all united and were coming on to Prattsville. So the people all got together and forted up the town and kept guard out day and night.

John O'Brien, of the Three Sisters, heard of it—somehow they had not attacked his ranch, and he with five others got into a whaleboat he had and took their guns and escaped from the Three Sisters. They thought they were all safe and got in front of Prattsville to land on the beach where the Indians could not molest them, but somehow the boat broached to in the breakers and upset and all six men were drowned, O'Brien being with them. Capt. Bill Tic[he]nor of Port Orford had a small sloop, the *Nelly Tich[e]nor*, named after his daughter; he ran his sloop into the mouth of Rogue River and took off some of the merchants and others who wished to leave Prattsville.[6]

One night the two Upton brothers, Mr. Pratt, and my brother, and Capt. Tich[e]nor, ran out with the tide without the Indians seeing them, and they all got safe to Crescent City. Then Charley came on to where I was. He had to leave his bakery and everything, even his clothes and bedding. Our house on the Pacific Ranch was burnt up, and I had left our bar outfit, cooking utensils and lots of potatoes, cabbage, beets, and everything in the house when I left, expecting to be back in three or four weeks

---

[6] Bancroft, *Oregon*, II, 395, states that Tichenor "was prevented by contrary winds from approaching the shore."

at the outside. Not a house was left standing from Prattsville to Port Orford (30 miles).[7]

The Indian War continued all summer, fall and next spring before they were conquered and subjected and all the miners on Rogue River, Gallus Creek, Slate Creek, Illinois River, Applegate Creek, had to go away and abandon their ranches and mining claims. The ball alley we had built on Sucker Creek was burnt up by the Indians. So we come to the conclusion we had done well in saving our lives and we would go back to Indian Creek City and buy claims and go to mining, because we could not go back to Prattsville or to our Pacific Ranch or Sucker Creek until the war was over.

So we bought a claim on the town site of one of the Lindsey boys. It had paid well and there was a water ditch belonging to the claim (and the tree that Phillips was hung on was on the same claim). We bought another creek claim and worked nearly two months with four others, the two Lindseys and John and Bill Buck, and myself. And at last could not make it pay, gave it up and commenced building a bakery and saloon in town. It was on our claim the summer of 1855 that Charley had baked and cooked for Joe Freez. He was an old friend from Sucker Creek, a German and a United States deserter, I afterwards found. He run a bakery and saloon and got married to a French Fancy. Charley had cooked, baked and kept bar for him while I was mining some the fall we put up the Eldorado Saloon and Bakery. We done tolerable well but too much credit among the miners. Before we opened up I took a trip up to Scott Bar on Scott River after some pictures for our saloon and a violin player. I engaged Chester Eastman who had played for us on Althouse and Sucker Creek in 1853.

When I engaged him he was just playing for a large gambling saloon on French Bar at $7 per night, but the night I got there, the city authorities had closed up the saloon, and left Eastman out of a job. So he came to us at quite low price, but we boarded him and give him about $4 per day and night. We had taken in a third partner named Edward Ryan; his mother, a widow, lived at Grand Harbor, Michigan. He and Charley whipsawed and cut lumber for the saloon and for sale. My partner in the bar claim was Harry Stone, an old sailor; he worked hard all fall and ground-sluiced with a big head of water, and ran off more dirt than I ever saw handled without a hydraulic, and got but little pay and had large expense fluming and repairing our ditch.

The winter of 1855 the Indians attacked and murdered a packer on the mountains between Indian Creek and Waldo (about 15 miles from us) and ran off the mules and horses. The other man got away. A com-

---

[7] All previous references to Prattsville indicate it was actually the earliest name for Gold Beach. In the previous chapter (see *note* 4), Reinhart stated that Prattsville was "a half a mile away on Gold Beach."

pany was raised to go after the Indians in our town, and the balance of the citizens built a large log fort, so that if the Indians did come we could fort up, and we kept guard out and had the place, or fort, well provisioned, and got 25 new rifles.

The company was gone some eight or ten days, and captured a lot of horses and mules, some of the train that had been robbed among them. They did not catch up to the Indians but caught up to the stock and some belonging to the Indians. The stock was sold, and after taking out the expense of provision, the balance was turned over to the widow of the man the Indians had killed. They lived up on the Klamath River.[8]

One day I waked up with a lame ankle, and thought it was wrenched, but that day the pain got up higher, and then shifted to the other knee, and was pronounced to be inflammatory rheumatism. When I took down I weighed about 185 lbs. I could not eat a mouthful for 22 days. Then I gave $2.50 for a small can of strawberries and I could eat a few at a time. I could not sleep or turn in bed, and had to be turned for I could not move. I begged for opium to put me to an everlasting sleep, but could get none. Everything that was thought of was used that anybody advised: brandy and pepper, cold ice water, wet sheets and everything, liniments of all kinds, also Mustang, till I got tired of it. Perry Davis's Pain Killer done me most good. I could not have anything touch me. My cover had to be propped up so not to lay on me. I wanted to die, but could not. And I could not let a light go out a night. I was sick about seven months and in the coldest of winter, and I kept thinking the Indians would take the town and burn it up, and I was perfectly helpless to move. After I got so I could get around on crutches to get weighed, I weighed 127 lbs.[9]

After I got well again, I went to ground sluicing, and felt better and could do a better day's work than before (so Harry Stone said) and kept working at our old bank claim.

After the Indian war was over, there was quite a rush again for Sucker Creek, where we had our bowling alley and bakery in 1853. The Indians had burnt up all the towns and miners' cabins and our bowling alley and bakery in 1855, but now some miners had made some discoveries on the right hand fork, where our party had prospected and worked in 1853, and now called Bowling Creek.

So I had got tired of my claim on Indian Creek, and I got Ed Ryan, Andrew Kelley and Samuel Hinks to go in with us to try Sucker Creek again. We struck acrost the mountains with our guns and two or three blankets, we thought we could get there next day, but we got lost and

---

[8] This incident was merely one of a series which led to the Humbug War of 1855. [Harry L.] Wells, *History of Siskiyou County California* . . . , pp. 131-141, gives details.

[9] Reinhart probably was suffering from beriberi, possibly scurvy (although he records no bleeding).

we liked to have starved and nearly froze to death, and came out thirty miles south of where we wanted to go. But we got to Sucker Creek at last and went to the new town at the forks of the creek at the mouth of Bowling Creek, and about three miles above the old town, (burnt up) where our Ball Alley and bakery was. We took up four claims (bank) that were vacant and had been worked three years ago. Two of us mined and two of us whipsawed. Ed Ryan took to drinking, and it was hard to get along with him.

The spring following, Charley rented out our Eldorado Saloon and Bakery on Indian Creek and come over to Sucker Creek again to us. He wanted to start a bakery, so I bought a lot in the new town right opposite the footbridge and main road up the creek, and put up the frame of a house, but after we got that far, concluded to build another bowling alley and bakery on a different lot, as it would take more ground for Ball Alley, and would have to be nearly a hundred feet long. So Charley took in a partner to whipsaw with him, and they took up some creek claims right in front of our bar claims, adjoining mine. My brother and Thomas Wilson (a Polander) whipsawed all winter and sold all they could, and sawed the lumber for an alley and saloon and bakery.

I had left Ed Ryan, Kelley and Samuel Hinks, and bought in with old Dutch John Schertz, one of the Dutch boys of Indian Creek. He had come to California in 1850—or 1849—I forget which, and was from our part of Lake County, Illinois. Charley and I knew his brothers, sister, and mother, and we had threshed their grain with our machine in 1850 on the Dusenberg farm, which the Schertz boys had rented. We found out this while we were on Indian Creek. We, John Schertz and I, had bank claims about a quarter of a mile below the mouth of Bowling Creek on the main Sucker, and they paid us well. Charley and Wilson lived in our cabin. Ryan did not like it because we would not keep him in company with us, but he drinked so we could not. But he went in with an Englishman named Thompson. (Ryan, Kelley and Thompson, and Freez and McPhearson and some others had once enlisted into the United States Service at Buffalo, New York, in 1850 or '51. They all deserted from The Dalles on the Columbia River, and had come to Althouse to mine.)

My brother and Wilson commenced the house in 1857 and opened up July 3rd, with the Bowling Alley and Bakery. We did not complete till fall. We had double alley, 78 feet long; the saloon was 30 by 40 and the bakery on behind.

I will here again go back to Indian Creek when I was just getting so that I could set up, and I was awful lonesome. There used to be some poker playing in our saloon and Charley would play to make up the game when he could get away from waiting on me while I was sick. There was a man called Captain Gunn, a sea-captain, whose family lived in Brooklyn, New York, and he drinked considerable, and would play poker.

One day he asked another man whom we knew for three years on Althouse Creek and at Gold Beach—his name was Henry Wiggans—Gunn proposed to Henry Wiggans to go into a poker game and play partners and get in my brother Charley, and if I would play, coax me to take a hand with them, and skin us out of what money we might have. Wiggans had not any money or no doubt it would have suited him, for he was a skinner.

But anyway he did not go in with Capt. Gunn. He may have had his doubts about him being able to skin us, for he knew us some time and could judge. So Capt. Gunn went to Bill Burk, another sort of a card sharp, and together they made up the plan spoken of. When they proposed the game, they said they would rather play in a private room where everyone could not look on, and my room was the one best suited. They were going to play against Charley three-handed, and asked me if I did not feel well enough to take a hand and make it four-handed. (Now Henry Wiggans had told me and Charley all about how he wanted him to play in with him, and we were prepared for Capt. Gunn and Wm. Burk.)

So when they proposed to me to play, I told them I would play a few hours, and if I got so I did not feel like playing any more, would quit the game. That it was company to me to look on and I would not get lonesome, so I was urged into the game. We played pretty high, and Captain Gunn kept calling for drinks, thinking that would affect us, but we only took mild wines or cordials. After a short time I had about all the money Capt. Gunn had, some $75 or $80, and he handed me a gold specimen ring worth $32. I loaned him $30 for it. He kept playing and borrowing from me or Charley, and never got any from Burk; in fact Burk had all he could do to hold his own.

After a little the Captain handed me a second specimen ring and I let him have $15 on it, and a bet or two took that. He pulled a third fine ring off his hand and told me he did not like to pawn it—it had been presented to him by Jim Covington, a gambler I knew, and his initials were on it. I gave him $25 for it. He said he would redeem it next day when he could see some of his partners (miners). So I had the three rings for $70. . . . And he kept borrowing and losing bets to me to the amount of $150. He gave me his note, and I saw that Bill Burk was nearly out of money . . . so I thought I had better quit the game while I had the money. . . . Capt. Gunn felt quite bad to think that he had lost $80 in money, $70 in rings, and a note for $150, at his own seeking "to skin us." Next morning he was still drinking, and he came and begged me to let him have his ring that he had got to remember John Covington by. He looked so pleading and reckless that I pitied him and give him back the ring, but he was to give me the $25 that I had loaned him on it as soon as he could raise it. I held good hands and played in good luck, and Charley and I never have been drunk or under the influence of liquors, although we have handled it and sold it many years.

In selling lumber of Sucker Creek, it was worth from $14 to $18 per hundred feet. In the winter of 1856–7, there was a tremendous excitement on Bowling Creek. Some six miles above town a company of three or four men, [James Dooling], Brown, Desmond, and Jim Hope, took out thousands of dollars per day, and one piece weighed 85 ounces, or seven pounds and one ounce. It looked the shape of a large Irish potato, and the miners made it ten times worse than it was. Such exaggerated stories as were told would nearly drive men crazy. I and Wilson went up to look, too, and saw the big piece and lots of other gold.

But report had it over half a bushel of gold a day. So Wilson and I went up about a mile above this fabulous claim, up a gulch and struck into a hill between two gulches, and got a good prospect and took up three hill claims. The gulch claims in front of our hill claims were being worked and had been over a year. In the company was William Reinhart, Cal Cooper, Henry Wilson, and it went under the title of Reinhart, Wilson and Company. Now our three hill claims were Reinhart, Wilson and Company, too.

The gulch claims were allowed by law, 45 feet wide in the bed of the gulch, and 50 yards to a claim. We claimed as Hill Claim[s] 50 feet front running back without limit. I put in a few sluices and I worked alone to hold our claim, for we had not water sufficient to wash, and my partner, Thomas Wilson, went down to our other claims. . . . I used to get from $4 to $20 per day when I could work. One piece weighed about $16 and all our gold was coarse.

Now when the Gulch Claim Company saw that I was getting good pay, they claimed the right to take their 45 feet all on our side of the gulch and take into our hill claims. And when I was not there, they commenced in our ground and were setting their sluices. I told them the law, but they would not listen. So Thomas Wilson came up and we two went up to our claims and found their sluices in our ground and they at work. Their cabin was but 50 yards off. We asked them to take their sluices and tools away but they said they would not. So Wilson and I just picked up their sluices and tools and threw them down into the gulch. There were four of them and but two of us, and at first it looked like they would show fight, but they did not.

They blustered and at last said they would dispose of us by law. We told them to pitch in, and so they entered suit at Kerbyville our county seat, and the papers were served on us—Reinhart, Wilson and Company vs. Reinhart, Wilson and Company. It was a puzzle to the lawyers which was which. They got the best lawyer in Kerbyville but there happened that one day General Preston, one of the best lawyers on the Pacific coast, came on Sucker Creek (he had mined on the Creek in 1853) and I engaged him to defend our case. He knew the laws and was an old miner.

We all ten went down to Kerbyville with our fourteen witnesses. They got the best of us in selecting the jury; they had six out of eight, but they hung. They thought they had us beat sure with their jury, but when they hung they became frightened of the next jury, which would have given us a verdict, and their lawyer, Sprague, found out, so they proposed to settle, and we did. They paid the cost and left us our claim as we had it. (We claimed that they must take 22½ feet each side of the centre of the bed of the gulch.) Our lawyer only charged us ten dollars, and $25 should have been the fee, if we had continued the trial. But we had our expense to pay, so it cost us $60 or $70, but the others must have paid over $200. And we went home good friends again.

William Reinhart said to me that I had better friends in Kerbyville than he had, or that he thought I had. I had got acquainted with him over a year ago. He must have been a cousin on my father's side, whose father had come to Virginia or Pennsylvania about the time of the Revolution, and his family were living in Indiana. He was raised in the family of Daniel Weaver, one of his partners in the claim, whose name I had forgot. William Reinhart in 1857 or 1858 was elected county clerk, and he volunteered during the Indian war, and in 1861-2 he was Major in the Regular Service, and commanded Fort Walla Walla (where we still were good friends).

The water gave out in the claim, and I went down home, and every nine days I would have to go up and see how our claim got along, and work a little to hold it. One day four Italians came in from California and wanted to buy some claims. I told them that I had three hill claims to sell that they could have for $600 if taken soon. They said they would go and prospect them two or three days. They went up on a Wednesday and took their blankets and tools and some provision, and on Saturday night got back, and showed me about $4 of fine gold which they said they got by working a day or two. They let on that they did not go much on the claims, but would give us $100 if it would pay $6 per day to the hand, and if not, nothing.

I mistrusted that they were not on the square, so told my brother not to appear to be anxious to sell, that they got more in the prospect and kept back the coarse gold. So that when they came again I told them that if they wanted to buy the claim that they must do so by Monday morning. My brother would start for Crescent City to buy liquors and cigars and balls and pins for our bowling alley, and that if they did not buy before he went, that we would keep them and work them when the water came. And we would take $300 in cash down and the other $300 in six months if it paid $6 per day per hand.

On Monday morning before we had breakfast one of the Italians came and paid me $300 in gold coin, and give me a note for the other $300 with the said conditions. (And it was a foolish condition, for no matter how much they might make out of the claim per day over the $6, we would have to take their word for it, and if they did not want to pay it, they could claim that they had not made the $6 per day, and we could not prove they had.) So Charley started for Crescent City that day.

Some two months after, the Italians abandoned the claim, saying they could not make anything out of it. The next summer one of them named Parodi Bartholomew worked for us in our creek claim, for he had had a falling-out with the three others and left them. He said when they all four went up to prospect the hill claim they dug out and washed out the dirt in the pan to see how it would average when worked with water and sluices. They worked about two hours and took out one piece of coarse rough gold and quartz, some over six ounces, or over $100. After getting out the quartz rock [they] weighed over four ounces of pure gold, and they must have got over a hundred dollars in that short time. They just quit work and done nothing and come down to buy the claims. [The claims should have] cost them $1000 [each]. But [they] were afraid if I saw the gold I would not sell to them at that price. (And that was lucky for me, for if I had got the gold they got in their prospecting, I would not have sold our claims for any price.) They worked for two months and never got back the amount they paid us, and we did not pity them one bit for their roguery. They deserved their punishment, loss of money and time.

In the fall of 1857 the Italian bought into my bank claim with John Schertz, and we hired a fourth man named Peter Kernan, an Irish young man. We made from $6 to $15 per day per hand. Wilson and my brother Charley found a good-paying hill claim away up the hill over 300 feet high, right acrost the creek from the town and our ball alley, which was called The Mayflower Saloon & Bakery & Bowling Alley. That fall and winter I used to be at the saloon a great deal when the weather was too bad to work. We had to put up a high aqueduct across the creek from the left-hand fork of Sucker above the town, and when the ditch got opposite our claims, crossed on our aqueduct and flume and brought the water high up on our high bar close to our house, and we could use it for ground sluicing and . . . [the water] was clearer than the main stream. But on the town bar above us were some thirty or forty Chinamen, and the bar they worked was awful deep. They worked in three changes, eight hours each, night and day, and when some were at work, others were smoking opium or asleep. The claim was from 18 to 25 feet deep. And no wonder that in 1853 we did not get good prospects, for we were prospecting this very claim and did not get down [but] 12 or 15 feet and so gave it up.

Now the Chinamen were called dirty, but they were too clean to suit us, for every day some of them would set in rows on the edge of our water ditch and wash their feet and legs above their knees; they would slip up their wide trouser legs up their bodies, and bathe off and wash in water we sometimes used for cooking or drinking. They had an interest in the ditch and used part of the water, so we could not object to their cleanness. They were called very frugal in their meals and considered close to their provision as to cheapness but these I knew once invited a lot of us storekeepers to a great dinner for the Americans, and they had a special table with the best of victuals, such as pies, cakes, roast pig, oysters in soup, or oyster pie, and all kinds of can goods and fresh meats the market afforded in great profusion. And only us white to the same table; they had their own table to themselves, and they waited on us as gentlemen; after eating they had wine and lemonade and nuts and oranges, figs and raisins and apples—in fact, as well got up as we could have done ourselfs.

Charley and Wilson had bought a dozen hens and a rooster and raised some 200 or 300 chickens on our bar, on the meat and offal of a butcher's slaughter-house. They sold their eggs at $2.50 to $3 per dozen, and chickens, young spring pullets at $1.50 apiece, and the Chinamen were [their] best customers, as they lived higher than I had any idea of. But they were sort of bound out to superiors, who had charge of them, as they were shipped out at the expense of their government, and they collected all their earnings until they had reimbursed themselfs and then after that they were at liberty to work for themselfs. If they died their bodies were temporarily buried and afterwards taken up and shipped back with a shipload to their homes in China. Their overseers are of a higher class, and some are overbearing and arbitrary with their employees, and sometimes hard feelings exist among them. I saw one of their higher class overseers died very suddenly; the Chinamen said he had smoked too much opium. No doubt. And it may have been foul play, for the young man was not well liked by them. But we took no stock in their doings and let them mind their own affairs.

One or two years before, we, a lot of miners, had run all the Chinamen off of Sucker Creek, and would not let them mine. But the county officers were making a good thing out of them collecting their licenses and $4 per month each head, and the sheriff brought back the Chinamen and told us the authorities would protect them as each Chinaman had paid $50 apiece to the United States government at San Francisco when they landed, and the government would have to protect them. So we had to let them alone, and they bought a great many mining claims on Sucker Creek and paid some big prices, too.

The fall and winter of 1857 I used to play some cards, and on the ball alley of ours to make up games. We got to playing Ten Pin Pool on the alleys. You draw a ball the same as billiard table pool and then try to make it up to 31, which is Pool. We would put up a stake of 25 or 50 cents apiece, and the one [who] made pool got all but ¼ or ante which was for the use of alley and pin setters. When you got as close to 31 as you thought would be safe, you would say "I plant," and then your turn in roll would count on the next man to you when you would endeavor to burst out. By chance sometimes you might make pool for him, not knowing what he had. If you bursted out, yourself or anyone could come in again with a new ball and the ante as many times as they please until someone made pool. Or when after planting no one makes pool, the nearest to 31 takes pot or the pool. Charley and I were quite good rollers and in practice but would have to play to make up the game. Sometimes Saturday nights and Sundays I would start a little Monta bank with $75 or $100 with a partner named French Joe Denny, and we sometime made some little winning. And I sometimes played poker and I usually came out winner.

I was the cause of two men making their fortunes. A boatman from New York (formerly) named James Daniels, and his mining partner Guss Hill, a Prussian, had claims on Sucker Creek a mile below town. They paid well but were nearly worked out. Jim Daniels used to be on Althouse Creek in 1854. He had a prize fight with a man named ———. I got acquainted with him and Guss Hill at Happy Camp in 1855. They used to come to our saloon and play ten pins and pool and sometimes poker. They had studied up some signs at the game so to play together so that one could pass a good hand and his partner would keep it from being passed out. So one Saturday night they both proposed for me to go into a game of poker with them, and another miner named Pickett come into the game and made it four-handed.

In the play I got away with all of them, and Jim Daniels got mad and excited and could not understand his partner's private signs, and he was very overbearing to his partner, Guss Hill, and laid all the blame on him, and kept quarreling with him till morning when they went home mad at each other. I had won about $150. Guss Hill owed me $30 or $35 and told me he would pay me next Saturday night. On Monday morning I came up town early to get some beef for breakfast when Charley called to me that Guss Hill had left some thirty-odd dollar for me and he had left Sunday for San Francisco, having sold out to James Daniels on account of the quarrel Saturday night.

In March, 1858, Jim Daniels got a letter from Guss Hill—he was at Fra[s]er River, British Columbia—and for him to come right out, that he had struck it rich. It happened when Guss Hill got to San Francisco there was a great excitement of a rich discovery on Fra[s]er River, and every steamer took from a thousand to 1800 passengers from California, and he, Guss Hill, being out of a claim, decided to go to Victoria and see. From there he found other friends he knew and they bought a small boat and provisioned her and tools, and went up the river. A lot of roughs from San Francisco, who had been run off from California by the Vigilantes, followed close behind. One of the roughs' name was Ned McGowan, very notorious.[10] Guss Hill saw a bar on his right hand side of the river; he landed and commenced to prospect and got a big prospect and staked off three claims. When the boat containing Ned McGowan came ashore and took up the next five or six claims adjoining Guss Hill's. The bar was named after the discoverer, as Hill's Bar, and turned out to be very rich. Jim Daniels went out there by steamer, right within a few miles of the bar to Fort Hope (an old Hudson Bay fur company trading post where the steamers made their terminus landing from Victoria, Vancouver's Island).[11]

When I came down Fra[s]er River in October [1858], I saw Jim Daniels just leaving Fort Hope for San Francisco. Guss Hill had left the day before. They had made eight or ten thousand dollars in less than eight months, and I the indirect cause of it!

The winter of 1857, Ed Ryan, our old partner, got killed by a saw-log. He had been on a spree for three or four days. His partner, Bill Thompson, got tired, and urged Ed Ryan to go to work at whipsawing that morning. Ed did not feel like working after his spree, got careless and cut a log or snag without taking the usual precaution to prop up the log at both ends, and when he cut a snag off the log started and rolled over him. It was over sixty feet long, two and a half or three feet through, and it mashed his breast and chest so inwardly, that he just told Thompson that he was a dead man, and died. When I heard of it, I took a scare. For that morning I had a very narrow escape from a frightfull death myself, and I will relate the circumstance.

I was mining in a tunnel and had put a blast in the bed-rock. It was in the water and I had to work in a hurry before the water would wet my blast. In tamping, the fuse got a kink in it, and in pulling it straighter, must have pulled it out of the powder, but did not know it then. So when I touched it off, the fuse did not go. So I lit it again and still it did not burn. So I took the candle in my hand and held it under the fuse to make it burn. All at once there was a hiss and it blew my candle out of my hand and I ran. But there was no explosion. So I went back, but I was awfully frightened, for if the blast had gone off, nothing could have saved me, for I was right over the blast.

And when I heard of Ed Ryan's death, I did not feel like working any more that day, but went and helped dig a grave on a high point and we carried him and buried him next day. His partner and others wrote to his mother in Michigan to let her know of her son's death.

[10] Carl I. Wheat, "Ned, the Ubiquitous," *California Historical Society Quarterly*, VI (1927), 3–36, has treated McGowan's career, including his 1856 brush with the Vigilance Committee and his British Columbia experiences. See Chapter VI, *note 47, post,* for Reinhart's reports on McGowan's Fraser River escapades.

[11] Established in the winter of 1848–1849, Fort Hope is situated on the left bank of the Fraser at the mouth of the Coquihalla. See Chapter VI, *note 34, post,* for Hill's Bar.

The following summer, Samuel Hinks, who was a partner of mine, Ryan's and Kelley's, was whipsawing with another young man. Sam took a handspike and raised the log, for the other man to change the head block. Sam's hand spike slipped and the young man was under the log and it came down on his head, killing him instantly. It happened near Kerbyville on the Illinois River.

# 6

## TO THE FRASER AND RETURN

(1858)

In February, 1858, John Schertz and I bought out the Italian out of one third of our bank claims and water privilege and aqueduct or flume across the creek, and the Italian and myself concluded to go to Fraser River in British Columbia where the great rich discoveries were made the fall and winter past, and some on Thompson River near Shuswa[p] Lake.[1] My old neighbors, the half-breeds of Gold Beach next to our Pacific Ranch, had been working out there the year before. Charley was to work half of our claims with John Schertz and employ some Chinamen, and still carry on our ball alley, saloon and bakery with Wilson. If I should find anything in British Columbia, [I would] take up claims for all and [then they would] sell out on Sucker Creek. . . .

We made our arrangements to start about May 8th, 1858. Charley went with us to Kerbyville and another of the Italians, named Sidney Bartholomew, no relative of the other, only they had been partners together in the gulch claim they bought of us. I had whipsawed with him some of the winter of 1856 and '57. He was a large rough ship carpenter —could speak French, German, Spanish, Russian, Portugee, Italian and Swedish. He had learned his German in Russia in a German colony. From his looks you would not think he knew one language, instead of seven or eight. So he made three of us to travel. We were going to buy some animals at Kerbyville, and strike through French Prairie to see the half-breeds, and on to The Dalles, Oregon, and join some company

---

[1] Gold has sporadically been discovered in various British Columbia areas from 1850. An effective strike on the Thompson near Nicommen, exaggerated by reports circulated in California, provoked the 1858 Fraser-Thompson River rush.

from there to the Interior of British Columbia.² So we got to Kerbyville and the two Italians bought two mules and one horse, and three riding saddles, one for each of us, and we could pack our blankets behind us.

At Kerbyville we found S. M. Charles, the carpenter, working at his trade; he had been a partner of my brother in the saloon at Gold Beach. He had worked for us a long time on our bowling alley and saloon and bakery on Sucker Creek, put up in 1857, for which we had paid him big wages, five dollars per day, and we considered him a great friend of ours. He came to us and wanted to sell me his new Navy Revolver, a very fine one, that he said he would not part with only to me, and that he had no especial use for it then, and said he would take anything for it from me, for a keepsake.

I had some fine gold specimens that I had bought from the Italian (my partner) and I gave double weight of choice pieces of smooth nice gold I had saved out of my claim. I sent twelve dollars in weight to Crescent City and had me a nice ring made and my name and date (1858) engraved inside, and I sent down two pieces of the same kind of gold, and had a pin put on each, with a fine linked gold chain to connect the two. The ring cost me $24 and the double pin I would not sell for $50. And I still had some of the specimens left. They were from the Jesus Maria mines in Cal[a]v[e]r[a]s County, California, and they were all as if in a molten state had been poured on some cedar branches, and the impress of the leaves was so natural.

So I gave S. M. Charles a fine specimen, weight about nine dollars, but costing me $18. Just the best shape for a watch seal for a gold watch chain; it was worth at least $20. Now S. M. Charles' pistol should have sold for from $27 to $30 with belt. (I needed a good pistol to carry on my horn of the saddle in a pair of canteens.) He seemed to be satisfied with the seal specimen; still he did not like to ask boot for the pistol.

So I through friendly feelings offered him a choice or selection of some other specimens. I had expected he would take one middling or medium piece, as I had some large and some small, but he surprised me by taking the largest of all, worth about $8. We (Charley and I) did not express ourselves, but thought it looked very hoggish. . . .

So I bid my brother goodbye and left Kerbyville about noon the tenth of May, 1858. We crossed Rogue River at Venoy's Ferry.³ We had for company a man named Thomas Mercer, a minister who lived at Seattle, Washington Territory, he had been in Illinois Valley to visit his brother-

---

² There were three known organized companies that set out overland from Oregon to the Fraser in 1858. They were captained by Joel Palmer, Archibald McKinlay, and David McLaughlin. Others might have followed except for active Indian hostility. Victoria *Gazette*, July 21, 1858; Leslie M. Scott, "The Pioneer Stimulus of Gold," *Oregon Historical Quarterly*, XVIII (1917), 153.

³ First established by Joel Perkins, later called Long's Ferry. James N. Vannoy operated the ferry subsequently.

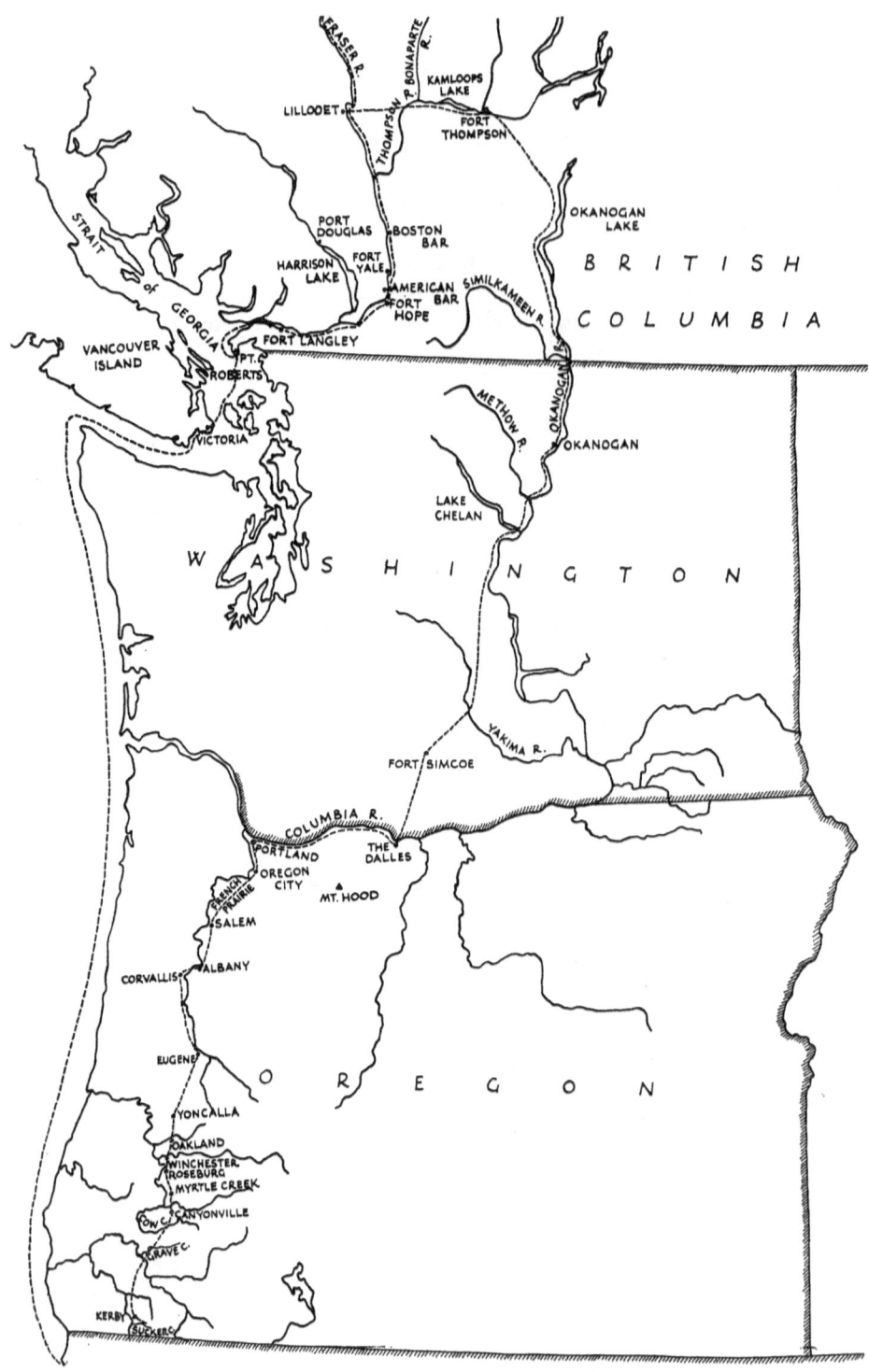

Map 4. The Fraser River Trip

in-law. He kept us company to Portland, Oregon.[4] He afterwards went to Boston and some Eastern cities and shipped by government steamers for the government some 700 young ladies for school teachers and servants, to get married in Oregon and Washington Territory.[5]

Our first day's ride was over forty miles. At Grave Creek Hotell I saw our old Chicago friend, Jimmy Twogood. He was a great stammerer and was well acquainted with Orson Breed and my sister Bertha, his wife, in Chicago. We knew him since 1852. He told me he had carried a letter for us from Orson Breed for several months, not knowing where we were, and had at last lost it.

The second day we passed through Canyonville and I saw many of our old friends and neighbors around Cow Creek and Umpqua. At noon we had dinner at old Lazarus Wright's at Myrtle Creek. A circus was just holding a performance there, and I saw more of our friends, and some I had crossed the Plains with.

That night we stopped at Roseburg, about fifty miles from where we started in the morning. Third day we passed Winchester at the forks of the North and South Umpqua River. A few miles farther we passed through Oakland. (I forgot to say we passed the old Joe Lane place—General Joseph Lane, the old Mexican and Oregon hero—he was senator and governor of Oregon, and the would-be Vice-President with John C. Breckenridge in 1860.) We passed through Yonc[al]la Valley next and crost the Cal[a]pooy[a] Mountain at Estes on the road to Eugene City. We got lost and did not get to Eugene City until a couple of hours after dark . . . but found the Hotell still open and got some supper, got our horses and mules fed and taken care of, and retired to rest after a long ride of about sixty miles.

And next day we traveled up the Willamette Valley (6th day) to Corvallis (or old Marysville), Benton County.[6] From Corvallis we passed through quite a fine country, and through Albany, Linn County, and on

---

[4] Born in Harrison County, Ohio, March 11, 1813, Mercer captained a wagon train west from Princeton, Illinois, in April 1852. On the trek west, he lost his wife. (He was visiting his dead wife's brother.) In 1859 he returned to Oregon and married Hester L. Ward. Although a staunch member of the Methodist Church, Mercer was not an ordained minister. He was one of the founders of Seattle and died on May 25, 1898. H[orace] S. Lyman (ed.), "Reminiscences of Daniel Knight Warren," *Oregon Historical Quarterly*, III (1902), 299–302; E. Ruth Rockwood (ed.), "The Letters of Charles Stevens," *ibid.*, XXXVII (1936), 340, 343. Clarence B. Bagley, *History of Seattle*, II, 701–703.

[5] Reinhart is in error. The reference should be to Thomas Mercer's younger brother, Asa S. Mercer, who came to Washington in 1861. The following year he became the first president of the Territorial University. Although Thomas Mercer was a strong advocate of bringing women to the territory, it was Asa who implemented the idea in two trips east in 1864 and 1865. Clarence B. Bagley, " 'The Mercer Immigration': Two Cargoes of Maidens for the Sound Country," *Oregon Historical Quarterly*, V (1904), 1–24.

[6] The change was made by the territorial legislature in 1853.

to Salem, the capital of Oregon, in Marion County, one of the best counties in Oregon.[7] The 7th day we passed through Oregon City and to Portland, the chief or largest in Oregon, on the Columbia River at the mouth of the Willamette. It is hard to say here which looks the largest. We put up at a hotell and stopped to buy some tools we wished to take with us, and here Thomas Mercer left us.

I will have to go back two days, for I forgot we struck from Salem for French Prairie, settled up by Frenchmen that had married squaws and were employees of the Hudson Bay Company. We had the address of our half-breed neighbors and friends, and it was Fairfield. We did not get there till one or two o'clock P.M. We passed a schoolhouse during recess, and saw a lot of half-breed and three-fourths white girls that were awful goodlooking of the dark style of beauty. When we got to where the half-breeds lived, we found only two of them at home, but one of them, Peter Grosluis, had been up on Thompson River the year before, and had prospected that part of the country quite thoroughly with a large party of French half-breed and old Hudson Bay employees, and he told me in good faith that the diggings he had worked on Thompson River and around Shuswa[p] Lake did not pay them over $2 to $3 and $4 a day, provision light and hard to get, no game to speak of, and short season, and he would not advise me to go there.

But there might have some other paying diggings been discovered since they had left. But taking all the disadvantages he had enumerated, it was unfavorable to go there, and he (Peter Grosluis) rather discouraged one of the Italians and myself, and we had a half notion to turn back. But I hated to do it, and [since I had] provided for my claims being worked without me, and the news we had from Lower Fra[s]er River was highly favorable, we concluded to go on to Portland and take along a whipsaw, cross-cut, and some other tools, so that if we did not like the mines, we could saw lumber to sell, and Sidolia could work at carpenter work anyway. We could make good wages in a new place, and we were anxious to travel awhile and see the country. So we went to Portland and hunted all the hardware stores to find a whipsaw, but could not. We tried to find some second-hand ones, but could not.

Then I heard that I might find some at Oregon City, twelve miles above, and I got on a little steamer that ran to the foot of the Falls Canema[h], and then it was about a half a mile to walk up past the falls and mill dam to Oregon City. I could find no whipsaws here, but I saw the finest water-power I ever saw any place. The whole Willamette River fell from 16 to 30 feet and good high banks, and power enough to run all the machinery that would ever be wanted.

So I went back by the next steamer to Portland and next day started

---

[7] The capital was established in Salem in 1851. In 1855 it was moved to Corvallis, only to be moved back that year to Salem.

*To the Fraser* 113

to go over the Cascade Mountains on the way to The Dalles in Wasco County, Oregon. We got near Forster's at the foot of the mountain, and we met Samuel Hinks; he was stopping on a farm close by; he was partner to Ryan, Kelley and myself on Sucker Creek, and was the one that let his handspike slip and killed his partner a year before at Kerbyville, Illinois River.

And while stopping at noon to let our horses rest and we eat a cold lunch, we concluded to try our pistols as we had not shot them off since we started. Mine was loaded when I got it from S. M. Charles, so I put up a target and fired a shot. And I thought I saw the smoke come out between the cylinder and the barrel, so I fired a second shot, and sure enough, the two Italians noticed it too. I looked at the pistol and saw that there was two short cracks in the barrel where it joined up to the cylinder, and they had been filled with some gum or glue to hide the flaws, or cracks. It was the biggest kind of luck that it did not explode the barrel of the pistol and hurt me or someone else. And I saw how mean and treacherous S. M. Charles was, for he had got rid of a dangerous pistol that was worthless and he did not care who got killed or hurt with it. And it looked so much worse, that he would profess friendship and only would part from it for me!

We fell in with some more company that night going to Fra[s]er by way of The Dalles, as we were going. They were Californians, and there was an old Frenchman camped where we did, too. On the 8th day out we got to a stream a few miles from Forster's. . . .

Next morning we took to the Cascade Mountains, passing quite close to Mount Hood, the deepest-snow-covered and highest peak on the continent of North America.[8] We had a splendid sight of Mt. St. Helens, Ranier, Jefferson, the Three Sisters, and several high mountain peaks in Oregon and Washington Territory. It took two days to cross the Cascade Mountains, and such jumping logs, and down trees and brush you never saw. We had to walk a great deal of the time, so that the mules and horses could jump the high logs, for if we had stayed on their backs, they could not have jumped them without the chance of being thrown off.

The 9th night out we stopped in Ty[gh] Valley at Ty[gh] Hotel. Next day we got to The Dalles in the afternoon, thirty miles. We passed some fine farms and stock ranches in the valley, and on the different creeks we passed many of them between Ty[gh] Valley and The Dalles. We had to stop at The Dalles for quite a while to buy more horses and to fit out for Fra[s]er [River], supposed to be five or six hundred miles by land, through wild, warlike and treacherous Indians.

The Dalles was quite a business town or city, three or four thousand inhabitants, with a government fort on the hill overlooking a city. A

---

[8] Mount McKinley, Alaska, has this distinction.

half a mile above, on the hill southeast, several companies of soldiers were in the fort, and Col. Steptoe and two or three companies of dragoons were up Snake River fighting the Spokane Indians.[9] Col. Steptoe whipped the Indians, and took 600 of their horses or ponies, put them in a corral and shot them all. He swore he would leave the Indians all on foot, and it whipped them the quickest way and made them know that the whites were in earnest and not fighting for ponies or plunder.

While we were at The Dalles, the Indians surprised and killed 60 dragoons out of Steptoe's command, and Col. Steptoe had to fort up and entrench himself for fear of a simultaneous attack from all the Indians, which was expected until he could be reinforced from Walla Walla and The Dalles, and it made quite a stir among the miners on their way to Fra[s]er River, as we had to pass right through the several Indian tribes that were fighting Col. Steptoe, the Yakimas, the Wenatchees, Spokanes, Snakes, Columbia River, [Ca]yuse, and some renegade Nez Perce. All were at war with the whites.[10]

Sometime in May, Major Robinson, with 175 miners, started from The Dalles by way of Fort Simcoe, and the Weewic[h] and Yakima and Wenatchee. At the Yakima River he was attacked by the Indians, and drove from over sixty miles northwest in Washington Territory to Fort Simcoe. He had lost seven or eight men killed by the Indians, thirty or forty horses and mules, and came near cutting them all off from getting back to the Fort Simcoe.[11] Some soldiers were sent out from Fort Simcoe,

[9] Colonel Edward J. Steptoe, a Virginian by birth, graduate of the United States Military Academy, was commandant at this time of Fort Dalles. He resigned his commission in November, 1861, and died in April, 1865. The fort alluded to by Reinhart was located on Mill Creek in the western part of the community.

[10] Reinhart gives an erroneous account here. Colonel Steptoe set out on May 6, 1858, for Colville to do something for the protection of miners who were on their way to the Fraser fields. His command consisted of 159 men. On May 17, in the Spokane country, Steptoe's troops were attacked by Palouse, Yakima, Spokane, and Coeur d'Alene Indians. He was compelled to retreat, sustaining casualties of eleven dead, fifteen wounded. With the aid of friendly Nez Percé, he was able to safely cross the Snake River and reach Fort Walla Walla.

In retaliation for this attack, General Newton S. Clarke, commander of the Pacific Department, decided upon a decisive conflict. In August a two-pronged expedition was sent out to chastise the Indians. The divisions were commanded by Colonel George Wright from Fort Walla Walla and Major R. S. Garnett from Fort Simcoe. The expeditions were successful: the expeditionary force consisted of one thousand men, the largest ever assembled in Oregon or Washington for hostile purposes. Some eight hundred horses were destroyed by the Americans to prevent recapture by the Indians. By September 12, the Indians surrendered. Thomas W. Prosch, "The Indian War in Washington Territory," *Oregon Historical Quarterly*, XVI (1915), 21-23; Prosch, "The Indian War of 1858," *Washington Historical Quarterly*, II (1908), 237-240; Elwood Evans, *History of the Pacific Northwest* . . . , I, 626-639, prints the official reports of Steptoe and Wright.

[11] Since the Indians in the region north and west of The Dalles were hostile, the overland trek to the Fraser gold fields was extremely hazardous. The formation of companies of miners for safety was a necessity. These companies traveled horseback. Joel

but the Indians kept out of their way up in the hills, so that the soldiers could not nor dare not follow. And they had not taken much provisions with them, so they returned to Fort Simcoe, and the Indians followed right after them, clear to the fort, and killed three or four soldiers and teamsters that were up the valley four or five miles, cutting and putting up hay. And would run in any party that would undertake to get out of the fort. A young chief and a war party came within 600 yards of the fort and dared the soldiers. One of the officers of the fort had a target gun and as the chief's son was patting his back at them, took an elevated shot at him. He was 800 yards off, and when he fired, the chief fell, and the Indians carried him further back, out of reach of the Boston gun. Soldiers were called "Bostons" by the northern Indians all through British Columbia. The chief's son died of the shot and his father swore revenge on all whites and "Bostons," so it looked rather gloomy about our going through the country to go to Fraser River.

Then every little while, some exaggerated report would come in, of the rich strikes being made on Thompson and Fraser Rivers, and at the Yakima, S[i]milk[a]meen and Shuswa[p] Lake, men making from six to twenty ounces per day, per hand. Everybody was excited and the government would not let small parties start.

After staying some four weeks at The Dalles, Maj. Robinson said if he could make up a company of 300 men with plenty of arms, ammunition, horses and mules and provisions, he would take us to Fraser River if we had to fight the Indians every day. When Major Robinson was out with the first party, he had along [a] Klick[i]tat Indian, [who] had been living and hunting with the Yakima and Wenatchee Indians, but did not like them, and Indian-like, would no doubt have robbed his own father. While the whites were fighting the Yakima Indians, this Klick[i]tat Indian by himself got in behind the Indians where they had their ponies in a corral with their young children to watch them, and while they were all fighting the whites, got into the corral and opened the bars and run out the herd and stampeded them, and came in with the whole herd into the road back of Major Robinson while he was making his retreat to Fort Simcoe. The miners heard the noise of the running herd, thought it was the Indians gaining on them, but happened to look and see that they were loose horses, stopped to let them pass, and to their great surprise, saw an Indian driving them as hard as they could go. And when the Indian saw the whites, he called out for them to help him drive the herd he had

---

Palmer took through a party of thirty-five in wagons from Portland to the Thompson River without hinderance, however.

Records fail to disclose a Major Robinson or Robertson (as Reinhart later calls him) on territorial military duty in 1858–1859. Most likely he was a civilian who served as captain for the overland party of miners here mentioned.

Fort Simcoe was established in August, 1856.

stolen from his old friends the Indians. He had rode until his horse would get tired, and take or catch another to ride, and he had killed several on the road when they give out or could not travel up with the herd. He brought them to Fort Simcoe and sold half of them to Fred White, a government herder.

He had given some horses to all the miners that had lost theirs by the Indians, and they all rode into Fort Simcoe. Then Fred White and the Klick[i]tat Indian came and drove them to The Dalles and sold some of them to the miners. But the Indian chiefs sent in a flag of truce to the commander of the fort at The Dalles by some Indians for him to deliver up the horses the Indian, the Klick[i]tat, had stolen from them, or they would attack the city of The Dalles and burn it up, and notified the mayor and city council of the same. The sheriff tried to take the horses from Fred White and the Klick[i]tat, but the commander of the fort no doubt had posted Fred White, and he run them over the Cascade Mountains, down to the Willamette Valley, and they sold them, so that the authorities could not find them. So [they] told the Indians with the flag of truce. They went off threatening, but the people at The Dalles kept a better watch than they had, and all passed over. (We had met the horses before they got to the Cascade Mountains at the head of Ty[gh] Valley; there were four or five persons with them.)

The government was buying up mules at the fort to work on wagons to haul supplies to Col. Steptoe and we got a good price, and bought cheaper Indian ponies, sold our nice riding saddles and bought pack-saddles. One of our riding saddles cost $45; the other $30 at Kerbyville, Oregon. We got about $60 for the two of Jack and Joe Crabb, who were keeping a livery stable at The Dalles. We kept our horses at Crabb's livery stable, and boarded at the Cushing Hotell.

Horse racing and foot racing and gambling was all the go among the miners, Californians and soldiers. Cutting and shooting, fighting all over town. The Dalles at that time, June 1858, was a lively place or city. It was in Wasco County, 160 miles up the Columbia River from Portland, above the Cascade Mountain range.

Parodi Bartholomew, my old partner, got a job cooking at the Cushing Hotell[12] to save expense (he was awful close). One day [a] friendly Indian brought some fresh fish to the hotell to sell. The fish were hanging to the back of the pony's saddle that the Indian rode. Parodi went to look and handle the fish hanging on the horse. The pony, a wild one, kicked Parodi in the side. He dropped as if he was killed, but he came to. He had a bad cut, and he had to lay down and keep quiet or mortification might set in and kill him. So he was laid up in bed, and when our company got ready to start, he did not dare go along yet, on account of his

---

[12] Milo M. Cushing erected a log hotel on Front Street in 1853, after his discharge from the army. William H. McNeal, *History of Wasco County, Oregon*, pp. 29, 83.

1. Yreka, California, in 1851. Courtesy of the Siskiyou County Historical Society.

2. Jacksonville, Oregon, in 1854. From *The Pictorial History of Southern Oregon and Northern California*, courtesy of the author, Jack Sutton, Grants Pass, Oregon.

3. Browntown, Oregon (c. 1852–1853). From *The Pictorial History of Southern Oregon and Northern California*, courtesy of the author, Jack Sutton, Grants Pass, Oregon.

4. A whipsaw-mill near the Oregon mines. From *The Pictorial History of Southern Oregon and Northern California*, courtesy of the author, Jack Sutton, Grants Pass, Oregon.

5. An early gold mine, with sluices and pans. Courtesy of Jack Sutton, Grants Pass, Oregon.

6. Fort Yale, British Columbia, in 1858. From *Harper's Weekly*. Reproduced by permission of The Huntington Library and Art Gallery.

7. Fort Langley, British Columbia, in 1858. From *Harper's Weekly*. Reproduced by permission of The Huntington Library and Art Gallery.

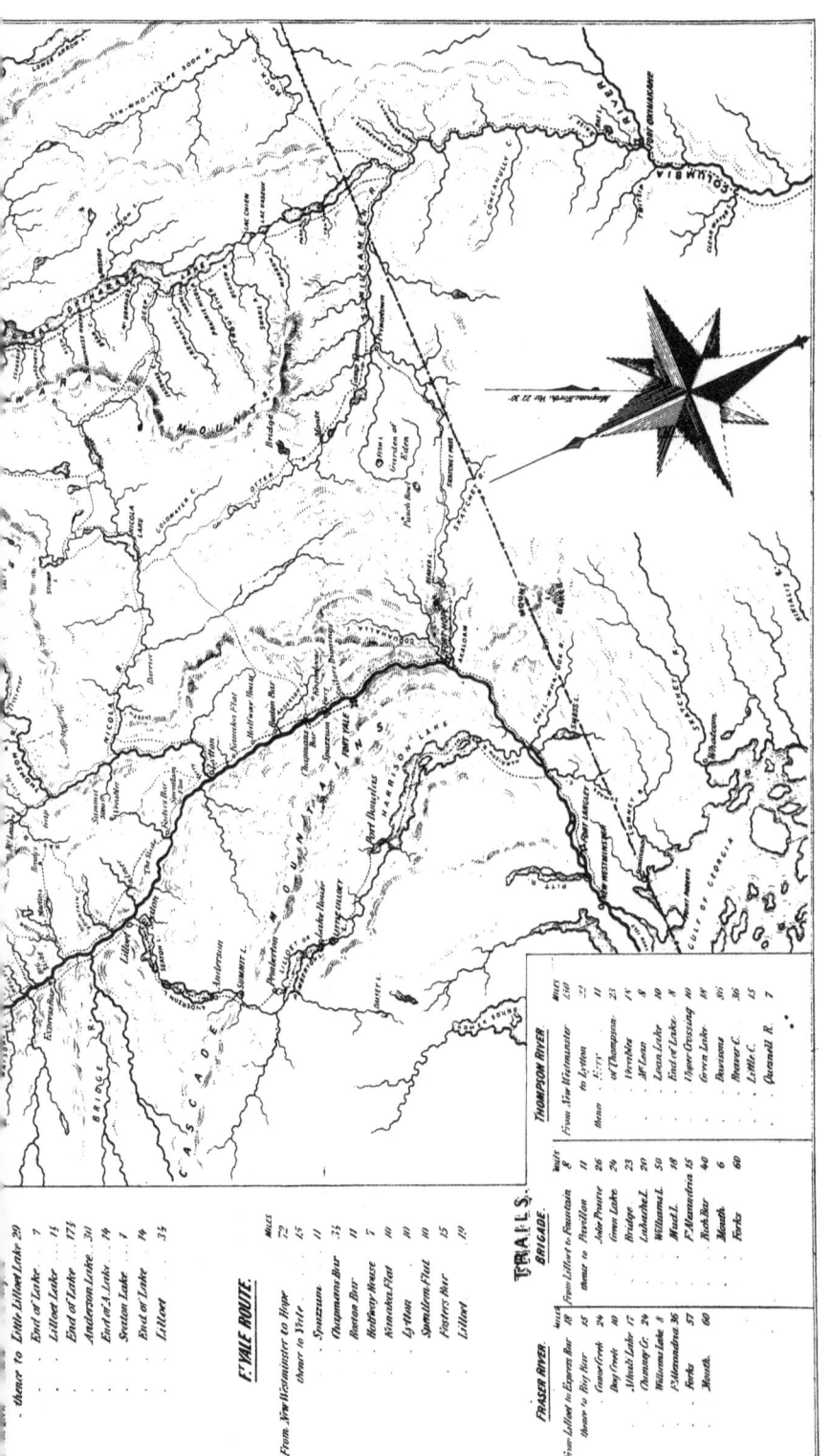

8. Map of the gold regions of British Columbia (1862) by Gustavus Epner. Reproduced by permission of The Huntington Library and Art Gallery.

9. Freighting to the Idaho mines. Reproduced by permission of The Huntington Library and Art Gallery.

10. Helena, Montana, in 1866. A Wells Fargo stagecoach on Main Street. Courtesy of the Montana State Historical Society.

11. Virginia City, Montana (c. 1866). Courtesy of the Montana State Historical Society.

12. Fort Benton, Montana (c. 1867–1868), showing the waterfront with a steamboat at the dock. Courtesy of the Montana State Historical Society.

13. Salt Lake City, Utah, in 1869, from the top of the tabernacle. From A. J. Russell, *The Great West Illustrated*. Reproduced by permission of The Huntington Library and Art Gallery.

14. Bear River City, Wyoming. From A. J. Russell, *The Great West Illustrated*. Reproduced by permission of The Huntington Library and Art Gallery.

15. Supply trains in 1869. From A. J. Russell, *The Great West Illustrated*. Reproduced by permission of The Huntington Library and Art Gallery.

wound, so he concluded to stay and let us two go on, and let him know after we got there, and if favorable, he would be well by that time, and he could come by water from Portland where the regular line of steamers touched on their way from San Francisco to Victoria, Vancouver's Island....

All persons wishing to go with the company of Major Robinson were to go to Fort Simcoe, sixty miles northwest, there to organize into companies, choosing captains and lieutenants. Sidolia, the Italian, and I, had four pack-horses. I had traded my pistol for a pony to a man named Gifford (I got acquainted with him afterwards at Clearwater and Walla Walla). We packed up our ponies with provisions, tools, a whipsaw and a crosscut saw (so to go to whipsawing out lumber and sluicers for sale) and a set of carpenter tools for Sidolia. Our horses were not used to being packed, and would stampede whenever they could, and it kept us very busy packing and repacking and running after them.

After crossing the Columbia River at The Dalles on a ferry,[13] after running and working so hard with the packhorses, I was hot and thirsty, and I nearly gave out. And no water to drink. I saw a sort of swamp or slough about a half mile off the trail; I went to it; the water was hot, green and slimy, and full of weeds, but I was suffering for water, so I waded in where I could stoop down and scoop some of that hot nasty stinking water into my mouth. I would have given five dollars or more for a glass of cool water, lemonade or beer. But we got to a camping place after dark, all tired out the first day, and only came eight miles from the ferry. The next day was nearly as bad as before. Our horses kept running away and kicking the packs off, and stampeding. (We had a gentle large gray American horse, brought with us from Kerbyville, and Sidolia had rode it to The Dalles. Now we packed it with some flour and our whip and crosscut saws, screwed down in between two long boards to keep the saws firm and stiff. We did not dare trust that part of the load on the other ponies.) But we made out to get twenty or twenty-five miles, and camped with more company.

The third day we done some better—not so many repackings and runaways—and we got to within six miles of Fort Simcoe, and camped. There were a great many miners camped about us, making up different companies. On the road to The Dalles I got acquainted with a young man from Yreka; his name was William Cochran, and he camped with us, and he and I were going up to Fort Simcoe to see when Maj. Robertson's [Robinson] company would be ready to start. Sidolia, the Italian, told me if a company was being organized, I should sign his name and mine, and what arms, ammunition and horses we had with us.

When we got to Fort Simcoe we found a nice place, several companies

---

[13] The ferry was operated by James Herman. *Ibid.*, p. 133.

of troops from New York, Boston, and eastern cities. I went to the baker's and we bought a lot of bread to take home to camp with us. We found a company being made up, Company A. I joined, so did Bill Cochran, and I signed for Sidolia. Our company was 56 men strong. Another was got up; Company B was mostly Californians from Redding's Springs and Colus[a], eighty members strong. Company C, a French company, organized, and Sidolia, the Italian, enrolled himself while I was at the fort. And when I told him I had put his name down in Company A, as he had told me to, he said he would rather go with the French company as he could talk to them best. I would not leave my Company A to go with the French, for I could not understand them. He said the French were the most experienced in prospecting and some had already been in British Columbia trapping, and they knew the country.

So he said he would take three of the pack horses and I could keep two, and I could go in my Company A with Bill Cochran; all the companies were to travel together anyway. I took the two poorest and lightest-loaded; he took the heaviest and strongest. He claimed the French, with him, were good packers, and would help him pack up and he had the big gray loaded with the flour, tools, and two saws in the long boards. Company D had some forty-five men, and Company F was a company with Dancing Bill (William Latham). I was acquainted with him at Yreka and Humbug Creeks, California, in 1851-52, and at Althouse Creek in 1853, at Gold Beach and Indian Creek in 1854-55. His company, all old Californians, [had] some hard cases. Several Spanish pack-trains [were] loaded with groceries and provisions and liquors and another with a set of blacksmith's tool, and stock of iron and steel. Some French pack-train[s] [had] groceries, provisions and liquors, boots and shoes, and kit of blacksmith's tools, . . . in all about 300 men, six companies, and about 700 head of mules and horses. We all had rifles and one or two large dragoon or navy revolvers to each man, and plenty of ammunition to last us three months. We had all selected our captains and lieutenants, and all were under Maj. Robinson. All started out from Fort Simcoe in good order.

The third day out got to the Yakima River, where the Indians had killed some miners out of the former Robinson company, and we found and buried two bodies (what was left of them—the wolves had nearly eat them up). About the fourth or fifth night, while standing guard, a German of Company B shot a man. He halted him, but the man would not stop, and he fired and wounded him in the side, quite dangerously. It transpired that some of the men in his company thought to try the Dutchman to see if he *would* shoot, and he made fun and said he would scare the damned Dutchman into camp. But he found to his sorrow that he would shoot. We laid by that day with the wounded man, and through medical advice he was taken back to Fort Simcoe. Some two of his comrades took him back to the government hospital at Fort Simcoe, and next day we started on our way.

We all had to stand guard half-night apiece, or each every four or five nights. Or about forty every night [because of] so many horses or mules to herd mornings and evenings. In traveling we had an advance guard of about 25 men, and a rear guard of 16 to 20. Every day a different company took the lead, and the next day it would fall back and the hind company would take the lead, and so on every day by turns. We had a doctor and if sick [we] could get medicines or an excuse from standing guard when sick.

Our major and advance guard would pick our camping places. We had a fearful bad road or trail—no road, just an Indian trail, over rocks, hills, mountains and streams both wide and deep and cold. Many of us had to walk, and all in a hurry to get to Fraser River. We were loaded too heavy, and drove too fast for our stock, and every day some ponies would give out and be left. Some would kill them, but mostly would just leave them to graze and shift for themselves. I had to leave my roan mare that I had traded my S. M. Charles pistol for, eight or nine days from Fort Simcoe. I had to put the most of his load on the other one, and some on Bill Cochran's. In a few days my bay horse commenced weakening with the heavy load, and I had to get [a] packer to take a hundred pounds of flour for me. Bill Cochran's commenced weakening and we took our only riding horse and put part of the load on him. Every day horses and mules gave out, until forty or fifty men would have to walk, or change with each other, half the time, riding.

Then we had many long stretches without water, road rocky and again so dusty. 700 horses and mules passing along raised an awful dust, and us foot-men had to walk fast behind each other's mules and horses, that sometimes we were nearly choked with dust and drouth. Our boots would hurt our feet, and some of us would have to walk in socks or rags, or make carpet-shoes to keep the rocks and prickly pears from our feet. Some of us suffered awful, but there was no stopping, we had to keep up with each other or be left behind in a dangerous Indian country, who were watching us to cut us off from each other, or attack us every opportunity.

I never suffered so much in my life as on that trip to Fraser River. My ankles [are] naturally weak, and I would sprain them every little while; they would swell up so that I could hardly get along, but I had to drag on anyway. I made me moccosins of carpet from a saddlecloth I had, and I would have to put on a new sole every night after I got into camp. How glad many of us were when we stopped to do something at the crossing of some stream, or to water all our stock at some lake, so that we could sit down and rest our sore feet. But the different companies, some wanted to go faster, others slower. Some would get up earlier and be ready to start sooner than others, and such a confusion and wrangle I never saw!

Some would threaten to go on, and accuse each other of cowardice and fear of Indians. The old Californian miners and Indian-fighters were the worst; they claimed they could travel in small parties and clean out all the Indians in the land, that others were all cowards, and it was all our major and captains could do to control their men, to stay all together, for if we had scattered out and separated from each other, the Indians would have completely annihilated us all, and all men of reason knew it.

When we got to a small stream emptying into the Columbia River, called the Weewich, a good many miners wished to lay by a couple of weeks to prospect the stream and its gulches, for some gold had been found, good prospects; but the majority wanted to go on. It was left to a vote and was carried to go on and not stop to prospect till we got to Thompson River or Shuswa[p] Lake. We got to Lake Chelan and had to construct rafts for ourselfs and goods and swim horses and mules over the river and lake.

One morning we had camped within a mile or two of a large stream that ran into the Columbia 10 or 12 miles below the mouth of the O[kanogan] River and Fort O[kanogan].[14] Some of the miners had lost sight of their horses, and before they had got them packed up the head of the train started on. Either they did not know that many were not ready—or they did not care. William Cochran's horse was one of the lost but we found him sooner than some others did theirs, and we loaded up and started to catch up to our company, just crossing the stream or river. Will and I had to get on our packs on the horses to cross. The water was very cold and waist deep. Our horses were very weak and our weights on them made them more so. So we had a hard time in crossing to the other side. Then we got off and straightened our packs. Men were passing and re-passing, belonging to our command. I could still see some pack-trains who had not finished loading, and some getting up their animals, so we pushed on slowly. My bay horse was nearly give out, and awful weak. I expected every hour to have to leave him. Our train was scattered for two or three miles along the bottom, on the side of the Columbia River. We could see the advance guard way ahead, over three miles. The bar next the river was narrow, with [a] high mountain on our left. We were on the bank of the creek we had just crossed and had started our three head of horses ahead of us, driving them in the trail, when all at once we heard some shooting across the creek we had just crossed.

We stopped and looked back and saw four horsemen coming out of the

[14] Reinhart consistently spells Okanogan, O'Kenakin. The route taken by his company is described as The Dalles-Okanogan-Kamloops route. *The Oregonian*, February 4; *Oregon Statesman*, February 14, 1860. William C. Brown, "Old Fort Okanogan and the Okanogan Trail," *Oregon Historical Quarterly*, XV (1914), 1–38, describes the trail north.

stream, running their horses as hard as they could go. We thought they were racing, and Will and I both stopped and rested on our guns to see who would beat. As they came closer I saw they were two Mexicans and two Frenchmen, and they were shouting "Indians! Indians!" and had their pistols in their hands and spurring their horses at their best speed. When near they called to us to run, that the Indians were killing some Frenchmen and Mexicans right behind us. They tried to help us drive our horses, but my bay, nearly-give-out horse, turned to one side, and Will Cochran's turned to the other side of the trail, and the third kept along the trail. So the Mexicans and Frenchmen told us to leave the horses and run ahead, and they went on and left us, and run their horses by our jaded animals.

So I called to Bill Cochran to get on his best horse and start on ahead in the trail and I would drive the two others, for they would be apt to follow better. He done so, and I put a fresh cap on my rifle and thinks I, "I cannot get away anyway if the Indians are a-horseback," so I drove as fast as my poor old bay horse could go, in a slow walk. Now all the money and gold specimens and my $24 ring and my fine double breastpin were in a sack with some clothing of mine, and it was made fast with a rope, with a sack of flour and a five-gallon keg of East Boston syrup. I could not take time to take the sack of clothes and my money out off the horse in the excitement. So I told Bill Cochran I would stay with the bay horse, for if I left him I would be broke anyway. I made up my mind to stay and fight it out and not try and run and leave everything I had in the world. Just then two or three shots were fired behind me, and the bay horse struck a trot for about fifty yards and weakened again.

I expected every minute the next shot would strike me, but we kept on and the first thing we saw, our Maj. Robertson [Robinson] and some of the advanced guards coming back toward us, not very fast but badly frightened. And they asked us where the Indians were. I told them to hurry back and go faster or they would not see any, and they clapped spurs to their horses, and some whipped them, and back they went until eighty or a hundred men had gone back. We kept on and found our train scattered for five miles along the Columbia River. . . . If there had been forty or fifty determined Indians, they could have commenced behind and killed half of our whole command whilst they were running to get on the next bar [of] the river to draw up in shape to fight. I never saw so many men so well armed and able to fight any amount of Indians, had they been rightly handled, and not so panic-stricken. Some laughable incidents took place; some of the French company jumped on their pack horses and rushed them by each other running over one another and knocking down all in their way. One old California miner named Pike was so excited that he said he saw at least fifteen hundred Indians on the side of the mountains [ready] to attack us. When we got up to where our company was, we stopped and waited to hear from the rear. As yet no news had been brought forward, but by the four men that gave the alarm.

After a while some of our men came back where we were ready to stand an attack if the Indians should defeat our Major and what men had gone back. And soon they returned with one dead man of Company B. Now I will relate the circumstances.

There was a French blacksmith in the French company; he had some nine or ten mules loaded with goods for a general store such as groceries, boots and shoes, clothing, a set of blacksmith tools all for himself; he had bellows, anvil and vise, and iron and steel. He was from Yreka, California, and had an old Mexican man helping him to pack his mules. Another Frenchman in the same company wanted to buy a pair of boots from him on time, or when he could get work, when he would get to Fraser River and make money to pay him for them. He was nearly barefooted. But the blacksmith refused to let him have them. . . . the French in his company did not like the blacksmith because he had plenty of means and was too close or mean for them. That morning the blacksmith's mules were hard to find, and it was late when he did find them, so that the French company got ready and just left him long before he got ready, for none would help him to load. So he and the old Mexican were working as hard as possible to get loaded and catch up to the balance of the train. They were way behind when they all got loaded. They noticed some eight or nine Indians on horseback coming toward them. There were some four or five Mexicans and French just starting out, and had got a few hundred yards from camp when they heard some firing of guns, and looking back, saw the Indians, shooting at the old Mexican and blacksmith and driving off their pack animals. But the four or five Mexicans and French, instead of going to help the blacksmith and old Mexican, came on after the train and to those ahead, with the result already stated.

Now if the four horsemen and one or two close by had gone back and helped the blacksmith and his Mexican (they all had big revolvers and two or three rifles or shotguns), they could have whipped or drove off the Indians, there being only eight or nine in number. But they got frightened and thought there was a large body of Indians, got excited and left them to give the alarm and get out of danger themselves. Or if they had told us, we both had rifles and revolvers, and altogether we would have been about the number of the Indians. But the old Mexican, when shot, fell as if dead, and crawled off into the brush. The blacksmith had to run off, the Indians shooting at him as he ran. Two or three Indians rode on after a white foot-man right ahead, and not over sixty yards behind me, and killed him. And then turned back and with the other Indians drove off the blacksmith's nine pack-mules and his one riding mule, and on the saddle of the riding mule a pair of leather cantinos containing $700 in gold coins, and revolver and a shotgun hanging to the

*To the Fraser* 123

riding saddle. The Indians just rushed the mules up the mountains, two of them too heavy-loaded to drive fast, had on the anvil, bellows and some iron. They just cut the ropes and cut the packs loose, and drove the mules right along with the others, and by the time the Major and our men got back where they had been, they were three or four miles up in the mountains, and the Major and men were afraid to follow for fear there might be an Indian ambush. So they let them go without a struggle, for Maj. Robertson's [Robinson] former bad luck did not encourage him to take the chances of being drawn into a fight on the mountains and maybe ambushed, and the balance of the train five miles off, and maybe watched by thousands of Indians. At least [he] was being cautious after the blunder of the morning.

The French blacksmith came out of his hiding-place to where our men was, and told them that the old Mexican was shot dead not far off, that he saw him fall. But just then the old man came out of the brush where he had been hiding unhurt. That man killed behind me belonged to Company B and had lost a horse and had been out across the stream, and Bill Cochran and I met him in the stream on his way back to camp. He told us of losing his horse (we were on the tops of our packs to keep from getting wet and cold), [when] he saw the Indians attack the blacksmith and come running after us. The two or three Indians saw him alone and that he had a pistol with him and came after him on horseback and shot him with their rifles; even after he lay on his face they shot him in the head. It was their shots we heard so close by [that] made my bay horse trot, they being but about sixty yards behind me. But there was a ridge between me and them and they did not see us two, nor we them, or if they seen us, saw that we both had rifles, and it might not have paid them to take equal chances with us.

Now the strangest part of the whole thing was that the man killed had his horse taken and no other; I thought it strange and it came to my mind. He was a teamster at The Dalles, an Irishman named McCandless; he joined at Fort Simcoe. One evening two or three days out from Simcoe an old Indian was caught prowling around our camp and stock, looking among the horses. He was arrested by the guard and brought to the Major's headquarters to see what should be done with him. The Irishman killed was very loudly in favor of shooting the Indian as a spy and was very persistent in trying to influence other miners to have him shot, and every Indian we should meet, friendly or hostile. The Indian understood the Irishman's threats (what he was in favor of and he would do). The Indian said he was a friendly, peaceable Indian, a Yakima, and did not mean any harm in coming to our camp, that some of his horses had been drove off by some Indian horse thief and taken to The Dalles, and he thought he knew the horse McCandless had was his horse, but by a vote the Indian was told to go and not be seen again or he would be shot. And

McCandless, the Irishman, told the Indian he had a good notion to shoot him. My surmise is that the Indian and his friends followed us for an opportunity to steal horses or to kill a lot of us, but could not get a good chance until then. They knew his horse and took it, knowing he would have to hunt it, and they could maybe kill him, which they did by chance, by the[re] being no rear guard that morning. . . . We [knew we] were close to Fort O[kanogan] . . . [and] thought there would be no danger of the Indians, although we knew we were watched and followed all the way from Fort Simcoe.

We camped on the bar and dug a grave close to the bank of the Columbia River and buried McCandless. No one knew anything about him, only [that] he had drove teams for the government at The Dalles. He may have been in the first Robertson [Robinson] company when drove back by the Indians two months before; it may have made him more bitter toward the Indians and [made him] wish to kill them whenever he could for revenge. So it ended.

Next morning about ten o'clock we got to Fort O[kanogan], an old Hudson Bay trading post.[15] A few half-breeds and a couple of Frenchmen with Indian wives kept the place, and plenty of Indians (they said they were friendly) around. They did not seem to know of the attack on our train and of the killing of the man. I think they did know, and maybe had a hand in it. If we had had any proof of it, we would soon have taken and burnt up the old fort and killed every one of them—our men were just in the humor for it.

. . . We had not gone but a few miles until [my bay horse] gave out, so I had to leave him. So my two horses had been left within ten or twelve days out from Fort Simcoe. . . .

We crossed the O[kanogan] River close to a high rocky pass canyon, and some of our men made remarks that some large body of men and horses had been running and tramping around—signs of Indians on the hills—and the men thought what a place it would be for the Indians to attack us, or to surround and ambush us in the canyon wh[ich] we had to pass through. Twenty-five well-armed men among the rocks could have defeated our whole train, being away up among the rocks which they could have rolled down on us and stampeded our horses and mules.

We felt a little squeamish when some of our men found a place where there were four or five new-made graves. On a headboard it says:

Captain David McLaughlin and 157 men passed on the 27th of June, 1858, and in an Indian attack and fight at this crossing of the River and Canyon five of their men were killed, and the company had to back out of the canyon and cross back over the river and go around this pass or

---

[15] Established 1811.

canyon. The Indians had defeated them, killing 28 or 30 horses.[16] But we got through all right, without trouble, and we had a guard of men on the top of the highest rocks in the canyon to keep any Indians from attempting to molest us. I knew Dave McLaughlin. He was a half-breed son of old Dr. McLaughlin of Oregon City.

The old doctor had been superintendent of the Hudson Bay Fur Company; he was very rich and Capt. Dave was his only son. He was considered a fast young man to drink, gamble and carouse, and a great Indian fighter and scout in several Indian wars on the Pacific coast.[17]

His company left The Dalles ahead of Maj. Robertson's [Robinson] first company, but they kept up the Columbia on the east side up to Fort Walla Walla, then on to Spokane and crossed the Columbia River just below Fort [Okanogan], and here was our first news of him or of his company.

For a few days we traveled along with great care, constantly on the lookout for an Indian attack. We crossed several nice streams and fine looking farming and grazing land, and got to the British line. Here about a hundred Californians out of our train concluded to go a different route, by way of the S[i]m[i]lk[a]meen, then on to Fort Hope, down low on the Fraser River. We tried to talk them out of going that way, but no, they were not afraid of Indians, and could travel where they wished to for all the Indians in British America. They were mostly from Northern California (I will speak of this command further on).

In a few days we got to O[kanogan] Lake. Our advance guards saw some Indians just leaving their camp and cross the lake in canoes for fear of us. The boys saw a couple of their dogs at their old camp ground, and shot them down, and they saw some old huts where the Indians had stored a lot of berries for the winter, blackberries and nuts, fifty or a hundred bushels. They helped themselfs to the berries and nuts, filling several sacks to take along, and the balance they just emptied into the

---

[16] McLaughlin's party rendezvoused at Walla Walla. The party was attacked first two or three days travel beyond the Columbia on the east side of the Okanogan River. They sustained several dead who were buried. (This is Reinhart's reference.) Two or three days later the McLaughlin party was again attacked on the west side of the Okanogan River without loss. Hubert H. Bancroft, *History of British Columbia*, pp. 367–368, dates the attack in mid-July and fixes the number of the company at 160. That number is supported by an item in the Victoria *Gazette*, August 24, 1858, which reported that a "Mr. Tucker, formerly of Tehama, Cal.," had arrived at the Forks, the confluence of the Fraser and Thompson rivers, in a company of 160 men with 400 animals, from The Dalles. The account (based on an August 17th letter) noted that the trip took thirty days and that the party was attacked at Fort Okanogan by Indians, sustaining three killed and six wounded.

Robert Frost, ("Fraser River Gold Rush Adventures," *Washington Historical Quarterly*, XXII (1931), 203–209, a member of the McLaughlin company, states that the attack took place in McLaughlin Canyon on the Okanogan River some twenty miles south of the Canadian boundary.

[17] Dr. John McLaughlin married Alexander McKay's half-breed widow. David's reputation is fairly stated by Reinhart.

lake, destroying them so that the Indians should not have them for provision for winter. I, and a great many others, expressed their opinion that it was very imprudent and uncalled for, and no doubt the Indians would retaliate. But they only laughed and thought it great fun to kill their dogs and destroy and rob them of their provisions. Most everyone but those who had done it disapproved of the whole affair.

The next night we camped on the bank of Lake O[kanogan], which is about 150 miles long and from one to six miles wide. Next morning a man named White, of Company B, could not find his horse. Some of his friends helped hunt for it, but as the train went on the men were coming down the hill, and someone fired a shot at White, and some men above him on the hill saw some Indians trying to cut White off from his companions. The men called to White to go down as the Indians were after him. So they gave up the horse, and did not look any more, for the train had already started on.

We traveled along the lake all day and camped on the banks at night. Every morning after we left camp some Indians would come across the lake in canoes and look over our camp grounds to look if we had left or thrown away anything (sometimes we threw away old clothes, hats, shoes, shirts or old blankets or crusts of bread or meat, and they would come and get them after we left. . . . That morning the advance guard planned to punish the Indians if they should come to camp as usual after we left. So right after breakfast some 25 men concealed themselves in a gulch close to camp, and the train went on as usual. We were passing along a high trail close to the lake and we soon saw three or four canoes start to come across from the other side, with seven or eight Indians in each canoe, to go to our camping place. I had gone with the train some one and one fourth to one and one half miles, when we heard some shooting. I stopped to listen and counted over fifty shots.

In the course of half an hour our advance guard that had formed the ambush came up to us and related how they were all lying down in the gulch, to be out of sight, and they got to talking to each other and forgot about the Indians to be ambushed, and they were surprised as well as the Indians, for the Indians had landed and were coming towards camp right to where the white men lay concealed. They had no idea of danger from the whites, so some whites happened to raise up to see if the Indians had landed yet, when behold! the Indians were within eight or ten feet from him, and they did not see the whites till they all raised and made a rush for the Indians with their guns and pistols all ready to shoot. As soon as the Indians saw the whites, they were so frightened that some turned back and ran towards their boat, some fell down on their knees and begged for [them] not to shoot, as they had no arms at all, and they threw up their hands and arms to show that they had nothing. But the whites all commenced to fire and shoot at them, and ran out to the lake after those who were getting in their canoes, and kept on shooting till the few that got into the [canoes] got out of reach of their guns and rifles. And lots jumped into the lake was shot in the water before they could swim out of reach of their murderers—for they were nothing else, for it was a great slaughter or massacre of what was killed, for they never made an effort to resist or fired a shot, either gun, pistol, or bow and arrows, and the men were not touched, no more than if they had shot at birds or fish. It was a brutal affair, but the perpetrators of the outrage thought they were heroes, and were victors in some well-fought battle. The Indians were completely dumbfounded to see a lot of armed men when they expected no one, and ran toward their canoes to get away, and the Indians knelt down and begged for life, saying they were friends. There must have been 10 or 12 killed and that many wounded, for very few got away unhurt. Some must have got drowned, and as I said before, it was like killing chickens or dogs or hogs, and a deed Californians should ever be ashamed of, without counting the after-consequence.

We traveled on, but many of us expected some revengeful attack. We could hear Indians, nights, and saw smoke and signals of lights and smoke on every hill and in every direction to each other in the mountains some forty or fifty miles away. About a week after the Indian slaughter, in the night (the guard had seen Indian tracks in the evening close to camp) the guard brought in two Indians. A mass meeting was called and the Indians were questioned by an interpreter. They were friendly Shuswa[p] Lake, British Columbia, Indians on their way to Colville, in Washington Territory (one of their wives lived there) and with the permission of the old chief Nick at the Fort Kamloops or Thompson, on Shuswa[p] Lake.[18] He was on the way to visit his wife; they had walked into camp without fear or evil intention. They said they had been at the Hudson Bay store at Fort Thompson, and old Nick's tribe were friends to the English, French and Scotch living there, trapping, and many were married to Indian squaws. At first our men were for taking them out and shooting them right off for spies, expecting we would be attacked, but they kept denying [it], and [said] they were good peaceable Indians. . . . At last we came to the conclusion to take them back with us as prisoners to Shuswa[p] Lake, and took their arms from them and always kept guard over them.

One morning Company F (Dancing Bill's) took leave and went ahead. They said we did not travel fast enough for them. Next day a part of the French company started on ahead. They thought they would do better by not traveling with the bloodthirsty Americans. They understood the In-

---

[18] Founded by David Thompson in 1810 at the junction of the two branches of the Thompson River with the eastern end of Kamloop Lake. The former name was in vogue in 1858.

dians better than us, and by their intermarriage with the Indians, expected the Indians on and around Thompson River would favor them with what they knew of the locality of the gold.

Some new discoveries had been made north of the Canoe Country, at or above the forks of Fraser River. Sidolia, the Italian, wanted me to go; he still had all three of our horses. I told him to go on, and after I got to Fraser River, I could come up to where he was. Next night the French company had only gained about one and a half miles, and after they had camped an old Frenchman that had traveled with us a day or two in the Cascade Mountains . . . had left a partner in our train, Company B, and he concluded to come back to his partner, stay all night, and catch up to the balance of his company early in the morning before they packed up, and then go on with them again. So at break of day he started ahead to catch up to the part of the French company he was going with, but after going about half way the Indians intercepted him and killed and shot him through the head, three or four shots, and his body was all shot full of holes. They stripped him and rolled him out of the trail into a gulch alongside of the trail. He had a shotgun; they took that, and no one, it seems, heard the firing at either ours or his camp. We started after breakfast and some of our advance guard saw the blood in the road, and Indian footprints or tracks, came to look close and followed the blood. A few yards below, they found the body, still quite warm; he could not have been dead twenty minutes. So the train stopped and we loaded his body, naked, across a riding saddle, and some men led the horse, and others held on the body, went on over the point of the hill where he was killed.

When we saw the body, we knew the old Frenchman and sent some horseback men ahead to hurry and stop the French train or company to bury their man. It took us three or four miles to catch up to where they had stopped, and we all stopped and dug a grave and buried him. He was perfectly helpless and harmless.

We kept on till we came to Fort Thompson. The Indians kept on the hills and making smoke signals all night, and kept speaking to each other in their own language. Our two prisoners said they were O[kanogan] Lake Indians, and had been following us ever since the slaughter of the Indians at the Lake. They had killed the old Frenchman and were trying to get the Indians on Thompson River to help them kill us all, but the Indians around the Fort were a sort of civilized, and under old Nicholas, and he was a good Catholic, and Capt. Mc[Lea]n of the Hudson Bay Company Fort was his friend.[19] The friendly Indians were all Catholics and had priests at the fort.

The next day at noon we camped right opposite the fort. There were

---

[19] Donald McLean was appointed Chief Trader in 1854, and succeeded Paul Fraser at Fort Thompson. A[drien] G. Morice, *The History of the Northern Interior of British Columbia*, pp. 109, 288.

lots of houses, the first we had seen after leaving Fort O[kanogan]. It made us feel more cheerful and more like civilization, and here the French company parted from us. We kept down the Thompson River to [Kamloops] Lake, where we had to cross over with rafts and canoes, and swim the horses and mules. Some would have to be held up by the heads from out of the canoes. It was a wide, rough place to cross. Some ten or twelve head of horses were drowned and strangled by not being held up properly at the crossing of the lake.

Old Nicholas the head chief of the Indians around that country, came to see us about the two prisoners we had brought back from Lake O[kanogan]. He was an old man about 65 or 70 years old, wore a stove pipe hat and citizen's clothes, and had a lot of medals of good character and official vouchers of good conduct for many years. He was quite angry and said he was surprised to see 300 men take two Indian prisoners and bring them back two or three hundred miles because we thought they were spies, and it was mighty little in us and did not show great bravery. And about the O[kanogan] Lake massacre, that it was brutal, and he could not think much of the Bostons, or Americans, that would do the like. Some of our boys were awful ashamed and some angry to hear an old man tell them so many truths, and some were mad enough to kill him for his boldness in his expressions to us all. But it was a fact none could deny, and Maj. Robertson [Robinson] let the two prisoners go. I think some of the men gave them some clothing and provisions, with some money to satisfy them for their loss of time and trouble.

We found horses were very high around that part of the country and a good pony was worth from a hundred to $250. Many had died and some had been eaten by the Indians a few years ago when the winter was so long and severe that they had eat up their provisions and fish laid up, and many Indians died by famine. The Hudson Bay company had to drive in a lot of ponies for them to eat and live on, for if the Indians had all starved, their fur trade in Fraser, Thompson and that whole country would have been no use to the Hudson Bay company or the British government.

So we traveled on, and in two days got to a stream called the Bon[a]part[e], and we found prospectors passing and re-passing.[20] Next day we passed through a divide in the mountains, a natural pass or canyon as high, but narrower, than the twenty-mile High Rock Canyon in Nevada. We met some Indians and they held out their hands shook them, and they all crossed themselves on their breasts and foreheads. They were all Catholic Indians and had priests among them once in awhile who belonged to the Hudson Bay forts.

The night before we got to the canyon, some of our guards played a

[20] The Bonaparte River flows between the Fraser and Thompson rivers and joins the Thompson north of the confluence of the latter two.

trick on the balance of the train by shooting off their guns and revolvers just before daybreak. And such a getting-up! Men ran out of bed, one boot on, some in their drawers, others nothing, hunting their guns and shaking and quaking with fear, some nervous with excitement, ready to hide—the alarm that the Indians had attacked us. After firing some 25 or 30 shots, it proved to be a hoax of the guards, and if it had been found out who done it, they might have fared bad, but the ones that started it had sense enough to let it not be known, and one blamed the other but kept still as to how it started.

The second night after entering the mountain pass or canyon we camped by a little chain of lakes, and found villages of Indians on the other side. A good many came over, all would cross themselves, would say "How! How!" and shake hands with everyone in a row as they came along the trail. They called us "Bostons" or "Soldiers" and were well disposed Indians. The squaws kept back and were shy of us, and looked innocent of any badness or boldness, not like some of our squaws in Oregon, California or Washington Territory. They lived mostly on fish.

It was late in August [1858] when we got to The Fountain on Fraser River (a large, high bar) where some miners were working along the river.[21] They had come by water from Victoria, up into the Gulf of Georgia, then into the mouth of Fraser River above Fort Hope. They ran into the mouth of Harrison River, up the last to Port Douglas, (named after Governor Douglas of British Columbia). At Port Douglas they struck a chain of three or four lakes which they now navigate with steamers, flatboats and canoes to within twelve miles of Fraser River.[22]

At The Fountain they got Indians to carry or "pack" their provisions and baggage. The first miners [had] gone in there at The Fountain in July. There had been an Indian War lower down Fraser River and the Indians had cut off the heads of many miners, "Bostons," or Americans, until the miners just went to work fighting them and killed all they came across. Since August there had been peace between them and the Indians carried big loads on their backs for the miners, and they paid them big while prospecting. The miners could not take horses around in the mountains and no grass to feed them. White men carried packs of 75 to 125 lbs for $5 to $6 per day; Indians got less.

We all camped at The Fountain and rested ourselves and horses, and we prospected some at the river. There was no flour or groceries of any

---

[21] The Fountain was situated a few miles below Bridge River, at the mouth of Fountain Creek, on the left bank of the Fraser, fourteen miles above the town of Lillooet. The Victoria *Gazette*, August 25, 1858, reported that a company of "some 300, with cattle and horses, had arrived at the Forks from Oregon." This was probably Reinhart's party.

[22] Port Douglas was located where the Harrison River empties into Harrison Lake. The trail described here by Reinhart was subsequently called the Harrison-Lillooet route. See *note 27, post*.

kind at The Fountain, only [w]hat our train had brought in by our packtrain. Major Robertson [Robinson] and a Dalles merchant started a store of groceries, provisions and liquors.[23] Flour sold for $1 to $1.20 per lb., sugar and coffee $1.50 per lb., bacon $2.50 per lb., brandy and whiskey 50 cts per drink or glass, and other things in proportion.

After staying at The Fountain four or five days, six or seven of us miners went up the river to prospect. We each took a saddle horse, blankets and provisions for five or six days, expecting to go to the Canoe Country, a new discovery at the forks of the Fraser River.[24] We heard of some rich strikes having been made.

(I had overlooked a part of the talk we had with old Nicholas the chief at Shuswap Lake. He blamed us for butchering the O[kanogan] Indians in cold blood and the O[kanogan] Indians had sent some messengers to him to help avenge the death of his people, but he said he had better teaching from good men and priests, and good advice from Capt. McL[ea]n, head of the Hudson Bay Company, and they advised him and his people to overlook the great crime. But . . . he had great trouble to quiet and calm down his young warriors, of which, with the Lake O[kanogan] tribe, he could have raised from 1800 to 2000 warriors, and could have surprised our command and cut them off to a man, utterly annihilating the whole of us, and taking all our animals and all our plunder. But he could not have told how it would have gone after, for he would have lost all control of his people, and the war chiefs would have usurped his power and carried on a general war against the whites, American and English. Being the massacre had taken place in British Columbia, it would be the duty of the English Queen Victoria to see justice done to her subjects, and he was right, no doubt.)

We thought we would take a few days prospecting and looking at the country farther north. There was a trail that took [us] up the river at the summit of the mountain range from the coast of the Gulf of Georgia, extending north five or six degrees of longitude, past what is now the Caribou River and country at the head of the Fraser River. (This trail was made by the Hudson Bay Fur Company, called the Brigade Trail, bearing north and south.)[25] By taking this trail in some places we could

---

[23] Efforts to identify Robinson (or Robertson) have been unsuccessful. It is possible that Reinhart is in error on the name of the leader of his Fraser company. The Victoria *Gazette*, July 29, 1858, notes a party of thirty-six Oregonians engaged in mining at Robinson's Bar some eight miles above the confluence of the Fraser and Thompson rivers. The account further notes that the company's leader, Wolff, "an old hunter and trapper," was on his way back to Oregon "with the intention of bringing another party of immigrants through."

[24] The Canoe Country was situated some fifty to sixty miles above Lillooet on the Fraser, and was named for Canoe Creek.

[25] This trail was cut across the mountains by the New Caledonia spring brigade in 1849, so that the Hudson's Bay Company could profitably tap the upper country. It was used until 1860 when a government road was constructed.

cut off bad places along the Fraser River where we wished to prospect as we went along.

It being September, we knew we would have to look around for a place to work for winter, as the season commences early and lasts long and late. We prospected on bars where we thought it looked favorable for gold, but got only light prospects, some places barely the color; some places the banks of the river were rocky and nearly impassable to horses or mules. . . . They said there was a great deal of trouble on the Brigade Trail by Indians and prospectors. Our French company had struck that trail ahead and went up north two or three hundred miles, and had not been heard of by us while we were there.

But the report about the new discoveries at the forks of Fraser River were all a hoax, and we did not find them or any one at the Forks. We talked to some Indians about gold; they would pick up pieces of gravel that would, had they been gold, have weighed from $10 to $15 apiece, and they said they would take us to where we could find them plenty, and some larger, say from $18 to $25 pieces, if we would give them a horse and some provisions. But when I took some of my gold specimens of California gold and showed it to them, they were astonished as to their size (And they were only $4 or $5 pieces). They made so much ado about mine, and some of them had a little bunch of fine gold done up in rags, and they would act as if they thought it wonderful and call each other across the river to show them their small bunches, and when they saw my California pieces they were all awe-struck, and I am satisfied they had seen none like them. So we came to the conclusion that the Indians were lying to us about their finding pieces as large as the gravel, and only wanted to beat us out of a horse and provisions, or would try and kill us if they could have got us to some place favorable, for our horses, guns, pistols, and what money and clothes we might have with us.

So we continued to prospect for five or six days, but did not find anything that would last but a few days, from $4 to $5 per day, and that would not pay in that cold climate where a miner could not work but a few months in the year in surface diggings. So after six days prospecting we were about a hundred miles from The Fountain, and were about out of provision, . . . we had to turn back and get to our camp at The Fountain as fast as possible, or we would have to starve, for there was no game to shoot, and in one place we found strawberries just in bloom (in September!). So you can judge the season. We were in about Latitude north 53 or 54th parallel, or about seven or eight degrees north of the forty-ninth parallel of the line between the United States and British Columbia.

The first day of starting back toward The Fountain, we run out of provisions, and I traded an old saddlecloth (the half of an empty 50 lb. sack) to an Indian for two dried salmon. He used the old cloth for a legging. The salmon was rather dry eating and had no salt in it, but it was better than nothing when a person was as hungry as we were. We did not stop to prospect any more, but kept right down the Brigade Trail, the nearest and quickest way to grub, or something to eat, and we traveled in two days what had taken us five days to come up the river. The third day after getting out of provisions, we came to some of our boys that were prospecting, and got some flour and bread to last us till we could get back to The Fountain, which we made the next day in the evening. We were all tired and half starved and discouraged in the "great, rich Fraser mines" and found most all the boys we had left at The Fountain the same way, and getting ready to start and return to California as soon as they could.[26]

William Cochran and I thought we would give it a few more days prospecting, which we did, and saw some miners close by, on a bar of the river, and we asked them how they were doing, and found that from $3 to $7 per day per hand was as much, or the most, they made. They had their own provision with them, or they could not have made board at the prices of provisions at The Fountain, and not much chance of getting it lower. It being September already, you might expect cold weather to set in, in October, and instead of provisions being lower, would be higher, because none could be got in again till next June by the Port Douglas or the Lilywhite [Lillooet] Route up Harrison River and by the Lake.[27]

After prospecting a few days longer with no success, we came to the conclusion to strike back to California. I had left good $8 and $10 per day diggings on Sucker Creek, where my brother Charley and my Partner Schertz were working my claim in company with them. So I sold some shirts, drawers, and books such as I could not carry with me. Will Cochran had sold his only horse at The Fountain, so we were both left on foot. I had a five-gallon keg of East Boston Syrup. When my last bay horse gave out, I got others in our train to carry it for me, and they had used

[26] The Fraser fields were a bust as far as the majority of the miners were concerned. Many called the whole thing a hoax on the part of the Hudson's Bay Company—which was not true. The San Francisco papers published many an exaggerated and misleading news report. If blame should be placed, those newspapers and San Francisco shipping concerns would be the proper culprits.

[27] After the successful upstream navigation of the *Umatilla* (see note 35, post), Governor Douglas laid plans to construct an access route to the upper country by way of the chain of lakes and rivers to Lillooet. He commissioned Alexander C. Anderson to head a group of miners who were employed for the task to launch the construction in August, 1858. In honor of the governor, Port Douglas, the southern terminus of the route at Harrison River and Lake confluence, was established.

The route indicated by Reinhart was formed by going up the Fraser into Harrison Lake, then up that river to Lillooet Lake, on into a series of lesser streams and lakes via the Douglas and Pemberton portages to Lillooet, situated at the head of Seton Lake on the Fraser. By this route the obstacle of Fraser Falls could be circumvented by the traveler to the upper gold fields. This trail was dubbed the Harrison-Lillooet Road. Margaret A. Ormsby, *British Columbia: A History*, pp. 158–160. See Robie L. Reid (ed.), "To the Fraser River Mines in 1858," *British Columbia Historical Quarterly*, I (1937), 243–253, for a contemporary travel account of the route.

the most of the syrup for their trouble. I took out a quart bottle of it to take with us, and I sold the balance, about five quarts, with the keg, for $20 gold piece, my shirts from $3 to $4 apiece, some undershirts and socks and the books—in all, I had some $75 or $80 left. And my breastpin, ring, rifle, pistol and blankets. I bought one pound of bacon for $2.50 of Robertson [Robinson] and Company's store; we had three or four pounds of flour left, and the bottle of syrup. We started with some five or six others for Victoria, right down Fraser River.

The first night out, four of our party concluded to go back and try and get work and stay all winter. They were discouraged and ashamed and hated to go back to Oregon or California broke, having spent all they had and lost their whole summer's work. They would be made fun of by their friends and acquaintances, so [they] turned back in the morning. One of them [was] named James Brown, of Corvallis, Benton County, Oregon. So Will and I were alone again. It was awful hard walking and carrying our guns, blankets, and what little clothes and grub we had, and our feet got quite sore. The whole country of upper Fraser River, the ground is thick with prickly pears, so that in lots of places you could hardly sit or lay down without getting into the stickers of the cactus. We crossed several bars where some miners were at work, but doing poorly.

The second day we got to the mouth of Thompson River, emptying in to Fraser River from the northwest. We had crossed it just below [Kamloops] Lake, below Fort Thompson, or Kamloops, with our whole train some one hundred or two hundred miles above. The river ran very swift and looked to be nearly as large as Fraser River, and Fraser River did not look to be any larger after Thompson River had run into it than above, but they run swift and deep, and are terrible dangerous rivers, so many rocks and points keep whorling the water, forming whirlpools that not one in a hundred men that get in, ever get out of, or are found after being drowned. So we had to cross Thompson River in a canoe, and swim our horses behind, and we had to pay $3 apiece.

We found some miners working American Bar, just above the mouth of Thompson River, but they did not seem to be doing much in the way of pay. After crossing Thompson River we got to a town.[28] There we saw some men of our command.

They said the lieutenant of our Company A had got killed and was buried close by. His name was John Ellrod, a Californian. A Mexican and some others were half drunk and the Mexican, Joe, flourished a revolver around and accidentally shot John Ellrod in the breast or side so that he died in a short time. The miners came near lynching the Mexican, Joe (he belonged to our train. I used to know him in 1853 on Alt-

---

[28] This was Thompson City, and American Bar was situated about four miles above it on the right bank of the Fraser. For an excellent contemporary description of these mines: Alfred Waddington, *The Fraser Mines Vindicated* . . ., pp. 10–14.

house and Sucker Creek). But he, Joe, swore and said he was sorry and he did not do it wilfully. So he was given 24 hours to leave the camp; if found after that time he would be hung or shot.

After we had traveled some distance from Thompson City, we came across some men that were in the Yreka, California company that had left us on the O[kanogan] River and went by the S[i]milk[a]meen route, one hundred men strong. They had come in on Fraser River below Thompson River some time in August. They told us that the Indians had followed them and cut off a man and killed him. His name was Reynolds, from Yreka. He got careless and dropped behind and was killed with another man. Then again they were attacked in front, a few men on ahead hunting, and two of the men were killed there. They said that Indians troubled them all the time, and at last killed four more men and wounded two more. One of the wounded had crossed the Cascade Mountains with us to The Dalles. He was badly shot in the side but then nearly well. We stopped to look at the miners at work along the river and bars, but as a general thing, they were not making high wages. At Boston Bar[29] we found quite a large camp and we stayed overnight one night and heard the miners tell what a time some of the prospectors had by the Indians, who would kill and rob them whenever they could find a man or two alone, and cut their heads off. As many as five bodies had been washed ashore at Yale's Bar[30] in one day, and it got so that a white man could not travel or work but the Indians would try to kill them. So the miners just quit work and organized into companies and went out to fight and kill all the Indians they could find, and found several camps of them, and just killed everything, men, women and children, so that the Indians were at last very glad to make terms of peace and promise not to molest the miners any more. That was in July, 1858.

But the Indians were not the only trouble; the river was awful dangerous and full of rocks and whirlpools, and more miners have lost their lives about the Fraser River—boats got swamped and whole boat-loads of men were drowned, and many never knew what became of them. Thousands of men have lost their lives by the Fraser, Harrison, Thompson, and other rivers, and by the Indians, that their friends and relations, both east and west, do not know whether they still live or not. A young man from Sucker and Althouse Creek named Joe Denny, "French Joe," they called him, and his partner Saml. Hurd, with a party of six or seven, bought a boat or canoe, loaded her with provisions, and several other boats with them left Yale's Bar and Hill's Bar to go through a long canyon up the river, and prospect bars along the river, which could not be got

---

[29] Boston Bar was located just above the confluence of the Anderson and Fraser rivers, considerably south of Thompson City.

[30] Yale's Bar lay one and a half miles above Hill's Bar on the left bank of the Fraser.

at from the shore, there being no trail close to the river near to where they wished to go, so the river was the best way to prospect. It took no horses or mules and they could land where they wished to prospect and have their provisions, arms and tools and bedding in the boat. Joe's boat was ahead and the other boats right after it, when Joe's boat got into a whirlpool which swamped it, and all were drowned.

Every day some boat would swamp or upset, and one or more would be drowned. And the Indians would kill them instead of saving their lives, or attempting to save their lives. Oh, what a great loss of life on and in Fraser River at that gold-excitement of 1858!

Will Cochran and I had often heard tell of the big huckleberries and the great huckleberry patch in the Cascade Mountains on the Brigade Trail. So when we nearly got there we inquired of some men we met how far it was, and if there were many berries there. They said it was not far off and there were millions of bushels of them. We thought they were just talking for talk's sake. One of the men told me he had left a mule on the mountain; it had give out and I might have it if I could drive it along. So we got near the top of the mountain to the huckleberry patch about 3 o'clock P.M., and we just sat down and eat and took some bread we had with us and just made a hearty meal on the huckleberries, as large as small black cherries back east, and we did not have to get up to fill ourselfs. And then we had a dinner pail and filled it, and came on to where I found the mule, and drove him ahead of us on the mountain next day.

Between the top and bottom, found two horses, and we drove all three head that day, but when we got to Harrison River I had to leave the mule—he could go no further. No grass, no hay or grain, so I just left him, as I had found him, and one of the horses with him. Will Cochran took his one along till he got to a ferry on Fraser River and sold him to the ferryman for $7.50.[31] At Yale's Bar I sold the chance of the mule I had left for $5 to some man who was picking up mules to winter on the Sound.

The second night after, we got to Yale's Bar, a city of about twelve or fifteen hundred inhabitants, in lumber and log houses and mostly tents. We both went to a barber shop and got shaved, at one dollar apiece, and haircut at $1.50, making $2.50 for what we now get here at Chanute, Kansas in 1885 for thirty-five cents. (!)

Next day we looked about town to see the sights. We found that Yale's, Texas, Emory, and Hill's Bars, and all the bars below the Cascade Mountain range, were quite rich in gold (surface diggings)[32] and the letters

---

[31] The ferry was at Spuzzum, six miles above Fraser Falls and ten miles above Fort Yale.
[32] Texas Bar was three miles below Fort Yale (which was established in 1848), on the right bank of the Fraser, opposite Strawberry Island, and was owned by King,

from brothers to brothers and partners, and to friends had been wrote in good faith, and truth as far as they knew. They had judged the Fraser River mines and streams and gulches like they did the California streams, and believed like in California, the higher up a river stream or gulch, the richer and coarser the gold. But it transpired that after you got above the Cascade Mountain Range, the poorer, and finer, and less pay there was. Until some two or three years after, some miners went up the river two or three hundred miles higher than our party had been, crossed the Caribou Range, and struck the great, rich Caribou Mines....[33]

While here at Yale Bar, I first heard of the death of French Joe and Saml. Hurd, and also heard that Guss Hill and James Daniels had sold out their claim on Hill's Bar and Guss Hill had started two days before to San Francisco, to wait till Daniels should come. Hill's Bar was only half a mile below Yale's Bar, on the other side, the richest bar of all.[34] We saw some old acquaintances at Yale, but we were anxious to get down to Victoria, so we did not look around much. We were in a hurry to get back to California before we would get broke or out of money, so we did not go over to Hill's Bar to see it. So we engaged our passage in a whaleboat to Fort Hope, for five dollars apiece, to take about two days. Some five or six of us took passage. It was quite a large boat.

Just below Yale's Bar we passed the Whirlpool where many had lost their lives right in sight of lots of men on both sides of the river.[35] But they were helpless, for the undercurrent in a whirlpool would take them under, and the best of swimmers could do nothing. We shuddered when we were told how many had got drowned there, and our boatman had to take care when in the undercurrent, and an experienced rower had many narrow escapes.

We got to Fort Hope, quite a large town, in the evening of the second day. A small river steamer lay just ready to go down to Victoria,[36] and

---

Severe, and Marshall. Emory Bar was situated on the same named creek four miles below Yale.

[33] The discoveries were made in the summer and autumn of 1861 on William, Towhee, and Lightning Creeks. In seven years the gold yield came to $25,000,000, a sharp contrast to the less than two million taken out of the Fraser in 1858–1859.

[34] Hill's Bar was a mile and a half below Yale Bar and *was* the richest strike of the Fraser rush. Hill made the discovery on May 11. *Daily Alta California*, August 19, 1858; James Moore, "The Discovery of Hill's Bar in 1858," *British Columbia Historical Quarterly*, III (1939), 215–220.

[35] Actually these were the rapids below the falls. The region above, Black Canyon, was extremely difficult to push through on foot. It proved a major handicap to miners seeking upstream placers.

[36] The Hudson's Bay Company licensed two American vessels, the *Enterprise* and *Maria*, to run between Fort Hope and Victoria. The *Surprise* and *Umatilla* also plied the same route. The latter reached Fort Hope on September 3, 1858, and continued the run until succeeded by the *Maria* in October, 1859. Bancroft, *British Columbia*, pp. 401, 444–445; Norman R. Hacking, " 'Steamboat Round the Bend': American Steamers on the Fraser River in 1858," *British Columbia Historical Quarterly*, VIII (1944), 255–280.

as I came close to it, James Daniels hailed me from off her deck. He came ashore as soon as I could land, and we shook hands. He told me that he and Guss Hill had sold their claims on Hill's Bar and Guss had left two days before for Frisco, and when he, Daniels, got down there, they were both going to South America (Rio de Janeiro) to go into merchandising; that Guss had made five thousand dollars, and he, Daniels, some three thousand, five hundred dollars. He left Sucker Creek in March, only two months ahead of me; he went by water to Victoria, and a little steamer clear to where he now was,[37] and no hardships or danger like me, by Indians or rivers. And here I was, only just got in, and he was just leaving, and had made a little fortune, in but two months difference in time. His boat soon put off. I had spent over $600 and all my time. I should have gone on the same boat, but it cost $25 apiece to go to Victoria, and some boatmen offered to take us for $8 apiece—that would be saving $17, and our funds were running low. So we engaged the $8 fare.

I bid Jim Daniels goodbye, and sent my regards to Guss Hill, and their steamer went on. We stopped in Fort Hope that night. There was a large Hudson Bay Company Store, and it was a sort of headquarters of the Hudson Bay Company, too. Their chief factor, Mc[Lea]n, lived here too—or superintendent, he was, and several of their head men of their company had their squaws living with them.[38] I saw plenty of children, quite white, red-haired, freckle-faced little Scotch fellows. Their mothers were squaws and could speak the Chinook, or Jargon language. Here at a bakery we bough[t] the first and best pumpkin pie we had eat for over ten years, about 3/4 inches thick, in square pans—it was good, and you got a big pie for 25 cents.

Next morning the man we contracted to take us to Victoria got ready. He had a large canoe about 45 feet long and from three to five feet wide, capable of carrying forty men and their baggage, and he had engaged about 22 passengers, and we all had to row or paddle. So we started down the river, and had lots of narrow escapes from upsetting and swamping the boat. The second day out we got to Fort Langley.[39] Here we found a United States war steamer with the English and American surveying party, surveying the line between British America and the United States of North America.[40] It was a great sight to some that had never seen a

[37] The *Umatilla* reached Yale Bar on July 21, 1858, the first steamer to go that far inland on the Fraser. Alexander Begg, *History of British Columbia*, p. 282.

[38] Donald McLean was chief trader and had been transferred from Fort Thompson. The chief factor of the region was Peter Ogden, the half-breed son of Peter Skene. Fort Hope had been built in 1848 on the left bank of the Fraser at the mouth of the Coquihalla (sometimes spelled Coquahalla) River.

[39] Fort Langley, completed in 1857, was the first sea fort in British Columbia and was built near the mouth of the Fraser on the left bank. James M. Yale was the resident chief factor in 1858.

[40] The Oregon settlement treaty of 1846 produced a boundary dispute, about the ownership of the islands situated in the Juan de Fuca and Georgia Straits. In 1856, the

ship or an ocean steamer, and the American flags were flying from her masts and stern. We all felt proud of the steamer and our nation.

At this place one of our boatmen gave us the slip. We had paid our fare in advance. It may have been that the boat did not belong to him. Anyhow he left us at Fort Langley, and the other man said he thought he would not be back. We were uneasy, because it required some knowledge of the river to make it safe boating, and the big canoe was hard to manage, for inexperienced men with paddles.

But the other boatman thought we could get along, so we started again for the mouth of Fraser River. Several times we took the wrong channel and had to come back, and I do not know how many times we came near swamping and getting into whirlpools. It was more by good luck than by good management that we got along at all.

One evening about four o'clock the boatman told us to take a left-hand channel and it would take us to Point Roberts, on the American side,[41] but we got into a wrong channel, and after going a couple of miles, found logs across, so that we could not go further, so had to turn back. When we got back into the main river, it was just getting dark, so we concluded to go ashore, and one of the men said he knew of a good camping-place not far away. And we kept going, and it got so very dark you could not see, and the first thing we knew we were in the breakers, and in the mouth of the Gulf of Georgia, and in the Pacific Ocean. And came near swamping. And if we had, no doubt every one of us would have been drowned. You could see nothing. So by hard work, cursing and swearing at the boatman for taking us out to sea, we at last managed to get back into the river. The man said he was lost, and looked for the place to camp on the wrong side of the river, and so it was with the other one—both good old sailors, and both lost their way. So as soon as we could, we run in toward the shore, and saw a fire not far off. It had been raining for the last half or three quarters of an hour, and we were all wet from rain, and the breakers kept coming into the boat before we turned back, so that we were cold and wet, and glad to see a fire. So we pulled in to shore and made for the fire, and found an Indian camp, and got some fire from

---

United States and Britain agreed to undertake a survey and negotiate the question. The two teams, the British headed by Captain Prevost of H.M.S. *Satellite;* the American headed by Archibald Campbell, accompanied by two topographical engineers and astronomers on board the survey steamer *Active* and the brig *Fauntleroy*, commenced their work in June, 1857. After two years of effort, the negotiations collapsed and the solution to the problem waited until 1871, when the question was solved by the arbitration of the Kaiser of Germany.

[41] The most northwestern extremity of Whatcom County, Point Roberts, was named by Captain George Vancouver in 1792 for Captain Henry Roberts of the British Navy. It lies directly south of Vancouver at the end of a peninsula across which runs the international boundary. Edmond S. Meany, *Origin of Washington Geographic Names,* pp. 222–223.

them to build us up a good fire, so that we could dry our clothes and blankets....

In the morning early, we inquired of the Indians, who were friendly, our way to Point Roberts. They said about seven miles by water, so we started a little after daylight, so as to make Point Roberts in two hours. But they proved to be the longest two hours I ever saw, and the most dangerous. When we started there was no wind, but when we got out into the Gulf of Georgia the wind raised and blew right from Point Roberts in our faces, so we had to keep in the breakers or the wind would blow us back, so that we would have lost headway and been beat back instead of forwards. We had to keep along in the edge of the breakers all along the coast, in shelter of the Point. And the surf would roll onto us and drench us every few seconds, so we had to wrap our blankets around us under the arms to keep from freezing, and the waves would strike us in the breast. It kept 12 to 14 men with paddles to keep the boat going across the breakers in such a way that if some one or two should not keep the boat in the right position she would broach to and swamp instantly. And if we headed too straight out, the headwind would blow us back further in five minutes than we could gain in an hour. We were all very tired and nearly give out, but it was life or death. If a man stopped, or even slacked, all would quit and curse him. It kept two or three men with tin wash gold pans to bail out the water to keep us from sinking or swamping. I remember two Oregonians that had never been on rough water; they got seasick and kept vomiting and bailing—they would not dare stop bailing or we would sink. Such cursing and threatening to anyone that would slack up was fearful, and we could see the shore all rocky and could see people and houses all along the coast....

At last we got in sight of Point Roberts town, and in trying to round the point, the wind would take us back, and we worked the harder to make the short distance. It was a life-struggle, and we succeeded at last. About 3:30 or 4 o'clock, we effected a landing. We were nearly perishing with cold, wet, tired, and hungry, and a gladder lot of boys were never seen. We struck right for a bakery and ordered good strong hot coffee, and something good to eat. The man was a German from San Francisco, California; he had some good French brandy, and he made us all some strong coffee, "coffee royell," with the brandy. And some good raisin pie, bread and butter, cheese, bologna, sardines, lobsters and oysters. We had our fill. We all had money, and would not have cared, if it had been our last dollar we were going to have a good time after our close escape from death. So we drinked coffee royell and eat good things as long as we could eat, and we dried our clothes and blankets and stayed all night. We were again on American soil, and we were proud of it!

Point Roberts was a town of three or four hundred inhabitants mostly Californians who started for Fraser River and Puget Sound. During the excitement several towns had been laid out on the American side for a big city of the future, and had gone down again, fortunes lost and large stocks of goods bought at high prices and sold for anything to get rid of them and get back to old California.

Now Will Cochran and I did not like to take to the old big canoe again if we could help it, and some few others said the same. We found a large life-boat at Point Roberts; the owner had but two passengers, merchants from Frisco, and he offered to take three or four more passengers at $3 apiece to Victoria, about 35 miles. But we had all paid our passage in the canoe. But rather than go back in her and take the chances of being foundered in her, we were willing to lose what we had paid, and Will Cochran and I and one other man took passage in the life-boat. It had sails and water-tight compartments, so if it was upset it could not sink. So we parted with the canoe at Point Roberts in the morning, and sailed away towards Victoria. We did not have to row or paddle as in the canoe. The two merchants were Germans, and they had lots of refreshments, and beer and brandy, bologna, ham, sardines, and we had cheese pies, crackers, lobsters, and so forth. So we lived fine and enjoyed our trip.

At noon we stopped at an island about half way and got our lunch, and got some fresh water from a spring on the island. While we were nooning, I was sitting in the boat and looked down in the clear water, saw something that looked awful to me like a crab, only the body was as large as a water-bucket and had eight or ten long legs, like a crab, only six or eight feet long, and would feel around with them long legs. I called the attention of the men and boatman; he said it was a devil-fish, and if he got a-holt of a man or animal in swimming, he could strangle it in the water. I took my rifle and shot into it, and we hauled it out with a boat-hook, and we jabbed it until I guess we killed or wounded it bad, for the water was covered with blood. We started soon after lunch and got to a point of the land, Vancouver Island, four miles from Victoria. We saw farmhouses and large flocks of sheep grazing.

Will and I were so tired of water that we said we would like to foot it to Victoria, as the boat had to go a long ways further around by water, and it would be late in getting in to Victoria. So Will and I took our guns and left our baggage and blankets in the boat to be brought to Victoria, and we struck out afoot. It did not take us long to get to town. We passed some fine farm houses and sheep and stock farms, and got to Victoria about four o'clock. We were surprised to find so large a city, and fine stores, hotels, saloons and dwellings there.[42]

We went to the American Hotell and found many old acquaintances from California and Oregon, waiting for a steamer to take them to San Francisco. Many had no money and made application to our consul

[42] Fort Victoria, the foundation of the future city, was established in 1843 by the Hudson's Bay Company. It became the center of their Pacific slope operations.

(agent for British Columbia) Edward Nugent.[43] He said he would try and make some arrangements with the company of the steamer *Pacific* to take a lot of American subjects to San Francisco, who had not the money to pay their own fare. It was the duty of the government to take its people to their homes if they were in a destitute condition on a foreign shore or land, and there were over one thousand men in that condition.

Will Cochran and I thought we would be entitled to go free too, for we were nearly out of money, but we waited till we were afraid we would not get to go, or get passage in that way, and our money would be spent if we had to wait longer. Our board at the hotell took our money quite fast. Our lifeboat got in at night. The big canoe did not get in for two or three days after, and they said they had a hard trip all the way. The canoe was unmanageable, so we were glad we did not go in her. We went every day to see the sights around Victoria. Governor Douglas had a fine palace or residence here, and the Hudson Bay Company have their headquarters and fort and stores, their soldiers and navy, some steamers and sail vessels.[44]

The English has several man-of-wars here, and at the government harbor at Esqimault, four miles north of Victoria they had several steam frigates and men-of-war.[45] While here at Victoria I went to church, the first I had entered for eight years, to witness a funeral service and see an English marine buried with military honors. He got accidentally shot and belonged to the frigate or Line of battleship, *Satellite*.[46] It was a grand sight; the marines, sailors and officers of the several ships and men-of-war marching to the church and to the grave, and the firing-off of guns by squads of soldiers over the grave. Some of the Americans and miners would scoff at the English soldiers, but they were half-drunk and done it for foolhardiness, like some drunken idiots will do to try and show off in another country, thinking to impress strangers with their bravery or daring, but it was uncalled for and only reflected discredit to the Americans, and many were ashamed of their countrymen.

There were many Indians at Victoria, some belonged there, others

---

[43] John Nugent was appointed commissioner and consular agent for the United States to handle the very problems Reinhart relates. Victoria *Gazette*, September 14, 22; November 16, 1858.

[44] James Douglas, long associated with the Canadian fur trade, served as governor of Vancouver Island, 1851–1863, and of British Columbia, 1858–1864, when that colony was established. He was knighted by the Crown for his distinguished record of service.

[45] Esquimalt started as a naval hospital during the Crimean War, 1855. In 1857, the area was taken over by the Admiralty. By 1864, a naval yard was fully established. John T. Walbran, *British Columbia Coast Names*, p. 171.

The man-of-war, *Ganges*, anchored in the harbor in early October along side the steamers *Satellite* and *Plumper*, the steam frigate *Tribune*, and the steam corvette *Pleiades*. Bancroft, *British Columbia*, p. 404.

[46] H.M.S. *Satellite*, commanded by Captain James C. Prevost, who was chief British negotiator on the 1857–1859 boundary survey, was Governor Douglas' chief law enforcement agency during the height of the Fraser gold rush.

from up north, where they were whiter and taller and goodlooking Indians. Most all of them could speak Chinook or Jargon, and all around their large huts, some built of lumber, there would be wooden figures carved to represent men, horses, dogs. And their graves, where their fences and posts were carved out of wood to represent the person dead. The squaws made nice matting out of reeds, and they were just the thing to spread on the ground under our blankets and roll them up in when we travel. We bought several, some colored and wove quite nice, and I saw a smoking pipe carved out of a piece of slate (black) rock by an Indian representing a locomotive and tender, baggage car and three or four passenger coaches. The smokestack of the locomotive formed the bowl of the pipe, and the end of the passenger car had the stem or mouthpiece. The windows, doors and wheels, all well got up, as if a fine mechanic had done it for his best, smooth and polished like black ivory, and it was a good smoking pipe.

A miner told me he was in Victoria in June 1858, and in the hight of the gold excitement for Fraser River, and that there were over ten thousand miners at Victoria, and the Indians from up north in their large war canoes, some of which held from 75 to 100 men, were trading with the Hudson Bay Company's stores, and the squaws got badly demoralized, and the miners had plenty of money to spend with them, and they gave them whiskey and there was an awful time among them, and they dressed up as fine as "White Soiled Doves" do in California. The Americans brought so much liquor into British Columbia that the English had men-of-war all up Fraser River and hailed and examined and overhauled all boats and vessels to find and search for liquors that was being smuggled and run into British Columbia by Puget Sound, Washington Territory, Oregon and California.

There were a lot of cutthroats, that had been run out of San Francisco by the Vigilantes, came to Victoria. They were watched so close at Victoria, . . . they went to Fraser River, and went up the same time that Guss Hill did. Their boats were just behind Hill's, and when he struck the prospect and took up his claim, they took up all above and below Guss Hill's claims. They were a hard lot from New York and the notorious Ned McGowan was in the crowd, and they used to do just as they pleased on the Bar or at Fort Yale or Yale's Bar, and the English authorities could do nothing with them up on Fraser River.[47] Justice Hicks tried at

[47] Edward McGowan early joined the rush to the Fraser, arriving in Victoria July 3, on board the steamer *Pacific*. His behavior at Yale's Bar, where he owned a rich claim, was, as usual, erratic and mercurial. He was wise enough to flee to the border in January, 1859, when the so-called "Ned McGowan war," an *opera bouffe*, came to a head. British justice was swift and stern.

The Fraser gold rush was well policed by the authorities. The judicial administration under tough Chief-Justice Matthew B. Begbie, enforced by the well-disciplined armed naval and army units, was commendable—a sharp contrast to the ruffianism that

Yale, but could do nothing.[48]

The American miners would take sides with the California roughs, but once in awhile if one should happen to come down to Victoria, the police and soldiers would nip them and punish them. Several had been caught and punished until they would give Victoria a wide berth.

There were some nice billiard halls and saloons in Victoria, one with six tables kept by a California gambler named Boston—here I saw the first 15-ball pool. Some old friends and acquaintances from Indian Creek, California came down on the steamer and took cheap passage on a boat for Puget Sound; they were going [to] Portland, Oregon, and to work their way back to California; one, Edward Hutchinson, a violinist, that used to play for us at our Eldorado Saloon on Indian Creek, and Pony (or John) Young, another from the same place, and some whose names I have forgotten.

Bill Cochran concluded to go the same way, and did. I concluded to wait for the little coast steamer, *Santa Cruz*, Captain Fauntleroy, that would touch at Crescent City, Scottsburg, Port Orford, and San Francisco.[49] Crescent City was my nearest seaport to Sucker and Althouse Creek. There were several that had come down the river in the canoe with us that wanted to go the same way. So we waited for the little steamer, and we still loafed about town.

One day a large three-masted ship came in from the Sandwich Islands, loaded with cattle and hogs. The cattle were very wild and had to be raised with slings and block-and-tacle and derricks, and when they struck the docks, rushed at anybody in their way or in sight. About 2000 miners were on the docks to see them, and got onto sheds and roofs of buildnigs and the fence around each side of the docks. The ship had some Kanakas[50] and Indians to lassoo the cattle on the dock, so that they could take the slings and rope off. Then you could look out for fun. . . .

---

marked the California mining scene. Ormsby, *British Columbia*, pp. 160–163, gives a good summary of the McGowan affair. Frederic W. Howay (ed.), *The Early History of the Fraser River Mines*, pp. 10–11, *et seq.*, contains the 1858–1859 documents touching on the matter.

[48] Richard Hicks served as assistant royal commissioner at Fort Yale, 1858–1859. His brief tenure of office was marked by constant strife with the miners. The miners in numerous resolutions condemned Hicks as "an unscrupulous man, as well as a corrupt public officer, and altogether unfit and unworthy the position" he held. P. B. Whannell was the resident magistrate at Fort Yale. Victoria *Gazette*, November 6, 1858; Howay, *Fraser Mines*, pp. 3–22.

The situation at Fort Yale did not, however, deteriorate into lawless disorder. Chief-Justice Begbie saw to that!

[49] The steamship *Santa Cruz*, which ran the northern route via Crescent City, commanded by Fauntleroy, was loaned to the United States government for a few months in 1861 as a revenue cutter, sailing under the name *General Sumner*. Later, she was purchased by some San Francisco businessmen for $40,000 and sold in China for $81,000. E[dgar] W. Wright (ed.), *Lewis & Dryden's Marine History of the Pacific Northwest*, p. 69.

[50] A "Kanaka" was a laborer from the Sandwich Islands (Hawaii).

When the steamer of the Pacific Company came, Consul Nugent got passage for some that had no money and there were still enough to load two or three steamers left to go to San Francisco. So when the little steamer, *Santa Cruz*, came, a lot of us engaged passage to Crescent City, California at $20 apiece. I believe I had not quite enough money to pay my passage, but a Yreka, Californian miner by the name of Kroney, a Prussian butcher, let me have about $40 until I could pay him back when we got to Kerbyville where I started from in May before. So I was fixed. We had to get our provisions, such as bread, crackers, bologna, cheese, ham, (and some whiskey or brandy for seasickness). We had to board ourselfs while on the boat. There were some three or four steamers ready to sail when we left Victoria, and when we started to leave the bay, the Puget Sound mail steamer and the Pacific steamers fired salutes to us, and away we started for our trip in return to California.

Kroney and I, and most all the passengers, stayed on the hurricane deck to keep from being taken with seasickness, but it was no use, many of us got deathly sick in two or three hours of the harbor of Victoria, and I took a blanket and went below and lay down on the floor. I was awful seasick, vomited fearful all night, and did not care whether we sunk or floated. We had to look out all night to keep others from vomiting all over us. Did not feel like eating but made out to go on deck and get a strong cup of coffee in the morning—the ship furnished it to all passengers.

After while I got up and watched the shore and saw the mouth of the Umpqua River, Scottsburg, Empire City, at the mouth of the Coquille, and after a little saw Point Blanco, then Port Orford and my old place, the Pacific Ranch. All looked familiar, way out at sea. Then we saw Prattsville, the mouth of Rogue River, and Ellensburg, just below. All along Gold Beach many new buildings had been put up, and whole villages started since I was there in 1854–55.

We got to old Crescent City about ten o'clock one morning. [I] went on shore in a whaleboat and stopped at a hotell.

# 7

## OREGON INTERLUDE

(1858–1859)

. . . I looked around and found lots of old acquaintances in the city. Next morning I had the choice to walk with a lot of our company, about 75 miles in three days, or pay $15 and go through by stage in one, and part of the night. I come to the conclusion with Kroney and a couple of others to go by stage, even if it did nearly break me. I thought I might as well spend $15 or $20 more, after spending $700 in seven months time. So we got our tickets and in the morning started in the stage. (The last time I had gone over the Coast mountains to Illinois Valley I was on horseback on my way to the Klamath River, California, to take up mining ground to work the Indians that broke out while I was gone in 1855.)

Now in 1858 everything looked peaceable, and the old settlers and some new ones had made considerable improvements, and Captain [M. M. Williams] and Lafayette Gates were still at the foot of the mountains in Illinois Valley and keeping public house. After a long ride we got to old Guthings' Hotell at Waldo, or Sailor Diggings, after night in time to get some supper.[1] It was Saturday night, and I sent word by the stage driver that if my brother was at Kerbyville to tell him I was there, and would wait to hear from him.

So Sunday morning about nine o'clock, here came my brother with two mules with saddles on, one for each of us. We were glad to see each other. I got four or five $20 gold pieces [from] him to pay the money I had borrowed from Kroney, the Prussian. I settled my hotell bill and bid our fellow travelers goodbye, they going back to Yreka. I strapped on our two mules the blankets and some clothes and my gun, and my

[1] This was the only hotel and had been built in 1858.

brother and myself rode to Kerbyville, about twelve miles, to where Charley was staying.

He was at present boarding at the Vining Hotell, kept by John Pringle, who used to keep the saloon and bowling alley on Indian Creek and was Justice of the Peace there.[2] He was Postmaster here, and Charley sometimes done some pastry cooking and baking for balls while at Kerbyville.

Kroney had taken the stage Monday morning for Yreka; some had gone to Althouse Creek, Oregon, sixteen or eighteen miles off. Charley told me all he had done with his mine, and John Schertz's claim. After I left they hired two or three Chinamen, one in my place, the others between us. It did not continue to pay as good as when I was there. So they, Charley and John, concluded to try and sell out to a large Chinee company for a big price, and they salted the claim some and the Chinamen working for us would see our cleaning up every night. At noon John Schertz would salt the sluice boxes—he superintended the work, and Charley was not there much of the time because he and Thomas Wilson were still running the bowling alley and saloon. So the Chinamen that worked for us told the others about our claim paying well, so the Chinese came to buy, and did at last after a great deal of haggling about the price to be paid. They only got $1500, half down and the other to be paid as it came out of the claim, so much every week until the other half, $750, was paid. Charley and John said it was well sold, if they never got the other half, which we never did, for after working awhile and getting work pay, the Chinamen let the claim fall back, rather than pay the balance (the $750). So after deducting expenses and dividing, there was but two or three hundred dollars apiece, and Charley had sold out his half of the ball alley and saloon for four or five hundred dollars more, and that was in July [1858]. And he was expecting to hear from me, for him to come to me, and at last he heard that I had been killed by the Indians.

He took a trip to Roseburg, Oregon, and while down there, there was a ball and he got up the supper, and all the cakes and pies. He stayed awhile among our old Cow Creek and Canyonville friends, where we used to live in 1852–53. Old William Riddle's girls had got married, so had Beckworth's, and Jess Roberts was keeping the hotell where old Knott kept in 1852, when we first came to the Umpqua Valley and Cow Creek.[3] Briggs, who crossed the plains in our company, had built a toll-bridge

---

[2] Kerbyville, established by James Kirby, was laid out in 1857. George T. Vining built the hotel for M. Ryder in the same year. It "was considered an extraordinary structure, indeed, it being really a large and commodious house." Ryder still owned the hotel in 1884. Most likely Pringle was an employee. A[lbert] G. Walling, *History of Southern Oregon . . .*, pp. 453–454, 509. Vining served in the 1860 legislature and Ryder later was elected sheriff of Josephine County.

[3] Jackson Reynolds and Joseph Roberts bought out Knott, and subsequently sold to Jesse Roberts. *Ibid.*, p. 425.

across the Umpqua (South) below Canyonville, and old Yokum, with his three boys, had cut a trail and wagon road around a rocky point and put in a toll-bridge one mile below Briggs's. Charley stayed among them a while and had just a few weeks before got back to Kerbyville. John Schertz, after selling out, was working at Capt. [M. M.] Williams' sawmill on the Illinois River, a few miles below Kerbyville. He came up to see me as soon as he heard of my getting back, and was awful glad to see me alive after hearing I had been killed by the Indians. I soon found out that Charley was doing considerable poker playing, and while I was there, he generally lost. I do not know how he had been doing, but I knew he was playing to get back even what he had lost, and one night he must have lost considerable, for I do not think we had over $100 or 150 left, and he said we had better see what we could do before the winter set in.

So we took a ride up to Althouse Creek, where we had a bakery and some claims in 1853–54, and we had some money owing to us, but we could not collect any.

So next we went to Sucker and Bowling Creek, eight or ten miles further, and saw old friends, but many had left and gone to other diggings and it did not look at all natural, and we did not feel like commencing again there on our old claim, which had been abandoned by the Chinese, and had been again sold, but did not pay.

So Charley and I concluded to go to The Dalles, Wasco County, Oregon. I told him it was a very lively place when I was there in June, last, 1858, and he, Charley, could work at his trade in the winter, and if it did not suit, we could go to Walla Walla City, or Valley, a lively fast place. There was a government post and forts at both The Dalles and Walla Walla, and there was winter quarters for six or eight companies of soldiers, and lots of miners wintered at both places in the cities. So we saw John Schertz, he said he would rather stay at the sawmill—he could get work all winter at $70 to $80 per month—until he could hear from us, and if favorable, would come to us. . . .

When we got ready to start for The Dalles, John Schertz, [who had decided to join us], brought a mule, a good one, to ride between us. He got it from Capt. Williams for $80 on his wages. We got a saddle and bridle, and bid all friends goodbye. . . . We stopped at hotells along the road. We were well acquainted with about every hotell keeper on the road to Canyonville. At James T[w]ogood's and Harkness, who kept the Grave Creek Hotell, I heard that Mr. T[w]ogood had carried a letter from Chicago for us over a year, but somehow had lost it. It must have been from my sister Emma Musgat, or the Breeds. We had not heard from home for four years at this time.

Our next noon stopping place was at Hardy Ellif[f]'s, at the south side of the canyon. He kept a good public house. He told us of some new gold discoveries on Coffee Creek, about thirty miles from Canyonville, putting into the South Umpqua River from the west. And there were three or four hundred men up there prospecting, and all the claims were already taken up.

The mountains [were] full of men prospecting more creeks and gulches, and a great many of our old acquaintances about Cow Creek and Canyonville [among them]. We stopped that night at Canyonville and old Jess Roberts, the landlord, was very sure that the mines would prove rich and extensive. We had a notion to stop, but Charley next thought it would be better wintering at The Dalles. So we kept on next morning on our way north. As we passed along the South Umpqua Valley, the old settlers all spoke favorable of the Coffee Creek mines. At noon we had dinner at old Lazarus Wright's on Myrtle Creek. He was glad to see us. He kept a hotell and had a store, mill, and the postoffice there.

That night we stopped at Roseburg, where Brad Robinson kept hotell, and [we] kept our mule at Col. [George] Ross's livery stable (We knew him at Jacksonville, Oregon). Charley had stayed at the hotell in July and got up some supper for John Pringle, who at that time kept, or worked for Brad Robinson, and he got well acquainted among the families that were at the hotell and some of their friends.

We saw some returned miners just down from The Dalles, and they said the place was overrun with miners and men looking for work to do until spring, and would be dull on that account. So Charley told me he had been offered a good place all winter and as long as he wished to stay, as Baker and Pastry Cook at the Roseburg Hotell [kept] by Brad Robinson.[4] He thought he had better stay, and I would go back to Coffee Creek and see the diggings, and see if I could either take up some mining claims or get into work for awhile.

So I started next morning with the mule and some blankets and what few clothes I had, and we divided what little money we had. I had some $15 or $20 and the mule—if I should get a chance to sell or trade it for a good claim, I was to do so.

I got to Canyonville at night and stayed there all night, and found some parties going up to Coffee Creek that knew the way. We passed some splendid farms above Canyonville on the South Umpqua River, and stopped overnight at John Bland's (a man who had married one of Charles W. Beckw[or]th's daughters, the oldest, and George Doty had married the youngest). I was well acquainted with both the Doty boys on Althouse Creek in 1853 and 1854. George had worked for us when we were building our bakery and our ball alley on Sucker Creek. Mrs. Bland was glad to see me. I had not seen her for over six years.

Next day we got to Coffee Creek, 18 miles northwest of the Umpqua

[4] Walling, *Southern Oregon*, p. 326, states that the hotel was built by Dr. Jesse Robinson, who "owned and conducted" the business. This was probably a brother, who assumed ownership on the death of Bradford Robinson.

River. We kept up the creek until we got to a store kept by David Ransom and Jerry McKay, and found Jim Yocum temporarily keeping store until Dave Ransom would get back from Roseburg where he had gone after goods. James Yocum had some claims with a partner named Williams; they were prospecting the claims. Ransom, McKay, and Major Cranmer had three creek claims together, which they were just opening.

I found many old friends and the creek was overrun with miners and it looked like an early California camp. I first concluded to go to whip-sawing out lumber, but the timber was not as favorable as at Althouse or Sucker or Indian Creeks. So I stayed a day or two with James Yocum at Ransom and McKay's store. The[y] kept groceries, whiskey, tobacco and clothing, miners tools—in fact, a general store and saloon, and lots of card-playing for whiskey or the drinks, and most anything you wanted, tobacco, cigars or clothing, and they got up meals, too, for transients who happened along, and somebody was there every night from four to twelve, just as it happened, and I helped James Yocum to cook for them. When Dave Ransom came back, he wanted to be at the claim some of the time, so he told me he would give me good wages to keep store for him and cook, and attend the store when he was gone.

He and McKay (who was deputy sheriff at Roseburg) had some five or six horses and mules, and he (Ransom) done his own packing and taking orders from miners to deliver. So he offered me $65 per month, and had the use of my mule until I wished to sell him. So I done so. I knew that was the best work for winter and not be out in the cold or snow, although it was a rough place and more whiskey drunk than any little place I ever saw. I would be waked up before daylight by the miners for their morning bitters, and always noticed that whiskey men were always the earliest risers; invariably, no matter how worthless they were, they were always the first up in the morning.

There were some rich creek and bank claims right in front of the store, owned by the discoverers of the Coffee Creek diggings. Their names were Jack Reynolds, John Danforth, Clement Glasgow, and two others. In the claim right above, which took in the forks of the creek, were Uncle Joe Southard, H. Wilson, and two others (one named Love—I knew him at Happy Camp, California, in 1855) and they were all on the drink.

Ransom and McKay done considerable of a credit business, but mostly to men that they knew, or had good claims, such as John Danforth or Uncle Joe Southard, or Major Cranmer (their own partner in the claim). I know at one time that Major Cranmer was owing over $350 but they bought him out of the claim, and I expect never got any of it back, for the claim did not pay as it prospected.

"Freeze-out Poker" was the game played for money or goods, and by the way things were carried on by Dave Ransom, every trip to Roseburg or Scottsburg, at the mouth of the Umpqua (afterward called Gard[i] ner's City) or at Canyonville, he, Ransom, always got on a spree and gambled, and his money went like water....

I commenced work about the middle of November (1858) and got some money laid up. I had nothing to pay out for clothes, and I won some little money [at cards].... Dave Ransom used to bring me word from Charley as to how he was doing well—he saw him at several dances and balls at the Robinson or Roseburg Hotell during the winter.

About the 18th of February, Dave Ransom took the train of mules and horses to Roseburg after a load of goods, and when he got back about the 25th of Feb., he gave me a letter that he said he forgot to give me before he went, and it was from my brother Charley to be sure and come down by the 22nd of February, that there would be a grand ball on Washington's Birthday and he was to be married the same evening before the ball to Miss Elizabeth Painter, a young lady friend of Mrs. Brad Robinson, the [wife of the] landlord at the hotell. Charley had give Dave Ransom the letter and told Dave, but [Dave] wanted to go, so would not tell me in time for if I had gone, he would [have] had to stay and keep store till I got back. I was mad about it, and no doubt Charley was disappointed, but what could I do about it? Dave told me to take the animals and make a trip after goods to Roseburg. At the same time I could pay Charley and his bride a visit, as soon as I liked, and it would cost me nothing, and my time would go on as usual.

So in a few days I got ready. It was quite cold and wet. Dave Ransom had a heavy beaver-skin coat, made out of eight beaverskins, not well-dressed, very heavy. He loaned me the coat to wear in the storm. The first night out I stopped at Mr. Bland's, the next night at Clark's, close by Canyonville. In the morning when I got all saddled and went to lead out the black riding-mule, out of the stable, I found him so lame that he could not walk. His stiffle-joint had slipped out of place, and I had to saddle one of the others and leave him until I could get back from Roseburg, which I hated to do, but had to.

I got to Roseburg that evening, and was made acquainted with my sister-in-law and liked her very much. She was short and heavy-set, about 21 years old, and had left home on account of her stepmother, and was staying with Mrs. Robinson at the hotell as companion, and helped to sew for Mrs. Robinson. After Charley's marriage she stayed a while as chambermaid [as] Charley was laid up with an awful felon on his left thumb. He said he had not slept for five or six nights, but got it cut open the day I come and he got better soon.

I stayed two or three days at Ross's Stable and got acquainted with some of the Roseburg folks, and Charley and his wife's friends, visiting and accepting invitations to dinners and suppers. Here I found Frank Smith keeping saloon. He used to be on Sucker Creek and Indian Creek gambling. He was married and doing well at Roseburg, and Vincent

Davis, the blacksmith from Jacksonville, Oregon—he too was married and doing well. So after my visit I packed up the mules and horses and bid the bride and friends goodbye, and got back to Clark's that night and found the black mule had got all right a short time after I had left him. So next morning I rode him. . . .

I got to Bland's [the third] night, and showed Mrs. Bland my new sister-in-law's photograph, which she had give me. Mr. and Mrs. Bland said they liked it, and she was a good wife for Charley.

The next night I had to stay on the Umpqua River at Ben and George Stout's ranch. There were several half-breed French had ranches along the river. Clement Glasgow, that used to keep the Cow Creek ferry with Thomas Smith, had a stock ranch on the river above the mouth of Coffee Creek. (The Coffee Creek got its name from a party of soldiers camping on it. They were after a lot of hostile Indians and were all tired out, and one of the soldiers cried because he had no coffee to drink. His companions made fun of him, and named the creek "Coffee Creek" and it kept its name after the discovery of gold.)[5] For several years the two brothers, Ben and George Stout, kept the ranch where I stayed all night. I will speak more of George Stout in the future. They both had gold claims on Coffee Creek.

I got to our store next day all right, and kept the store while David Ransom and his partners mined their claims. There were many carousals there at night, and I was not to refuse certain customers their whiskey, even when I knew they had enough. If I did, they would complain to Ransom, and he would tell me to let them have it. There was one man named John Danforth, that owned an interest in the claim in front of the store, who drinked a great deal but was harmless and when I refused him whiskey and told him he had enough, he would curse and swear, then beg and cry for drinks, until someone would let him have a drink. A neighbor of his, named Joe Crutchfield, a Texas ranger, used to drink the same way, but he was more vicious and desperate when drunk, and he always carried his rifle or pistol with him. He was a partner of old Joe Southard and Jim Wilson. In Danforth's claim were Clement Glasgow, Jack Reynolds, Sy Gilbert and Thomas Love, and they had the discovery claims. (I used to know Love at Happy Camp, California, and at Fraser River; he was considered a very lazy young, stout six-footer, and we all thought he had no energy at all until after a few years he went to San Francisco and had a tumor taken out of his side, weighing some 18 lbs., and of which he died.)

One day John Danforth told me a long story, how he had left his home in Illinois, where he and his brother had a large machine manufactory, and had got up the first threshing machines ever invented, before there

---

[5] The creek reputedly got its name by some miners because of some unexplained joke. Here, perhaps, is the real explanation. Walling, *Southern Oregon*, p. 442.

# Oregon Interlude 153

was any patent on them, and they had been trying to perfect the whole so as to thresh clean and separate, stationary.[6] They had one completed to travel, and were at reapers and threshers on one when he, John, had a girl he was engaged to, and did not know she was also engaged to his brother, and he got so that he could not stay, and he took to drinking, and his girl got married to his brother. He said if he had stayed he should surely have killed his brother or his sister-in-law, so he just left for Oregon in 1846, at the earliest settling of Oregon. He did not let his brother know where he was until he got to Oregon City, and he never let him know the cause of his leaving, and his brother thought it was on a drunken spree that he left. When he got a little liquor in him he would cry and grief about it to me.

One day John Danforth was quite stupid from drink and sitting on a bench half asleep, and Joe Crutchfield was drunk and had his gun in his hands. I had taken it away from him and put it behind the counter, but he said he was ready to go home and wanted his rifle. So I give it to him, but he did not go right off, but sat down on a bench. I happened to look up and he had raised his rifle to the level of John Danforth's head, and was just a-going to pull trigger when I shouted and threw up the muzzle of the rifle, and the ball went into the chimney. It sobered John Danforth right up and he did not want to spree any more with Crutchfield. He said he was only trying to see how near he could shoot to John Danforth's head and scare him out of his sleep—that he was too dull for him. But if I had not thrown up the muzzle he would have shot him through the head, for he was too drunk to hold his rifle steady, and could not see that he had the muzzle so close to his head.

Dave Ransom come home from Roseburg and was telling of a joke that a girl had played on a young man named Coom. He had been trying to keep company with her, and she could not get rid of him, so she told Brad Robinson, the hotell keeper, about him. My brother was paying attention to her, too, and was at the time engaged to marry her, so between four or five boys and some three or four girls and Mrs. Robinson, they got Mr. Coom to believe that he could get her (Miss Elizabeth Painter) and he proposed and she accepted him.

[They] told him Mrs. Robinson would oppose the marriage, but if he would come on such a night, she would go with him to the squire's and

---

[6] The claim of inventing the *first* threshing machine is hyperbolous. F. Hal Higgins, Davis, California, distinguished collector and historian of agricultural history, writing to me under cover of July 24, 1961, noted that in *The Union Agriculturist*, August, 1841, there was an advertisement for Churchill & Danford. Their machine "was a strip combined harvester. It was of course a 'threshing machine' pulled through the field by horses. Since the patent was taken out by A. Churchill of Geneva, Ill., it is likely that Danford was the local threshing machine manufacturer." The advertisement refers to "E. Danford." This might be the brother of the Danforth (possibly misspelled) mentioned here by Reinhart.

get married, but to let no one know about it. So Mr. Coom got ready and come after his intended wife. She met him, and when they got to the Justice of the Peace, he saw that his companion was a young man dressed in Elizabeth Painter's dress. His name Albert Chase, a young artist and painter about the height of Miss Painter, and it was all got up to fool Mr. Coom, the butcher of Roseburg, and Mr. Coom got quite mad over it and threatened to prosecute the parties that had played the joke on him. But at last he got so badly plagued that he left his business and Roseburg. It was shortly before Charley, my brother, got married to Miss Painter, and he had encouraged the joke.

Some people did not think it was a good joke, and spoke of any girl that would do that could not be much. There always are some that such things do not suit, and they will talk. When I heard of it, I wrote to Charley that I did not approve of it, and I acted quite foolish and said I would not own such a sister-in-law. But Charley told me it was a joke got up by Mr. and Mrs. Brad Robinson, Mr. and Mrs. Vince Davis, and several young ladies that had got tired of Mr. Coom's attention to them. But Charley did not let his wife see my letter, and it was best he did not, for it would have caused hard feelings between her and me. I was prejudiced against her on that and other accounts which I will mention further on.

While I was clerking for Ransom and McKay, there were several miners stopping at the store, and one an Irishman named John McClancy. He had a fight with some other Irishman at Canyonville, and was arrested for it, but was cleared. I think he must have cut someone with a knife in self-defense. Dave Ransom got acquainted with him and he came up to Coffeeville with Dave to go to work on the creek, and he stayed several days at our store.

Two Scotchmen came up from a creek running into Grave Creek, Douglas County, Oregon, and they had made some little money in the mines there, and one, named Andrew Robertson, got in with John McClancy, the Irishman, and they concluded to go on a prospecting tour on some creek 75 or 80 miles off in the mountains. They both took guns and blankets, and a pack of provisions to last them two weeks, and took along a pick and shovel.

Now McClancy had little or no money, and Robertson had some $400 or $500 in gold dust with him. There was some snow in the hills and mountains, it being February. The two went by themselfs afoot. In about eight days John McClancy came in to Stout's ranch on the Umpqua River one morning about daylight and told of his getting lost from Robertson, that Robertson had give out and he had to leave him, nearly dead, to come and try to find help to go after him and bring him to Coffeeville. But he got lost and out of provisions and got his feet froze, so that he could hardly travel. He was given something to eat, and the Stout brothers brought him to Coffee Creek on a mule and as soon as the miners heard of it they proposed to raise a party of 15 or 20 men, take provisions, blankets and guns, and go out to find Robertson. McClancy said it could not be over sixty miles off to where Andrew Robertson had give out.

John McClancy had got fed up and rested, so he could go along and try and show them the way, But Jack Reynolds, a mountaineer, said they could track by the snow, and McClancy was too lame to travel. So the party left, well-armed and provisioned, and McClancy stayed at the store with me until he could go to work chopping wood. Some of the miners that went after Robertson suspected foul play of John McClancy, for Robertson was the stoutest and toughest and hardiest of the two. The story of McClancy saying that Robertson gave out after only three days out did not look plausible to many. So it was thought best to go without McClancy, for he would plead being lost and try and take them off the right track.

In five or six days the men came back and had found Robertson's body. The wolves had eat up his flesh but they saw a bullet mark in head, and his gun some distance away. His clothes were torn, and if he had money it was gone now, for none could be found. So they buried the skeleton and bones and brought his gun back with them. Some thought he might have shot himself because he could get no further, and no assistance came. Others thought, and I do today, that John McClancy killed him for his money, for I will have more of John McClancy to write about in after years. But the most of the miners would not act too hastily on circumstantial evidence for they had no positive proof, for some wanted to lynch McClancy, but concluded to let it go. Robertson was dead, and they might kill an innocent man.

I will now give another tragedy that proved quite serious for several persons, most all I was acquainted with. While on Sucker Creek, Josephine County, Oregon, in 1856 and 1857, I got acquainted with a small, light-complexioned Irishman named McPherson. He had been acquainted with Ed Ryan and Andrew Kelley, my old partners in 1852. Ryan told me McPherson was a [military] deserter from The Dalles or Vancouver on the Columbia. So was Ryan, Kelley, Freeze, and several others they pointed out to me. So was Thompson. My brother Charley knew him too there.

The winter of 1858 McPherson came to Roseburg and worked around and got drunk considerable. Now Vincent Davis, a blacksmith that Charley and I knew at Jacksonville, Oregon, and at Cow Creek, was another great whiskey-head, and he got married at Roseburg to a blacksmith's daughter named Louisa Bowen. She had a sister married to a farmer living close to Roseburg. McPherson got to work for the farmer, and Vince Davis's wife, then not married, used to come out to see her sister. She, Louisa, was living then as a hired girl at Col. George Ross's, that kept a livery stable at Roseburg. This farmer's wife made McPherson believe that he could marry her sister Louisa, and so when Vince Davis married her, he was awful abusive and got drunk and kept on sprees until the farmer turned him off. He went to town, got drunk, and tried to kill the farmer that he had been working for by shooting at him out of his coat pocket while they were in a grocery store, but he missed his man. He claimed that it was done accidentally, but the farmer knew better, for he had threatened it to someone else, and the farmer went to a Justice of the Peace and swore out a warrant.

In the meantime McPherson walked down the street to Vince Davis's house. Davis was home, and McPherson asked Vince if he had any navy-size revolver bullets. Vince saw that he was drunk, but did not know that he had shot at his brother-in-law, and to get rid of him, he said he would go and get him some, thinking McPherson would leave if he did not come back. But when Vince left, he told Louisa that he was going to shoot and kill her husband, Vince, when he came back. As soon as Louisa heard that, she ran out into the street to meet Vince Davis to let him know that McPherson was going to kill him and she met the constable looking for McPherson with a warrant got out by her brother-in-law. McPherson was running after Louisa to keep her from telling Vince when he met the constable, who called to him to halt, that he was under arrest. But McPherson ran back out of town with his navy revolver in his hand and swore he would not be arrested alive. So the constable come back to get help and guns to keep from getting too close to his pistols.

McPherson kept on down the street, crossed the bridge over Deer Creek and kept retreating into a field towards a farm house. Five or six men followed to help arrest him, among them Brad Robinson, the Justice of the Peace. He called on McPherson to surrender, but he would not, and they commenced firing at each other.

I had just come down from Coffee Creek and rode into Col. Ross's livery stable, and had not yet unsaddled when I saw the crowd run down the field. The men in the stable told me about it, and who it was, but just as we were looking, the firing commenced. McPherson first shot Dunsmith in the hand, and knocked the pistol out of it; he next shot Brad Robinson in the side. And they kept shooting at him, but no one hit him. A man with a rifle could not get a chance to shoot, for the constable's posse was in his way. At last, after McPherson had shot all of his shots out of his pistol, hitting three men, he threw his empty pistol at them, and drew a big knife; still retreating and the crowd following up, till a man named Singleton threw a large rock and knocked the knife out of his hand, and the crowd closed in on him, got him down and tied his hands, and were bringing him up to town.

When I met them, two or three men afoot and one on horseback came running ahead as hard as his horse could run for a doctor, that Brad

Robinson had been badly shot. (My brother and his wife were both at the Robinson hotell) Mrs. Robinson had two children; when they heard the news they were nearly frantic, and the wife fainted. The crowd brought up Mr. Robinson to the hotell, and over five hundred men had gathered aound McPherson and threw a rope around his neck and tied the other end to the horn or pommel of saddle on a horses back, and commenced to drag him, when the United States Judge Deady,[7] who happened to be at Roseburg, called on the citizens and Sheriff [John] Fullerton and Deputy McKay and Constable to assist, and take the prisoner from the crowd.

So we all rushed in and got the rope away from the saddle and off of McPherson's neck, and Sheriff Fullerton and posse got possession of the prisoner and took him to jail for safe-keeping. Judge Deady then addressed the crowd and asked them to wait and see how bad Brad Robinson was hurt before they hung the prisoner, and the sheriff and constable quelled the disturbance for the time and the jail was well guarded by the sheriff's posse. If I could have seen McPherson before the crowd commenced to attack him, I could have reasoned him into allowing himself to be arrested, because they could not prove the shot at the farmer was not an accident, and could have done nothing but to fine him a few dollars for disturbing the peace, and he would give himself up by my advice.....

Brad Robinson died on that night, and McPherson was to have his examination before a Justice of the Peace in the morning. The sheriff took McPherson upstairs to the justice's office—the stairs were on the outside. Some of the bystanders went up to McPherson in the courtroom and threw a slip noose around his neck, and the crowd at the end of the rope jerked him off the stairs to the ground about twenty feet below; they then dragged him to the Deer Creek bridge. There was a crowd of four or five hundred men and they were determined on his hanging right then in spite of judge or sheriff. They got on the bridge and fastened the rope high up on a crossbeam, and shoved him off and kicked him when he held on to everything he could get ahold of, and he begged and prayed, but it was of no use, he had to die. The fall from the bridge either broke his neck or he strangled to death, and he was guarded by the crowd to keep Sheriff John Fullerton from cutting him down. Judge Deady knew that it was of no use trying to keep him from being lynched, for the crowd was so exasperated about the killing of Bradford Robinson that he or the sheriffs would have fared hard had they attempted to take the prisoner from them.

[7] Matthew P. Deady, an 1849 emigrant, was appointed federal territorial judge for southern Oregon by President Franklin Pierce in 1853. In 1859, he became federal district judge, a post he held until his death in 1893. Moreover, he served as chairman of Oregon's state constitutional convention. He died March 24, 1893. Harvey W. Scott, *History of the Oregon Country*, V, 188–190.

The widow and Fred Robinson, a brother of Brad's, carried on the hotell a short time after the death of Brad. I was down to see them once in a while. On one of the trips there was a minstrel theatrical troupe. I knew one of the performers named Bill Shepherd—he was barkeeper for Spanish Mary at Port Orford—when I was there and when I kept the Pacific Ranch. Vince Davis was acquainted with a couple of them, one named Watson, called Old Eph, and after Charley and his wife, Vince Davis and wife, and several of our party had been to the theatre that night, we invited Watson and Shepherd to an oyster supper at Vince Davis's, and they had good music. Vince Davis was an old performer himself, and sang and played on the guitar, and we had wine and cake and a good time until two or three in the morning. . . .

The Widow Robinson concluded to go out of the hotel business, and so Charley and his wife talked of going on to a farm and persuaded me to go in business with them. I did not know how soon I would be out of a place, so said I would consider the plan, and try and find a place suitable for us. So I made inquiries and found a farm and hotell eleven miles east of Roseburg, right on the road to Canyonville, and the main road to and from Portland Oregon, to Calif. and through the big canyon to Rogue River Valley. It was the Old Burnett House, owned by old Col. Burnett. It was a postoffice called Round Prairie, and it had a great deal of travel pass it, and some fine farming land across the Umpqua River, and about one half mile to the river.[8] The old Colonel wanted to rent it for two years, including some crop already in of wheat and oats, so we rented the place for two years, and paid some down in advance. . . .

When we moved out on the Burnett place, we had to buy furniture, bedding, dishes and provisions. Some seed and feed we got from Judge Burnett, and it took my money pretty fast, and I found that Charley had not saved a cent, but run in debt when he married, and the one hundred dollars or more of mine he took . . . went to pay off [his] debts. He told me his felon on his hand throwed him so much behind that he could not work with it for a long time.

When Judge Burnett give us possession of the house and farm, his daughter Molly, a dashing young brunette lady of 18 years lived with us awhile, and kept the postoffice till I could be qualified as deputy postmaster, and until my appointment could come from Washington, which it did not come till late in the summer, while I was out running a threshing machine.

The postoffice business was a nuisance, and amounted to only five or six dollars per quarter year, and only paid a per cent, but there was always lots of reading matter, sample copies, papers and magazines to read, and the mail a-Friday weekly. I would have to be there when the

---

[8] Round Prairie is on the South Umpqua River between Roseburg and Myrtle Creek. James D. Burnett was the owner and settled there in Douglas County, 1852.

mail come in, when I should be at other business.

So we put in some spring crop—the fall wheat was in already—and I got certain shares to cut and thresh it out. We moved to the Burnett House the last of May, 1859. Col. Joe Hooker (afterward Fighting Joe) had a government contract to build and make a military road from Scottsburg through the canyon at Canyonville, and he was a frequent visitor at old Judge Burnett's when they were there, and he would come after we got there and stay overnight, and he paid considerable attention to Miss Molly Burnett.[9] He used to take her to festivals and balls at Roseburg, or Canyonville, and Molly used to be plagued about old Col. Hooker. He was gentlemanly, military bearing, goodlooking and fine-appearing, elderly —he was quite gray—but ruddy and fresh, and as upright as a dart. He was an Englishman and had won military renown in Mexico.[10] He had sometimes from twenty to fifty men working on the road, shoveling, picking, and chopping trees, and he done the overseeing of it. He was a great friend of Ed Baker of California, who came to Oregon and was elected to the United States Senate in less than seven months.[11] Old Joe was a warm supporter of his, and old Gen. Joe Lane was another of Hooker's Mexican comrades; it must have been Gen. Lane that got Hooker his government contract. Old General Joseph Lane had been Governor of Oregon and United States Senator with Delazon Smith, and they were both stumping the state of Oregon that year of 1859.[12] Col. Stout, a Democrat, ran against David Logan, a Republican,[13] and both Joe Lane and Delazon Smith stopped at our house, and preferred sleeping together in a double bed.

The bedbugs were awful bad at that time, all through Oregon, and not a hotell or private house but what was overrun with them, no matter

[9] When twenty-three years old, Joseph Hooker graduated from the U. S. Military Academy, serving with distinction in the Mexican War, and finally resigning his commission in 1853. He moved to a farm in Sonoma, California, remaining there until 1858. In that year he accepted the superintendency for military road construction in southern Oregon. He played a major role as a Union general in the Civil War.

[10] Hooker was born in Hadley, Massachusetts, in 1814.

[11] Edward D. Baker, an Illinois friend of Abraham Lincoln (whom he defeated for nomination to Congress in 1844), emigrated to California in 1852. His oratorical brilliance made him well known. In the winter of 1859, he moved to Oregon and was elected to the United States Senate by the legislature, September, 1860. With the advent of the Civil War, he raised a regiment and was killed at the head of his command at Ball's Bluff, Virginia, October 21, 1861.

[12] Delazon Smith was born in New Berlin, New York, October 15, 1816. He was a graduate of Oberlin College Institute, Ohio; studied law, and later worked as an editor on several newspapers. After three terms in the legislature, and one as a delegate to the state constitutional convention, he was elected United States senator in July, 1859, for the short term—Lane receiving the longer one. He died in Portland, November 18, 1860.

[13] Lansing Stout was elected to Congress over Logan in the 1859 election, even though Logan had the personal endorsement of Abraham Lincoln. For particulars see Walter C. Woodward, "The Rise and Early History of Political Parties in Oregon," *Oregon Historical Quarterly*, XII (1911), 245–263.

what pains they took to try and keep them away, not a house was without them, and they, Lane and Smith, next morning were joking about the lively time they had with the bedbugs during the night. I have known Col. Joe Hooker to take some bedding, or blankets, and go to the barn and sleep in the hay rather than be molested with the bugs in the house. Charley's wife took down the beds every day, and washed and scrubbed, to no purpose; the walls and paper were full of them, and we kept things as neat and clean as any house could be kept.

Old Joe Hooker used to tell of the hardships they went through in Mexico (he distinguished himself at Buena Vista). I heard him say that he had slept on the top of dead soldiers that were killed in the swampy land and they piled up the dead bodies to sleep on them to keep out of the water. He took Molly Burnett to a grand ball and reception at Roseburg, but Molly said the colonel was too precise, and did not dance (she liked to dance). But Hooker was too ceremonious and precise to suit Molly, who was pretty wild and full of fun, and old Aaron Rose's girls (Roseburg was named after Aaron Rose) and he had two nice, fine-looking girls, one named Emma, the other, I think Lissy [Elizabeth?]. They and some more young ladies plagued Molly about old Joe Hooker and teased her so that she did not like to go with him. He was a great friend of her father, the old judge Burnett, an old Kentuckian of good family. Col. Joe Hooker, Ed Baker (afterwards killed at Ball's Bluff) and old Joe Lane had all fought and won great notoriety in the Mexican War, 1846–1847. Joe Lane was born in North Carolina, but had an Indiana regiment and fought bravely.

Col. Joe Hooker was one of the most sociable and pleasant gentlemen I ever met, the best-behaved and good-mannered; in fact, he was a perfect ladies' man, so polite. He took a great fancy to a mulatto who used to chop trees for him on his road contract. He was a good ax-man, named Collins; he could throw an ax with the handle farther than any man I ever saw. (I have sometimes wondered in after years, in the War of the Rebellion, when Gen. Joe Hooker got to be a major-general, if he ever had Collins with him or not.)[14]

After harvesting our wheat and oats we concluded to buy a second-hand threshing machine, and we threshed in our neighborhood, because we did not do business enough at the hotell to engage both our attentions. One of us could be home part of the time. I was generally with the threshing machine, and left Charley to help his wife run the house, and we kept a hired man. The machine was not much account, and we had some few breaks and had to send to Yreka, California, some 200 miles, for castings or pinions, and we had to send back to Jacksonville by mail, and I had to go to Jacksonville, about 70 miles, with a team, and the delay was too long, and we could not make it pay. We had only bought on

[14] Hooker became a major general during the Civil War.

trial. Our own wheat and oats were in shock as most people in Oregon hardly ever stack it; no danger of rain, and they save time and expense by threshing it out of the shock.

While off threshing, Col. Burnett lived nearby with one of his sons, Josiah, a surveyor, and he found fault (about not threshing out our own grain before doing so for others) to Charley's wife, and made some remarks of her husband's carelessness. She took offense and scolded him and they had a war of words and she ordered him to leave the place and not come there again while she was there, that he was no gentleman to run down her husband in his absence. He got mad and left the house, but Elizabeth concluded she did not want to live on his place any longer, and when Charley got home, said if they did not move, she would, and got Charley into the notion of going on to a large farm belonging to Captain Gordon, and rented it before I knew of it.

Now I did not approve of giving up the Burnett House and place or farm, for we had rented and contracted it for two years, and paid part of the rent in advance, and we were holding for the two years rent when I went off threshing. While I was gone Charley moved all his effects away, and give up the key and place sometime in September, and moved to the Capt. Gordon place, six miles up the South Umpqua River, above Canyonville, and Judge Burnett took possession and sued for the balance of the two years' rent, claiming we had abandoned the hotell and farm. I was awful angry about it—if they had but consulted me about it, I could have given up the place to Burnett and got my lease canceled without more pay, but as they had given up the keys and abandoned the place, Judge Burnett could recover the balance of the rent. We had paid over one year, and had not been on the place over four or five months. The judge got judgment for the full two years and costs. Charley had engaged an old lawyer (Ab Gazeley, afterwards Representative)[15] whom he had paid in advance in threshing, and he no doubt sold Charley out, or he was careless in the case, and I expect he got a fee from Burnett, for he was tricky, I found out afterwards. Dave Ransom had crossed the plains with him; they were from Tioga County, New York, from the city of Oswego. He had been a school teacher and shyster lawyer, and had a spot in his eye (and I never saw a man that had, but he would bear watching mighty close). So Elizabeth Painter's temper and Charley's dullness in not seeing what he was doing when he left the place came so near breaking me up that I concluded to go to California and stay the coming winter.

[15] Reinhart's reference should be to James F. Gazley, who emigrated to Douglas County in 1851 from California. He served in the territorial legislature for Douglas County in 1854 and 1858; in the state legislature (House), 1860 and 1868. He was elected state prosecuting attorney for the first district in 1862. Reinhart's caustic comments should be taken lightly. *Transactions of the Oregon Pioneer Association, 1919*, p. 170; *The Oregon Journal*, November 12, 1937.

# 8

## PEOPLE, PLACES, AND THINGS

(1859–1861)

I had some other business in Kerbyville to attend to, so I told Charley to take what we had, and make the best of it. I took my gray horse and left for California, and Charley let the threshing machine fall back into the owner's hands, and he went on to putting in a big crop of wheat, rye and oats on the Gordon place. Capt. Gordon had been keeping a Dry Goods and Grocery store at Roseburg before he married the widow of Pat Day, whose brother, George Day, owned a large body of land and had plenty of stock on their farm, and were very popular men and well-off. So when Pat Day died, the widow got quite a large fortune in farms and stock. She had but one daughter, the only child, Dora, and the widow thought she had better marry again and have a manager for her farm and other property. So she married Capt. Saml. Gordon, and he gave up his store-keeping at Roseburg, and went to farming on a large scale.

We had threshed out his crop that summer, and he had more land than he could handle, and some houses on another farm, and Charley moved into the other buildings while Capt. Gordon still lived in the old Day homestead with his family. Charley got two teams with the place, and had over 200 acres in cultivation, and his four horses with it, some cows and hogs. He hired a man to work one of the teams.

I got to Jacksonville the first day, where Charley had built the second house put up in town in February 1852, when the Rich Gulch was discovered. Now in 1859 it was a nice town of about 2000 or 2500 inhabitants, well-built, and a good farming valley [in] Rogue River Valley, where there was not a house for twenty miles in September 1851 when we first got to Rogue River Valley off the plains.

The second day I crossed the Siskiyou Mountains, near old Applegate's.[1] I had not been over the road since I came from Humbug and Yreka, California, in 1852. All now was strange to me, and improved; fine hotels in the mountains, and good farms on Cottonwood Creek and Rich Diggings.

Mount Shasta showed fine from the mountains; it was a splendid view, and looked finer, but not quite as high, as Mount Hood on the Columbia River in Oregon. I crossed over the Klamath River on a bridge where I had crossed the ferry on my trip from Humbug Creek, and well I remembered how near I lost my life that day, twice, as I related that time in May, 1852.

I got to Yreka the evening of the second day; I had not been in Yreka since September, 1851, and [in] over eight years it had made a fine city of the then Shasta Butte City of clapboards, logs, and tents for houses. Now [it had] nice brick and stone two-story stores, and over 8000 inhabitants; whole streets lined with Chinese wash and fancy houses full of Chinee women.

I had not been in town but a few minutes when I ran across William Cochran, my old Fraser River partner. He was clerking in a book and stationery and notion store. He was awful glad to see me. The owner of the store was named Romain. I will have something to say of him further on.[2] He put me in mind of John Musgat, my brother-in-law of Fond du Lac, Wisconsin....

I stayed all night in Yreka to see the Hon. John D. Cosbey; he had been senator from that district, and was one of the best lawyers in California.

---

[1] Reference is to Jesse Applegate, who with his two brothers, Lindsey and Charles, emigrated to Oregon from Missouri in 1843.

[2] Anton A. Roman, a native of Bavaria, went to Scott Bar, Siskiyou County, in 1851. In 1857, after several enriching partnership-mining ventures, he left for San Francisco carrying a hundred ounces of gold. As he relates:

> I incidentally stepped into the book store of Burgess, Gilbert & Still on the old plaza. Then I had no possible intention of purchasing even a single copy of a book, and so informed a clerk. However, he seemed quite interested in what I told him about the mines—likewise observed my fondness for books—and easily persuaded me to exchange my gold dust for them.

This launched Roman's career as a bookman. Returning to Shasta City, later Yreka, he opened up "Roman's Book Store," a bit later taking in as his partner A. E. Raynes.

In September, 1858, he moved to San Francisco. He records: "I established myself permanently with a large stock of bound books on the west side of Montgomery Street north of California Street. It was from this store, nine years later, that the first issue of *Overland Monthly* was published." He died at the age of seventy-five, a casualty of a railroad accident. San Francisco *Chronicle*, June 22, 1903; San Francisco *Examiner*, June 22, 23, 1903; Madeleine B. Stern, "Anton Roman," *California Historical Society Quarterly*, XXVIII (1949), 1–18.

Anton Roman should not be confused, as Reinhart later confuses him, with James P. Romaine, who was hung in 1864 for murder and theft. (See Chapter XII.)

The firm of Crook and Cosbey was one of the best in the state.[3] Now Cosbey was to defend Abel George for the murder of Hough McCaslin in Jacksonville in 1857 or 1858. The Masons had employed Gen. Cosbey and I knew it. I wanted him to attend a case for myself and brother, so went to see him and he promised to prosecute our case at the same term of court in November, at Kerbyville, Josephine County, Oregon.

Next morning I concluded to have a look at old Humbug Creek, where I worked the fall of 1851. I passed through Greenhorn and several other mining creeks, all worked out, and I could not recognize any place. The face of the whole country looked the same, only worse; the hills all washed down and the creek beds all filled up. At Humbug City—the Forks—some few buildings were left, but the creek bars and banks were all leveled up, the timber all off the hills, and they looked so low that I could not find one familiar landmark, house or man, on the whole creek. Nearly eight years' absence in a mining camp or creek is worse than fifty years anywhere else but California. So I stopped my horse and made some few inquiries, and kept on over the mountain range, and came down on Indian Creek where some rich quartz had been struck, and several quartz mills were at work.

I first thought of trying to get work, but I had no experience in that kind of work. So [I] concluded to go to Scott Valley and I got to Fort Jones,[4] where I heard of a man who had kept [a hotel] there a long time, named Joel Sherman. He was from Fort Hill, Illinois, where I used to live and where our parents had our home. He was suspected of killing a peddler at a dance at Goodale's Hotell, three or four miles from our place, in 1847 or 1848. He, Sherman, left there because he was suspicioned of the murder, and he came to California and had lived at Fort Jones, keeping a hotell most of the time. I did not stop to make myself known, or that I was from Fort Hill.

I went on down to Scott Valley where I heard James Cranston (who used to play the violin for us, with Dick Jac[obs] at Althouse Creek, and afterward tended bar for us at the bakery and ball alley on Sucker Creek) had a farm, so I thought I would call on him. But when I got nearly to the place I heard from a neighbor of his that he had died that summer,

---

[3] John D. Cosby launched his legal practice in Yreka in 1851. The following year he ran for county judge but lost. Later he served as state senator for one term. His first legal partnership was with John D. Cook, also a native of Kentucky, who came to Yreka in 1852. That partnership was dissolved in 1855. (Cook died in Sacramento of consumption in 1857 or 1858.) In 1857 Cosby formed his second partnership with Elijah Steele, which terminated in 1859.

During the Modoc conflict of 1885–1856, Cosby served as major-general of the Siskiyou Militia. He died on May 5, 1861. [Harry L.] Wells, *History of Siskiyou County, California* . . ., pp. 72, 91–92, 92a (for Steele's biography); *California Blue Book, or State Roster, 1909*, p. 570.

[4] Fort Jones was established October 16, 1852, by Major Edward H. Fitzgerald of the First United States Dragoons. J. Roy Jones, *Saddle Bags in Siskiyou*, p. 25, *et seq.*, gives a brief history.

*People, Places, and Things* 165

so I went by his place. He had a partner on Althouse Creek, mining. His name was Horace Spears. He lived close to James Cranston in Illinois and he was in delicate health and went back to Illinois in 1854, and I heard he died there.

I thought I might get to run and feed a threshing machine in Scott Valley, but work was scarce. I then thought I would work on some farm a month, until the November term of court at Kerbyville, and I would have to be there. So I stopped at a farm one night, and the farmer said he would give me work for a while at, I think, $25 per month, board and washing, and my horse kept. I was to chop wood, plow, drive up the cows, and feed a lot of cattle or stock. There was a nice girl or two, and the farmer seemed to be well-off. So I thought I would try it awhile. I was well-dressed, had my big gold ring on my finger, and my fine specimen gold double breast-pin with a gold chain between the two pins. I had a good horse, saddle and bridle, nice saddle-bags, some clean clothes and shirts, and a nice pair of canteenas with holster, with a navy revolver. They rather suspicioned me working out in the fix I was in, but I was nearly out of money. I do not think I had over $2.50 left, or I would not have tried for work.

So I got up in the morning awful early. Everybody got waked up before daylight. I went out and fed the cows and then chopped wood till breakfast, an awful long time, I thought, for me to get up before day. After breakfast I yoked up two yoke of cattle and went to plowing and harrowing, and I was not used to that kind of work, and it came hard to me. I thought noon would never come, I was so leg-weary. After dinner I worked a while again till night, quite dark, and come to the conclusion it would not pay me to work that way at that wages. So at night I told the farmer I would quit. I stayed all night and left next morning.

I looked around for some surface diggings where I might get work, but I could find none. So I concluded to go back to Yreka and look around there for some kind of work, and to sell my horse, for he was a great expense, and I could walk where I wished to go. At the stable I put up at, at Yreka, a man told me a German butcher in Yreka had a farm out some six or seven miles on Little Shasta Creek, in Shasta Valley, and he was digging a large water-ditch or canal, and maybe I could get work out there. Next morning I was nearly out of money, so I took my big ring to a pawnbroker and pawned it for ten dollars (It cost me $24) for a few days until I could sell my horse. Next morning I went out to see Mr. Jullian, the butcher, on his farm. He had a fine place and he was very wealthy, and had everything fixed up to carry on a stock farm of some fine stock. He was a Swiss, and had just got married to his ward, a young lady of seventeen years old (he was over fifty years old).[5] I asked about

[5] Reinhart's reference is probably to Julien Neuschwander. Wells, *Siskiyou County*, p. 106c.

work, but he had all the help he wanted for the whole winter, some 18 or 20 men.

So I asked him to buy my horse. He was awful cautious, because I was a stranger to him. He found out I was German, [and] he made me a low offer. I at last sold him the horse for $65. He was worth a hundred, but it was a bad time to sell, at that time of year. So I rode the horse to Yreka, and sold my saddle and bridle, and left the horse for Mr. Jullian at the livery stable. I went and redeemed my ring, in two days after I had pawned it, and paid 25 cents for the use of the ten dollars. I did not stop over two or three days in Yreka. I saw John D. Cosbey before I left, so to be sure that I could depend on him to take our case; bid William Cochran goodbye, and left Yreka on my way back to Kerbyville. I had my saddlebags and holster to carry. I sometimes got chances to ride in a wagon or on horseback, and got back to Rogue River Valley and stopped one night at a hotell kept by Lorenzo Oatman and his sister Olive, who had been a prisoner so long among the Apache Indians, after they had murdered her father, mother, some brothers and sister, and the balance of the whole train. I saw her once before near Evans Bridge where she lived with some doctor's family. She had her face, mouth and chin all tattooed, and her features had a regular Indian cast, and not knowing who she was, would have taken her for a half-breed squaw; her complexion, form, and features had become so by their habits and diet and mode of living.[6]

I found some old friends along the road, and passed William Ishe's place, one of Molly Burnett's old beaus.[7] He lived close to Jacksonville and had a stage line and big lot of cattle. One of Miss Molly's sisters had got married to a wealthy farmer close to Jacksonville. I did not stop long at Jacksonville, but I stopped near Evans on the Rogue River, and saw James Wilson; he had sold out on Coffee Creek and come down to Roseburg and stayed with us a few days at the Burnett house, and he tried to get me to go down on the coast below Port Orford, where I used to live

---

[6] Royce Oatman and his family left Independence, Missouri, in a caravan of Brewsterites (a Mormon sect) for Yuma, Arizona, in August, 1850. Eighty miles above the mouth of the Gila, the Apaches fell on them and captured Olive and her sister, Mary Ann. Lorenzo, left for dead, survived and later made his way to Los Angeles. Through negotiation, Olive was delivered up to Fort Yuma, February 22, 1855. (Mary Ann died in captivity.) A cousin, Harrison B. Oatman, came to Los Angeles several months later to claim guardianship of Olive and Lorenzo, and took them home to the Rogue River.

Several years later, Olive and Lorenzo returned to California, spending six months in the Santa Clara Valley. In March, 1858, they sailed from San Francisco to New York. Olive married John B. Fairchild in the fall of 1865, and died at Sherman, Texas, March 20, 1903. Los Angeles *Star*, March 8; April 19; June 21, 1856; Alice B. Maloney (ed.), "Some Oatman Documents," *California Historical Society Quarterly*, XXI (1941), 107–111 (on facts leading to the captivity); William B. Rice, "The Captivity of Olive Oatman," *ibid.*, pp. 97–106. Royal B. Stratton, *Captivity of the Oatman Girls . . ., passim.*

[7] Jacob and George Ish lived in this vicinity as did William K. Ish.

and had our Pacific Ranch in 1854, when the Indians burnt us out. I had not been back there since, and did not know if anybody was on it. If there was, it would have been hard to get it back after five years.

Wilson said he had an aunt somewhere down the Willamette Valley, and she had lots of cows and cattle, and we, Charley and his wife, should take up a large stock farm and ranch and go into raising cattle. But he was going to see his aunt first and let us know if his aunt was agreeable, [and] I had not seen him since until I saw him here. He said he had a racehorse, and had made up a race and was training him for the race. He, Wilson, stopped on a ranch belonging to Thomas Kav[a]naugh, a gambler, who had lots of horses and cattle on the ranch seven or eight miles south of Jacksonville. Wilson was just loafing around the ranch. Jim Kav[a] naugh had a herder with the stock, and he himself stayed most of the time in Jacksonville, gambling. (I will speak again of these parties further on.)

Next day I got to Kerbyville and court was about to commence—the murder trial of Abel George for killing Hough McCaslin. The sheriff, James Hendershott, had selected a jury of whom nine or ten out of the twelve were Masons.[8] The prisoner had been a good Mason in good standing, but of late years had got to drinking and become reckless and a dead beat and a dangerous man when he got drunk. But he let his dues to the lodge run until he was expelled, or not recognized by them, but he had been a first class Mason in good repute, and well thought of when he was captain of a company of volunteer miners from Scott's Valley, and had done good service in the Indian War of 1853-54-55. So the Masons concluded to defend him, for he had a nice family, a wife and two or three children.[9]

Now I will relate the circumstances of the murder: A young Irishman miner named Hough McCaslin lived on one of the creeks near Jacksonville, mining. He was in town, and his horse at a livery stable. He got ready to go home, went to the stable, had him saddled, and just as he got on the horse, Abel George asked him if his horse would carry double. McCaslin said no. Then Abel George said he would get on behind anyhow, and took a-holt of the stirrup to get on behind. There were several persons present in the stable, and they saw that Abel George was drunk, and advised and tried to persuade him to let the horse alone, but he would not. So McCaslin said if he would get on, he would get off, and stepped down off his horse, when Abel George stabbed and cut him in the side and McCaslin died on the spot. Abel George escaped to home before the

[8] See Chapter XI, *note* 3, *post*, for James Hendershott.
[9] Abel George had also served on the first board of county commissioners for Jackson in 1853. The Jacksonville Masonic Lodge was established March 15, 1855. George was vehemently anti-Mormon. He proposed raising an army of volunteers to wipe them out. Shasta *Courier*, October 21, 1854; Red Bluff *Beacon*, January 1, 1858; A[lbert] G. Walling, *History of Southern Oregon* . . ., pp. 340, 367.

men realized what he had done, then a crowd went to his house to take him. But the city marshal or sheriff and posse got in ahead and took him to jail, or the crowd of McCaslin's friends that came to lynch him would have done so.

But the citizens and miners at last concluded to let him have a fair trial, feeling sure of convicting him. Abel George had many friends in town, so had his brother John George, and his brother Masons tried all they could to clear George, and by postponing the trial from one court to another, had at last after two years got a change of venue to Josephine County by swearing that they did not believe they could have justice, that the Jackson County people were prejudiced against Abel George. The prosecuting witnesses, some had left, and others believed it would be a farce, so took no interest in the trial, and no doubt the prosecuting attorney from Jackson County was feed or bribed not to work hard against him, the prisoner, and John D. Cosbey, the best criminal lawyer in California, was hired by his friends and the Masons, and nine or ten out of 12 jurymen were Masons. (One of the Masons told me himself—Mr. Wesley Pleasands of the firm of R. B. Morford and Pleasands, that kept a general store on Sucker Creek. R. B. Morford was son-in-law of old Dr. Pleasands of Jacksonville, and was our county clerk of Josephine County. He afterwards moved to Umatilla Landing, just below Wallula on the Columbia River, and was the county seat of Umatilla county, where I lived after, and had my ranch and farm.)

The trial of Abel George lasted three or four days and five or six doctors were brought from Jacksonville by the defense to prove that George was crazy when he killed Hough McCaslin, but that it was temporary lunacy, and he recovered soon after, and thought that hard drinking had brought the crazy spell on; that he had been so afflicted before; and [the defense] had the prisoner's wife and children [in court] to work on the sympathy of the jurors, which they did to perfection, and when Gen. John D. Cosbey got nearly through pleading, every man, woman and child in the courtroom was in tears (jurors, bar, and all), and it continued to the end. It was hardly necessary for the jury to leave their seats, for everyone in the room knew that he would be cleared, not but that he was guilty, but because they pitied the wife and two children, one a girl of 8 years old, the boy about 9 or 10, and I think they had a younger one at home. Anyway it did not take long for the jury to find a verdict of "Not Guilty" for the defendant on the plea of temporary insanity, and he was cleared by great shouts, and shaking of hands and so forth.

As I came out of the courtroom I saw Schertz had been in to the trial, and like me, had tears in his eyes when I went up to shake hands with him. He was still working at the mill, but was about to quit work there to go to Sucker or Althouse Creek to try to mine again. I saw Judge Cosbey after the trial (he was pretty full already) and Sheriff Jim Hendershot[t], an old Althouse friend of 1853.

He was holding a payment for us from the Chinamen for our claim, and a man named Ben Norton, with whom we left our accounts and books and the renting of our saloon, the Eldorado, on Indian Creek, attached it in the hands of the sheriff for some debt Charley owed him, and thought he had collected enough off our books and rents to pay his claim, but he said not, and I, being gone to Fraser River, Charley could do nothing. We agreed at last to divide, each take $75 rather than make more costs (took it, but not at this time).

I told John Schertz about Charley's farming, and Charley had told me to get John Schertz to go in with him, but John said he had no money to go. I told him I would give him [some] and did, and paid his way down, and both went back together. John was a good farmer, and he went in with Charley, and I helped awhile till we got most of the crop in, and I got tired because there would be no money to come in till next harvest or fall, and the two could get along, and less expense to the two men than three would be.

So I took some $15 or $20 and left some with Charley and John, but I had paid out a good deal of the $65 that I got for the horse, [which] did not leave but little. For fear I would run short, I borrowed $15 on my ring of a saloon-keeper at Canyonville till I could redeem it. I went to Wolf Creek, about 60 miles off, where there had been some new surface gold diggings discovered. It was in Jackson County and only 7 or 8 miles from Grave Creek, afterward named Leland Creek, where my old friend from Chicago kept the hotel with his partner Harkness.[10] They were doing a good business.

When I got to Wolf Creek I found several old friends, William Woosley, and John McClancy, and two others, had a creek claim just about to open. They had a cabin on their claim and the two, Woosley and McClancy, invited me to stay with them until I should make other arrangements to stop.

There was a man named Dan Levens, and he kept a public house not far from Cow Creek, on the east side of the canyon on the military road. He brought a small stock of groceries and clothing and built a log cabin for a store near Woosley's claim. He had a little Irishman clerking for him that I used to know when he was working for Jess Roberts on his place on Deer Creek, while we were at the Burnett place. So I stopped with Woosley and McClancy a few days till I looked around.

I did not think their claim paid over two or three dollars a day to the hand. They had done considerable work at a great expense to open and get lumber to make sluices, and buy tools. I found wages low, and winter

---

[10] Martha Leland Crowley died in 1846 on what is now called Grave Creek. At first the legislature sought to honor her memory by dubbing the creek, Leland. But popular usage won out. Lewis A. MacArthur, *Oregon Geographic Names*, p. 273.

coming on. It was the last of November, 1859, and it was getting quite chilly nights, and the water cold to work in, and I knew it would soon freeze, so that a person could not get in much work in the week, and the board would still go on.

So I thought I might get a whipsaw and get me a partner to saw lumber, and I could work under shelter of a tree and shed, and make more than at mining, without working in the cold water. I already had a twinge of rheumatism in my right shoulder and I was afraid working in cold water, getting wet and exposed to cold, it might come back on me, for I had it in 1855 on Indian Creek, California. So I made inquiries and it was some ten miles off, on another creek, where the sawmill was that supplied Wolf Creek with lumber at $8 per hundred, and when the road was bad or the mill stream low, they would charge more, or could not get in with their teams. It was rather low to run opposition to a sawmill by Armstrong Power [elbow grease?], but I could make more at it, and I would rather do it than mining, and be always with dry feet. So I went over to Grave Creek and found that Jimmy T[w]ogood and Harkness had a whipsaw to sell. I bought it, and paid the biggest part down, for I could not spare the whole amount, $15. So I paid about $9 down, and was to pay the other whenever I could....

I found a young man that wanted to learn to whipsaw, got ready, and several miners said they would want sluices, and give me orders rather than patronize the sawmill. Some said I could sell all I could saw, and I soon made my partner understand how to saw and when the road got bad so that the teams could not haul any more lumber into Wolf Creek, I raised the price to $10 per hundred feet and so made treble good wages, and we built sheds over our heads with some of the lumber and worked in rain and snow. The young man I took in as a partner was named George Shepherd, from the state of Massachusetts. He told me he come out to San Francisco with his father when he was but 14 years old, and had been out five years, and he ran away from his father and went to Fraser River, but I believe he enlisted in some eastern city and deserted when he got on the Pacific coast at Puget Sound. He was a good worker when he felt like it, but would rather play cards and gamble than work. But I kept him straight. He had no money to spree on and I kept the purse, so that he could not get off while we had work to do. But when we did not work I would divide out a certain portion and he could do whatever he wished to do with his. So did I.

We played cards a great deal, mostly freeze-out poker, in Dan Levens' store, and Irish Jimmy generally took a hand. I won considerable underclothes and provisions, sugar, coffee, bacon, socks, or anything we wanted, but we had to pay cash for what we lost. We did not run in debt. There were two gamblers come in from Jacksonville and started a saloon right opposite Dan Levens' store. We sawed out a bill of lumber for them. One of them was named Hiram Long, the other Bob Steward. Long did not stop much, but traveled, and Robert Steward run the saloon. Some gamblers from other places used to come and want to play, and could get no one. Steward would tell them he had no cards, but I had a deck and they could play with me. I still had my deck that I had over three years, and William Cochran gave me two decks of common stamp cards, and I had learned them pretty well. I would play with the sharpest, and they did not detect, although they mistrusted the cards, and they would not want to play with me long, for I always beat them with them old cards.

During the winter Woosley and John McClancy sold out of their claim. Woosley stayed a short time after selling out, then went off. John McClancy prospected and worked some for another company named Donehough [Donahue?] and Ferguson.

We had a game of freeze-out poker one day, and I put up my breastpin for $25. If I had not got it broke by Dave Ransom, I would not have parted with it for less than $75. I had been offered $65 for it at The Dalles on my way to Fraser River, by James Brickley, a gambler whom both Charley and myself knew at Sucker Creek in 1856–57. (I will speak of him further on.) The game was made up, five of us at $5 apiece and Donehough[?] won it. He was to let me have it by redeeming it whenever I wished to. It did not seem to be worth much to me after it got broke, and I would not have give $25 for it again. One day Donehough[?] lost it out of his shirt-bosom into the sluices, and after hunting a half a day, only found the pin. So I would not care to own it. All the novelty had worn off.

A man came from Jacksonville—he was a carpenter, and had gone on a spree and spent the money he had, and was ashamed to be seen at Jacksonville. He worked in a furniture and cabinet shop for Burpee and Linn.[11] I was acquainted with both men while I was there. This man, named John Larkin, was broke and nearly barefooted, and John McClancy run acrost him and made his acquaintance.

He had a brother, a New York alderman, a contractor, well-off, but he had disowned his brother John Larkin on account of his spreeing so much. John got on a spree and enlisted into a company of soldiers sent out to Washington Territory, and between Seattle and Olympia he had deserted, and got away and worked his way to Jacksonville. But after a while he would get drunk and so lost his place with Burpee and Linn. John McClancy done what he could for him....

The first Monday in March my suit, that had been postponed by the Abel George trial, was to come off, and Charley wrote to me that he could not go, that his wife was so that he could not get off and stay any

[11] David Linn, a carpenter, was associated with James S. Burpee in the furniture business.

length of time.... I started the last of February [1860]. It took me about three days to get to Kerbyville....

When I got to Kerbyville I found I would have to wait, and was uncertain of my case coming up. There were several criminal cases first on the docket. At last Ben Norton offered to settle and divide the money, for no telling how much longer we would have to wait, and fee our lawyers, and the cost would come to half the amount in question, and by looking at it right it was so, and I took the $75, but after paying all my expense, I did not have over $50 left.

I left Kerbyville on my way back to Wolf Creek next morning. I got company as far as Slate Creek, Jefferson Howell, a young man, that first come to where I was mining on a gulch on Cow Creek not far from Canyonville. He and five or six others before wrote about, [had just come] acrost the plains [by] the northern route—I spoke of him at Althouse and some of them at Gold Beach and through Oregon. Alexander Fuller was his partner on Althouse. He was from Missouri, near Creek 101.

During the Indian outbreak of 1855, a company of volunteers from Althouse chose Capt. Bob Williams their captain, and Jack Driscoll their 2d lieutenant and Jefferson Howell 1st lieutenant, and the war lasted a year or more. After the war was over and the men discharged they were still at Jacksonville and the old grudge between Col. Bob Williams and Jack Driscoll got some way revived (about the whipping of McCloud and McCoy on Althouse in 1853 when Bob Williams struck at Jack Driscoll with a knife). Bob had friends and so had Jack and both sides egged on both their friends. They told Jack Driscoll that Col. Bob said that Driscoll belonged to a band of horse thieves and Driscoll threatened to kill Col. Bob Williams on sight, and some of Bob's friends told him of Jack Driscoll's threat.

So one day in Jacksonville they saw each other, one on one side of Main Street and one on the other side. Jack Driscoll hauled out his big Navy revolver and got on one knee to aim and shot at Bob Williams, who had a shotgun in his hand. Jack shot first and after two shots Bob Williams shot Jack Driscoll and he fell, but still kept shooting until he had emptied his pistol. Col. Bob Williams fired the second and he killed Jack Driscoll, who died shortly. Williams went down the street, got on his horse and left town. He was not even arrested at the time, but I believe he afterwards gave himself up, was tried and discharged. He had acted in self defense. (This was the result and outcome of the Salmon River miner that got drunk and thought he got robbed in 1853 on Althouse, and two men whipped and run off the creek, and 1857 or 1858 some men who had claims were mining up the old town bar and found a buckskin sack with 3 or 4 hundred dollars of very fine dust, all in a pile in a prospect hole. The purse was rotten and the gold fell through and it proved to be the purse of gold dust. The man was sober enough to think someone might take his purse while he was asleep, and before laying down he got into a prospect hole, buried the purse of dust, and got out and went to sleep against some houses, and when he awoke next morning he missed his gold and he had forgotten that he had buried it, and so accused everyone he could. And he might as well have accused me or Jim Downer, the owner of the alley; we had all of us handled his purse, and the miners were just crazy enough to believe all he said, and we might have been the victims as well as McCloud and McCoy, but no doubt they both deserved what they got, for their agreement with each other to rob or get the money from the man.)

Jeff Howell after the war was deputy sheriff awhile of Josephine County and was once elected to the legislature, and he mixed up and become quite a popular politician and worked for the election of Stout against Dave Logan, a Republican. Stout and Howell were Douglas Democrats....[12]

When we got to Slate Creek Jeff Howell stopped, and I went on to the Vannoy Ferry on Rogue River and stayed all night. Next day to Jimmy T[w]ogood's and stayed over night and paid my bill and the balance due on the whipsaw, and next day got to as far as Dan Levens' place and stayed over night, and next day got to Charley my brother's, where he was on the Gordon place.

His wife had a little boy two or three weeks old; they named him George Ellwood. John Schertz was doing well and not drinking much; he was too far away to get it, and he worked well and had a good crop. I divided the money I had got at Kerbyville with John and Charley, stayed over night, and started back to Wolf Creek....

I got back to Kyoty [Coyote] or Wolf Creek the second day after leaving Brother Charley.[13] Shep[h]erd was doing something else and prospecting. So I took in John Larkin and we run the whipsaw for a contract of lumber. He made a good sawyer, and John McClancy went and got on a spree and abused an old man of family named Wallace, and the old man stabbed and cut John in the side. For some time it was thought he would die from it. The old man sloped and left his family nearly unprovided. So when McClancy got so there was no danger of his dying from the effects of his cuts, the old man came back and there was nothing ever done about it. McClancy stayed with Larkin and I and cooked our meals, for he was not able to work.

One day someone put up a horse at freeze-out, and John McClancy paid half the entry and I won the horse in company with him. A few days after Robert Steward, the saloonkeeper, put up his big gray at freeze-

---

[12] Howell was a delegate to the Democratic National Convention, 1864. Records fail to show him a member of the legislature, but he was a Douglas Democrat.

[13] Coyote and Wolf creeks flow more or less parallel, but are separated by several miles at their closest point.

out at $10 a chance, and he took a chance in, and there were six besides he and I. We froze out the other six and he proposed to give or take so much money, and not play against each other any more. I done so and sold John Larkin and McClancy my half of the other horse, a bay, and bought out Steward's half of his gray horse. I got a saddle and bridle and in May sold out my saw and what tools I had, and concluded to go down to the Willamette Valley, and go to work or get into something....

Now I will go back to Henry, or James Wilson, he sometimes called himself one or the other. When he stopped with the herder of Tom Kav[a]naugh on the ranch seven or eight miles south of Jacksonville, one morning he went out walking after his horse, and in passing through a gulch or ravine, picked up a piece of gold-bearing quartz. He found several pieces, and they looked favorable for gold, so he brought the pieces to the cabin, and laid them on a shelf. A few days after, he went out while the herder was gone, took a saddle-horse belonging to Tom Kav[a]naugh, and skipped out for parts unknown. He must have got in debt or borrowed some money. He was gone a few days before they missed him, thinking he was up to Jacksonville, but when the herder asked Tom Kav[a]naugh, he said he expected he was on the ranch with him, as he had not seen him for several days. So they knew he was gone, and so was Tom's horse.

A few days after, old Col. Ross of Jacksonville (he was father of Col. George Ross of the Roseburg livery stable—he was a gambler and well-off) came out to Tom Kav[a]naugh['s] ranch, and he happened to see the quartz rock on the shelf, and examined it close and saw free gold in it. He asked Tom where he got that. Tom did not know and he called to the herder if he knew, and he told him Wilson had got it down in the pasture, and at that time had said he thought there was some good-paying quartz close by, but he, the herder, had thought no more of it. Col. Ross asked the herder if he could show him the place where Wilson had said he got it. He said yes, and they all got up their horses that were close by, and rode down to the gulch they found it. And Col. Ross and Tom commenced looking around and found where the rock had come out of a small ledge or quartz lead, decomposed, and found some pieces nearly half gold, and very rich. So Kav[a]naugh and Ross covered it up and told the herder to say nothing about it to no one, and they staked off some ground that would contain the lode, and went on to Jacksonville, recorded their discovery and claims, and made up a company to work the mines, called Gold Hill Mining Company.

Tom Kav[a]naugh sold one half fifth of a share of the whole for $60,000 to Statler of Crescent City, Col. Ross kept one fifth, Jack Pope bought one tenth, Charley Williams one tenth, and Tom Kav[a]naugh kept half of the fifth—he sold the other to Statler. Fowler, and all the moneyed men around Jacksonville, bought certain interests, mostly all gamblers too. It proved to be the richest discovery of gold and quartz ever found up to that time on record, and supposed to be worth millions of dollars. The people around run nearly crazy, and thousands of men or miners had walked over, rode over, and passed by, and never saw it until found by Ross, Kav[a]naugh and Company. Only Wilson suspected that there was paying quartz, and he by thieving, stealing, meanness and dishonesty, had lost a big fortune, and he would have been widely notorious as the discoverer of the celebrated Gold Hill, Jackson County, Oregon. (I have more of this same James or Henry Wilson to write about further on).[14]

So in May, 1860, I concluded to go to Willamette Valley to get work, and go to Salem, the capital of the state, to collect some money, and [then] go to [The] Dalles and get the note or due-bill cashed that I won on Wolf Creek. McClancy and Larkin packed up their bay horse and struck out for California by way of Jacksonville....

I left the same day, rode my gray to Charley's, and stayed overnight with them and started the next day for the Willamette.... Charley's and John's crops looked well, and Elizabeth and the baby George were both well. They had plenty of chickens, pigs, and cows to make butter, and all were doing well when I left. I saw many old friends as I passed Canyonville, Myrtle Creek, Round Prairie, and Roseburg, and the second day got to Oakland.

A nice large grist mill is located here at Oakland and is quite a little town. Charley's wife's father lived close by. So did John Ladd, a neighbor who had married a minister's daughter, some relation of the Painters, of whom I will [say] more hereafter.

Passed the North Umpqua at Winchester.... I enjoyed my trip very much and passed through all the towns and cities on the way, crossed the Cal[a]pooya Mountains ... and went by Eugene City, the head of navigation at high water, a nice city, but it tried several times to be made the capital of Oregon, but the people that favored Salem, the then and now still capital, beat it at every election, for the southern part of Oregon was not so well populated as the northern and older part of Oregon. From Eugene City I went on to Corvallis (old Marysville of 1852).

After I left Corvallis on my way to Albany I met some farmers and I was inquiring for work. I was referred to William McCoy, about four miles northeast of Corvallis, in Benton County. I stopped at McCoy's and he told me of a man that wanted help, at a little town named Peoria, Linn County, a few miles from McCoy's. The man's name was James Martin, he was a large heavy man, weighed about 300 lbs, and he gave me work for a while on his farm until haying commenced. They were Seceders of

[14] Colonel John E. Ross was one of the leading men of Jacksonville and was instrumental in bringing Gold Hill into production. A variation of this gold strike is given in Walling, *Southern Oregon*, pp. 328–329. Thomas Chavner and Charles Williams were members of the original partnership. Walling also includes George Ish, James Hayes, and Long John in the initial agreement. Subsequently, Henry Klippel, John McLaughlin, and Ross bought out several members of the first partnership.

the Methodist Church, and attended church in Peoria (named by Illinois folks who settled around it).

Old James Martin was a great hog raiser, and he made money on them. His wife was a little old body that would not weigh ninety pounds. She used to shave the old man, and he was so fat that he could not tie or untie his shoes, and she would have to do it, or a little girl of about ten or eleven years they had adopted. Their own were grown and married, some in Illinois, some in Oregon, but none at home. I stayed with old Jimmy Martin a couple of months, and then got work from James P. Hogue. He had a large farm and large harvest.

I got to binding for his McCormick reaper. I tried raking, but I could not go it—I would rather bind. He had a son-in-law named Silas Storey, from Ohio. He was a schoolmaster and had a farm adjoining Hogue's. The Hogues were a good family from Monmouth, Illinois. He was a carpenter, and had crossed the plains in 1850 or 1852 to Oregon. He had a large body of land, a band of mixed Spanish horses, and a new threshing machine, which I was to run as soon as the season commenced. I worked through harvest, and went to church regular every Sunday morning. So did the family. They had three daughters married, one to Storey, one to Col. Davis Leighton, a rich man, and one to Jed Powers, a get-up stock man, the fourth, a younger one at home, about 12 years old, named Ida; one son at Portland, Oregon, in the lumber business (he was a lawyer). Their firm was Harvey, Hogue, and Solomon Jacobs.

Then there were two boys at home, going to school, one 14, Charles P., and one about 9, Montgomery. Mrs. Emma Leighton, with a little blind daughter, lived with her father then, for the colonel was off somewhere on business and for his health (consumption). It was a large family, and I was treated like a lord, went wherever I pleased on my own horse, and I kept on with Hogue till the threshing season commenced. The machine belonged half to James P. Hogue, and the other half to John Leighton, brother of the colonel, a widower with a large family, with a large farm near Albany, where Col. Davis Leighton had some farms adjoining his brother John.

I run the thresher during harvest at $2 per day, during threshing at $1.50 per day, and washing, and my horse kept at Hogue's. While running the thresher, we threshed for lots of rich farmers, and they had plenty of fruit. Apples are as plenty in Oregon as any place I ever saw; the trees bear younger, and never grow very large, but the apples were some of the largest I ever saw. Some weighed from 20 to 39 ounces (the Gloria Mundy). We lived well and had good meals and nice victuals. While threshing, Saturday nights I would sometimes go home to the Hogues to stay over Sunday, and get clean clothes.

During the last few weeks of threshing I got sore eyes and I had to quit the dust of the machine, and came near being laid up all winter. At first

I thought it was but some beards of barley that got into my eyes; some thought wild hair had grown through the eyelids or winkers. So I went to see a celebrated eye doctor named Dr. Alexander, a Kentuckian. He asked me if I ever had the rheumatism. I told him I had in 1855. He said my eyes were affected by the rheumatic inflammation of the eyeballs or pupils and I would have to be quiet and stay in a dark place, out of dust and strong sunlight, and to wear green shades. He did not say if he could effect a cure, but gave me a lotion or liniment to wash the eye. He put in a few drops and it gave me ease right off. I think it was part syrup, and it kept the friction from the pupils. [My eyes] first felt to me like there was sand in them, then the eyeball would contract, then expand, until I thought they would burst out of my head....

So I stayed at Hogue's two or three weeks, and felt better, but I could not stand it to be in a dark room all the time. So I concluded to take my horse and ride around and see the country, and some of the cities not too far off. I had a lot of old accounts from Althouse and Sucker Creek and Gold Beach, and George Eads, the sheriff of Marion County, lived at Salem, the state capital, and the best county in the state. He owed us a pie and whiskey bill from Prattsville, Gold Beach, and I concluded to call on him for it. Another man, Joe Davenport, was owing me at Althouse Creek, and he paid me $65 in a violin when he left, but still owed a balance. He, too lived at Salem.

So I took . . . [a] trip to Albany and on to Salem. I stayed a few days and collected some little, but not enough to pay my expense for my horse and myself. But it done me good, and I saw all the finest land in the Willamette Valley, crossed the river at Salem, and I went to Dall[a]s [in Polk County], and Lafayette and many other towns in Yamhill County, and came back by way of Benton County, on the west side of the river. I came by the Wheeler and Pyburn farms. I was somewhat acquainted with the men that were concerned in this affair I here relate, of a case of Vigilantes in the heart of a well-settled valley.

A man named Wheeler, a school teacher by profession, came from one of the eastern states with his wife, and took up a half-section of good land, not far from the city of Corvallis, some ten years before. Wheeler made money by teaching, and whenever he could, bought stock and cows, and his wife was a good worker. They had accumulated a lot of cattle, cows and hogs and horses, and were worth 12 or 15 thousand dollars. Mrs. Wheeler had a brother-in-law and sister living but a short distance, named Pyburn, not of very good repute.

A Californian came along to buy up stock cattle, and sometimes made his home with [the] Wheelers, and boarded with them. He had plenty of money and spent it free, and Squire Wheeler never suspected him or his wife of anything improper. They used to be together a great deal; he would bring home oranges and candies and apples or fruit. One day while

Mrs. Pyburn, her sister, was on a visit, she told her (Mrs. Pyburn) if she would not tell, she would tell her something, and made her promise not to tell that she was expecting the Californian to come, and run away with him. When he got ready, he would write a letter to Mrs. Pyburn and she was to give it to Mrs. Wheeler, and the place she was to meet him at would be stated in the letter.

Now the Wheelers had four children, the oldest a girl ten or eleven years old, all well raised, and the whole family were highly respected and looked-up-to as a model family. After Mrs. Pyburn got home, she concluded her sister would be acting very imprudent to elope with the cattleman, and she at last told her husband. So when the expected letter came to her, Mr. Pyburn took it to Squire Wheeler and they opened the letter and made the man that brought the letter tell who sent it, and made the man take Mr. Wheeler and Pyburn, and five or six of the near neighbors to the Californian (he at first refused, but they threatened to rawhide him if he would not, so he took them to him) and they took the California cattleman to a tree, tied him, and gave him a fearful cowhiding, and told him if he ever was caught in the neighborhood he would be shot.

The men were all disguised in black masks, so that the Californian should not be able to identify the whippers, and they let him go. He went into Linn County and had all of them arrested, and at the trial came near convicting them. Squire Wheeler paid all their fines and costs, for he said that it was on his account, and it must have cost Wheeler four or five thousand dollars for lawyers and postponing the trial so often.

His wife at first wanted to divide up the property and part, but her sister and friends at last persuaded her to stay with her four children and not break up the family, and she said she would stay and be a faithful wife if her husband would promise not to throw it up to her. She claimed that the man had drugged or doped her with some medicines to make her like him and run-off from her husband and children. She was always looked on with suspicion, and . . . her own sex were worse than the men in their slurring remarks about her, so I was told where I stayed overnight.

My trip done me good, although it cost me considerable, but I enjoyed it. I got back to Hogue's in October. While at Mr. Hogue's I bought me fine clothes and I was always well-dressed and went to church, lectures, parties and some dances, and I never enjoyed myself better in my life. . . .

When I got back to James P. Hogue's, I did not know what to go at. Mr. Hogue said he had a band of mares, colts, and horses, Spanish, and they had got domesticated to half and ¾ breed with American, but still very wild. J[ed] Powers, a son-in-law, had been living at The Dalles, and had a lot of sheep with a partner named Brownlee, and Brownlee had a ferry on John Day River with Walter Cushman, an old acquaintance of

Hogue's Brownlee had some trouble with Bill or Elijah Bunt[o]n about a land claim, and Bunt[o]n got the drop on Brownlee, but he got over it and was still with the sheep.

Powers came down with his wife and family and stopped with Hogue's. Powers talked to Hogue about breaking his horses, and in the spring to drive them to The Dalles and sell them. A young man named James Johnson came along and wanted work breaking wild horses, and he asked me to go in with him and take the contract of Mr. Hogue to break all that he wanted broke, at one for every five broke. So we started in.

The winter was wet and the ground soft and muddy. We would just drive the herd into a corral lassoo two, blind them, saddle them, let them flirt and buck or jump a little, get into the saddle, take the blinds off, and let someone open the big gate and let the band out. The bell mare would take the lead, and we would let our wild horses follow on the run. They would scarcely ever buck, most of them were stubborn and had to be whipped along; others would roll around and beat their heads and break their saddles or bruise their heads awful, some so bad that we had to let them go a few weeks, and then have to go over them again. We took big chances of being killed or some of our limbs broke every day. . . .

We broke some 40 or 45 head during the winter, and we got two horses apiece and some money, and we had to pay for one that died by Johnson's brutality. He should have replaced it with one of his own, but he would not, and Mr. Hogue was always so good to me that I would not let Johnson off without his paying his half, and I the other half. I broke one saddle-tree at first by a horse falling over backwards, and I bought me a strong, stout, heavy Sacramento tree at Corvallis, and it lasted me until I parted with it.

I have still some more to relate that happened in 1860: While on Coffee Creek I got acquainted with a man named George Green. He and his wife and family were living about two miles below the store where I was clerking, keeping a boarding house, and he had some mining claims with other partners, which he was working. (He had been a widower or bachelor and his wife had been a widow with some children. They both had parents at or near each other at Looking Glass Valley, not far from Burnett's, or Round Prairie Post office.)

Green was a whiskey man, and so were his partners, and his boarders, and he himself kept whiskey to sell, or a sort of saloon with his boarding house, and there was always a great deal of drunkenness around his place. One of the Stout brothers, George, was either a partner or only a boarder, but they were drunk a good deal of the time, and they quarreled a great deal. I expect that Green was jealous of George Stout in regards his (Green's) wife.

One Sunday morning, during some drunken brawl, Green stabbed George Stout in the side and abdomen, and after a few days Stout died.

The miners at first came near lynching Green but his family was mostly small children and they concluded to let him be tried by law at Roseburg. He got a hearing and got some one of his friends to go his bail, was bound over until July for trial. They had left Coffee Creek and went back to their own folks at home, and in July, while we were at Burnett's keeping the hotel, they came and took dinner with us as they were on their way to attend court at Roseburg. I do not think the case came up for trial till I had left, but I heard after, he only got a few years in the penitentiary. They were both a hard couple.

I heard afterwards that Ben Stout, [George's] brother, and one of the Boyles and some other of the cattlemen that lived on the South Umpqua River drove a large lot of cattle to Nevada to sell, and while down there, the great Comstock mines were discovered, and other rich places, and our Umpqua River cattlemen got into some rich claims and some got rich and got town lots in Virginia City and other places, while I was up at Walla Walla, Washington Territory.[15]

[15] The Comstock Lode was discovered in June, 1859.

# 9

## WINTER IN WALLA WALLA

(1861–1862)

So in about May, 1861, J[ed] Powers got ready to drive his and Mr. James P. Hogue's horses to The Dalles. (I had got into the notion to go to the new-discovered Nez Perce mines—they were on a branch of Snake River, some distance above a town just laid off, called Lewiston, at the mouth of Clearwater, and the Or[o]fino, or Clearwater Mines were some one hundred and twenty miles further in the mountains.)[1] I would have to go by way of The Dalles, so Powers and Hogue said they would pay all expense to The Dalles if I would go with Powers and help drive; they made the same offer to James Johnson and he concluded to go along too. I thought I could not do better, so got all ready.

We started out with about 55 or 60 head including our own. We were to go slow, as the Cascade Mountains were still quite deep with snow, and we took blankets for bedding and clothes, and packed two or three horses with them, and provisions from Mr. Hogue's, and we had some little help from Mr. Hogue, Charley Hogue and William Winn, a young

---

[1] The Nez Percé-Clearwater mines, located in and about the mouth of the Clearwater River and Orofino Creek, were discovered in 1860. Lewiston was founded by a group of men who were trying to bring in supplies via river steamboat. As one of that party relates:

> It was soon seen that the Clearwater was not practicable for navigation, and that its junction with the Snake River was the logical location for a town to supply the mines in the inland country. . . . The names Lewisville and Lewistown were suggested but finally they agreed on the present name, Lewiston.

Since the site was on an Indian reservation, permission had to be obtained and William Craig helped in negotiating a lease. Henry L. Talkington, "Story of the River: Its Place in Northwest History," *Oregon Historical Quarterly*, XVI (1916), 190, quoting the recollections of George E. Cole; *The Oregonian*, May 6, 20; June 4, 1861.

man stopping at Hogue's, to help us start the band from its old grazing ground. We had all we could do to keep them from making a break to go back, for we had left a few old mares and a lot of young colts at home, and ours kept trying to get back. We did not get very far the first day, and the Hogues and Williams left us as soon as we could do without them, so they could get back home to do their work for the night. It was like leaving home to me, but I bid them all goodbye.

We passed through the best parts of the Willamette Valley, from Albany, Linn County, to Salem, Marion County, on to other towns, until we got to Oregon City, [C]lackamas County, one of the earliest settled in Oregon, and through French Prairie.² We kept changing our saddle-horses two or three times a day, to make them gentle and tame, but still some would be stubborn and wild, and it made us quite sore and tired the first few days a-riding.

From Oregon City we left the river to strike in toward Foster's where we were to cross the Cascade Mountains by the Emigrant Trail,³ and when we got to near where we had some creek to ferry, we heard that it might not do to go over the mountains, the snow was so deep, that it might be June before we could cross. So we camped, and J[ed] Powers made some inquiries, and found it so. He concluded to go to Portland and see how he could drive by the river trail up the Columbia River to above the mouth of Salmon Creek at the foot of the mountains, on the west side of the Cascade Mountains. Powers asked me to go with him to Portland—it was not over 12 or 14 miles off—and we left Jim Johnson to keep camp and watch the horses.

We took a look about town, and saw Mr. Henry Hogue, the lumber dealer, son of James P. Hogue. We found we had to keep up on the south side of the Columbia River and 6 or 8 miles below the Cascade's Falls, cross on boats or swim our horses to the other side, and then drive them up to Upper Cascade and ship them on a steamboat to The Dalles, for there was no trail on the north side.

So we went back to camp, and next morning started up the river on the trail. It was very rocky along the trail, and after crossing Salmon Creek, about swimming, and ice cold, we drove up the river, and it was very bushy, and through timber.... We ferried our horses over, just six or eight miles below the Cascade, and kept up the other side to the town, and here we had to catch all of the horses, put halters on them to lead them on the steamboat, and a fearful time we had, for there was no corral, but a large field, and if run, [the horses] would run out of the fence, or

---

² Oregon City, Clackamas County, was laid out in 1842 by John McLoughlin. French Prairie was settled earlier by former employees of the Hudson's Bay Company.

³ The reference is to the northern route of the Oregon Trail into Willamette Valley. Philip Foster came to Oregon in 1842, and his farm at Eagle Creek, Multnomah County, was the first on the westbound trail.

*Winter in Walla Walla* 183

over it. Powers hired a couple of Indians to lassoo them in the field as we drove them past. Their lasso ropes were too light and they threw too small a noose, but one of them was good at throwing them small loops as swift and straight as an arrow.

When we had them all caught we had to lead them down to the docks to where the steamboat lay, and then they did not like to go up the gangplank, and we had 12 more than we could get on, so left them to some man to put on the next time the boat went up, next evening. Powers paid all our expense of shipping mine and Johnson's horses. We had two apiece left.

It was a splendid sight to go up the river below and above the Cascades. The mountains on each side from 200 to 1000 feet perpendicular, and Mount Hood 14 or 16000 feet, all snow peaks and other mountains on the Washington Territory side, such as Mount Ranier and St. Helens. It was pleasant steamboating on the Columbia and we passed many large rivers emptying into the Columbia, such as Dog River, and others I do not know, or now remember.

We got to The Dalles in the night, and Brownlee came on board and we unloaded the horses and took them into a livery stableyard and went and got some drinks and something to eat, and turned in awhile and slept until after daylight.

I looked around The Dalles a day or two, and concluded to push on up to Walla Walla. There had not been but little changes in The Dalles since I left in 1858 for Fraser, when I had stayed there four or six weeks and had got tired of it. The big hotel, the Umatilla, had been built since then, and some few business houses.[4]

Powers took his horses out to grass awhile before he was to sell, and Johnson concluded to stay at The Dalles awhile, so in a couple of days I packed my half-wild gray mare with some blankets, clothes, saddlebags, and a few provisions, and I had a riding saddle on my big gray, and my pistol and holster, or canteenas on him, and I got to the De[s]ch[u]tes River ferry, where I had to go over.[5]

Next day I got to the John Day River where Brownlee and Cushman kept ferry. Ferried over and stayed overnight. Next day went to Willow Creek, where Tom Richmond kept a large cattle ranch and public house, and kept a great deal of travel[ers]. I was acquainted with Tom Richmond at Jacksonville and Kerbyville. He was doing well and his cattle were all fat on grass, and he never fed or cut hay, or had any grain, but they were as fat as if fed on corn and fattened. And in fact, they were too fat for beef for many.

---

[4] The Umatilla House was built in 1857 by A. J. Nixon. It burned in 1878, but was rebuilt the following year. It had 120 rooms with a dining room that could accomodate 200. William H. McNeal, *History of Wasco County, Oregon*, p. 25.

[5] The Deschutes River ferry was established by Nathan Olney in 1852.

Next day I got to the Umatilla River, and went up the river to where a man named Marshall was keeping hotell or public house. There were some fine bottom farms on the Umatilla River, and I crossed at ———— Ferry, some distance below where I stayed overnight. Mr. Marshall was either sheriff or deputy, and off on business, and a man was stopping at the house. It was a double log house, and the man's name was Isaac Stonebraker, and his wife and girl seven years old were stopping at Marshall's that time. In the morning I heard that Mrs. Stonebraker had brought a little boy, and I had slept in the next room, and must have slept soundly, for I did not hear when he came.

I had a talk with Mr. Stonebraker the evening before. He was from Iowa, a Millerite[6] and a carpenter, but was farming at Marshall's that season. He told me that some distance down below the ferry, some men, one named Guss Hill, and the other James Daniels, had both taken up farms and were making improvements. They were both miners and had worked on the S[i]milk[a]meen River, mining, and then come and located their land and built houses, stables and fences, and broke up some ground and put in a crop, and they had some horses and considerable young stock, and cows. Both had got married to some emigrant girls. Just acrost the plains the emigrant road crossed the Blue Mountains not far above on the Umatilla River.

When the Clearwater excitement broke out, they went and prospected and located some mining creek claims on Rhodes Creek, one of the branches of the Orofin[o], a short distance above the town of Pierce City, named after Captain Pierce, an old California prospector.[7] He used to be around the Hogue neighborhood. I had heard of him but had not seen him. Hill and Daniels had got good prospects and took up claims and went back home, done some work, and Guss Hill stayed home to take care of both the farms, and the families, and James loaded up some horses with provisions and went to work their claim on Rhodes Creek. I knew by the names they were our two old Althouse, Happy Camp, and Sucker Creek and Fraser River acquaintances that had gone to San Francisco to go to South America or Rio de Janeiro, but here they were, three years later. I did not go back to where Guss Hill lived, for I knew I would come acrost him after a while.

[6] A follower of William Miller, who in 1831 prophesied the second coming of Christ and the end of the world in 1843.

[7] Elias Davidson Pierce, a former California prospector, joined a group of retired Hudson's Bay Company employees in trading with the Nez Percé on the Clearwater at Lapwai in the fall of 1852. In April, 1857, he heard from the Indians the incredible story that there was gold on their reservation lands. As the rumors persisted, he set out on February 12, 1860, with Seth Ferrel, to prospect the Clearwater and its tributaries. On February 20, he discovered placer gold on the north fork of the Clearwater. Pierce City was named for him and was situated twenty-five miles above the mouth of the Clearwater. *The Oregonian*, May 30, 1861; M[errill] D. Beal and Merle W. Wells, *History of Idaho*, I, 282–289.

So in the morning I got ready to start, and Mr. Marshall got home and I started for Walla Walla. At the place just above Marshall's was the commencement of the Umatilla Indian Reservation, and they had the Upper Umatilla and the east side of Wild Horse Creek, and they had their Indian agency, twelve or fifteen miles from Marshall's.[8] When I got to Wild Horse Creek, there were some fine-looking farms and farming lands, mostly on the Indian side of the creek. On the road where the wagon road leaves the creek there was a hotell, and farm acrost the west side of Wild Horse Creek. It was 21 miles to Walla Walla.

I next got to Pine Creek, where there was a settlement, and the creek was wooded with pines (how it took its name) six miles from Wild Horse Creek. From Pine Creek it was five miles to Dry Creek, dry only from July until fall, and the water sunk and run under the ground and then would come up when the soil was suitable for it rising.

I saw quite a body of land just below the crossing of Dry Creek covered with willows, thorns, and large cottonwood trees. The creek was quite high, sometimes too high to ford. This was about 12 miles from Walla Walla City, but only two miles over the hill to the Walla Walla River. At the crossing of the Walla Walla River, a store and hotell was kept by John Broughton, and right acrost the other side of the road a claim had been taken up by George Anderson, who was living with old Mrs. Bunt[on], and had married her daughter Minerva. They were the same Bunt[on]s from Umpqua Valley near Winchester, and near old Gen. [Joseph] Lane's place, and the same boys that had shot Brownlee on Eightmile Creek, near The Dalles in Wasco County. Broughton's claim run up the Walla Walla River, Anderson's or the Bunt[on]s ran down the river. The stream was usually quite high and rocky and hard to ford.

I next came to some fine farms in the valley, well-tilled and in good cultivation, belonging to Walter Davis and several others. Then come to another branch of the Walla Walla River, but called the Tumlum, deeper than the other branch. This was on the line of Oregon and Washington

---

[8] This reservation had been established in 1855. Through negotiation led by Joel Palmer, the Walla Wallas, Cayuse, and Umatillas received eight hundred square miles. Wild Horse Creek, in Walla Walla County, is called Mud Creek today. By 1870 there were only a thousand Indians remaining on the reservation. Frances F. Victor, *All Over Oregon and Washington*, p. 102. Elwood Evans, *History of The Pacific Northwest*..., I, 486, defines the boundary of the reservation as:

> Commencing in the middle of the channel of the Umatilla river opposite the mouth of Wild Horse creek; thence up the middle of the said creek to its source; thence southerly to a point in the Blue Mountains, known as Lee's encampment; thence on a line to the waters of the How-tome creek; thence west to the divide between the How-tome and Birch creeks; thence northerly along said divide to a point due west of the southwest corner of Wm. C. McKay's claim; thence east along his line to the southeast corner; thence in a line to the place of beginning.

By this description, Reinhart's memory probably has substituted Marshall for McKay.

Territory, only six miles from Walla Walla City. Good farms all along on both sides, well-timbered; so were all the streams that came down the Blue Mountains and emptied into the Walla Walla or Tumlum, then into the Columbia, just below the mouth of Snake River, ten miles above Wallula, a old Hudson Bay fort.

The next creek I got to was Yellowhawk, then one or two more, the names I have forgotten. The next was called Garrison Creek because it ran by Fort Walla Walla, and was the garrison for five or six companies of soldiers, who had fine buildings and a large government farm and garden, corralls, barns, sheds, and the sawmill.[9] It was on a high sort of a hill and laid off very nice with parade grounds in the center and officers' quarters and commissary and Sut[l]er's store [a licensed army post trader], made a large town, and they had long stables to stable several thousand head of cavalry horses and teams of mules to haul supplies. It was a general distributing post for that whole upper country of the Columbia, Pend l'Oreille and Coeur d'Al[e]ne Mission on to the British possessions on the north and west.

I got in sight of the City of Walla Walla—it was a lively little place of about two or three thousand inhabitants, mostly transient miners, teamsters, packers and land-seekers. There were two or three hotells and a lot of boarding houses, some four or five restaurants, eight or ten saloons, three or four bakeries, three drug stores, four or five butcher shops, three or four barbers; at least 8 or 10 general stores, dry goods and groceries, boots and shoes, doctors, lawyers, two breweries, one nice theatre, and several churches (not very costly buildings) and some five or six livery, feed, and sale stables, some nice billiard rooms, and several large gambling saloons with all kinds of games and gaming tables where they had music and singing every night free, and one or two dance houses of evil repute, and one or two hurdy-gurdy dance houses of fair repute—taken altogether Walla Walla was a fast place, and a great outfitting place for the mines all over, just then the Clearwater, Nez Perces, or Pierce City, and Orofin[o], and towns were building up in the new mines, and the rush was from California. Sacramento and San Francisco boats to Portland were crowded for the Nez Perces mines.

I found a man in Walla Walla that wanted someone to cook and keep house on his cattle ranch for awhile, and I heard it was too early yet to go to the mines. All the streams were high and provisions in the mines yet scarce, so I concluded to get into something for three or four weeks, until

---

[9] Earlier forts had been destroyed. In 1857, it was rebuilt by Colonel George Wright, and is situated on the east bank of the Columbia, half a mile above the mouth of the Walla Walla River. Frank H. Woody, "From Missoula to Walla Walla in 1857 on Horseback," *Washington Historical Quarterly, III* (1912), 285-286.

The nearby city of Walla Walla was first called Steptoe by its founder, A. J. Cain. When incorporated, January 1, 1862, it chose the former name. *The Oregonian*, June 18, 1862; Frank T. Gilbert, *Historic Sketches of Walla Walla . . .*, pp. 214, 297-298.

the waters went down and a person could go so much quicker and at less expense, and live cheaper after the pack trains had taken in stocks of goods, groceries and provisions.

The man I spoke of lived out some ten miles on Dry Creek in Umatilla County, Oregon, where I had crossed Dry Creek at the wagon crossing. Some four miles above, he had a log house, and something over 100 fine milk cows and four and five-year-old steers, and from calves up to cows and steers, all rolling fat, about 600 head. He had wintered this stock two years on Willow Creek, near where Thomas Richmond had his cattle, and he had moved them up to where he now was, and bought the cabin and claim to ranch on. His name was Caleb Grover, a large heavy robust man of about fifty years old.[10] He had a sort of partner still living on Willow Creek—his name was William Todd—and his Indian, a boy, was with Grover, who had some eight head of pretty good Nez Perces horses, quite fat, to herd his cattle on. Todd only came up once in awhile but there were many stock men around, and buyers for fat cattle for the government to butcher, that there were from four to seven or eight every day to cook for, which was my job, and to keep ranch when Grover was away.

The Indian had to go out amongst the cattle and cows to attend their calves, and keep all the herd from scattering too far off, and the fresh cows with their calves were brought up to the corral, and milked a few times, and what the calves could not suck was milked out on the ground, or some calves would suck some cows not their mother, and they run with their mothers until fall, and it made the largest and fattest calves to be fed nothing but grass I ever saw. I have seen yearlings butchered and dressed weigh 640 lbs., 2-year-olds 900 lbs., and most too fat to be eatable, and government buyers for contracts calling for so many head of 3-year-olds and so many of 4-year-olds, would make up their contract of 2-year-olds, and then their aggregate weight would overrun the amount calculated for four-year-olds.

Caleb Grover told me he had commenced on ten cows, and had been selling some of his four or five-year-old steers, but kept all heifers. He had sold several thousand dollars worth of steers, and now had over one hundred cows, and all from calves up to his oldest (now four-year-old) steers, had over 600 head. He offered to sell the lot, cows, calves, yearlings and all, for $6000. That was in June, but he did not sell but only some of the largest four-year-olds and three-year-olds, and some twos. . . .

I got tired at Grover's. I went up the creek, up to the wagon crossing, four miles above, and took up the claim I spoke of as I passed on my way to Walla Walla. I cut some foundation logs for a house, and drove some stakes and put up a notice that I had taken up a donation, or pre-emption claim of a half-mile square, from the crossing of the wagon road of the

[10] Grover had moved from Oregon. He was Douglas County's first coroner in 1852. Evans, *Pacific Northwest*, I, 398.

creek to the lower end of a grove of cottonwoods, and the large willow patch was included in the claim. I laid a few of the logs up for a foundation, and left it until I could build on it, as it was customary in that part of the country. . . . I sold [my mare] for $45 in Walla Walla and next day started for Nez Perces mines.

While at Walla Walla I heard that Jim Buckley, whom I used to know on Sucker Creek and at The Dalles, was elected sheriff of Walla Walla County, Washington Territory,[11] but he was away up in the Bitter Root Valley, and he was living with a young squaw that was acquainted in that country, and he took along a white man and another Indian and his wife, or squaw, and taken a lot of whiskey or alcohol and when he got to Bitter Root Valley, made his alcohol into whiskey and bought and traded for Indian ponies. When he got a lot, some five or six hundred head, had lit out in the night and got away, and he was stopping on some creek between Patuka [Pataha] and Snake River, where he had them cached or hid, and kept his Indian and squaw and white man with them, but he owned them and sold some on the sly, for it was against the law to sell Indians liquor, or trade it to them, and he was afraid the officers at Walla Walla would molest him, but they did not. I did not see him at that time, as I was in a hurry to get to the mines.

I traveled quite fast, and first day got to the Touchet River, about 28 miles from Walla Walla.[12] I had crossed Dry Creek, eight miles; Whiskey Creek, twelve; Coffee, about sixteen, and stayed all night at a store on the Touchet, kept by a man named Stump, a packer who had a hard name then, and got it worse a few years after.[13] Next day I got to T[u]can-[n]on Creek, stayed overnight, and found some nice rich ranches and farms on T[u]can[n]on Creek or River.

Next day nearly noon, I overtook a lot of men afoot, and heard one called James Brady. I inquired who he was, and found one of our old Centre Market New York Louffers whom I knew for years, and he had been in our bakery hundreds of times in New York when we kept op-

[11] James A. Buckley was elected sheriff in July, 1860, and reappointed again to the office in 1862. Gilbert, *Historic Sketches*, pp. 223, 229, 345.

[12] Reinhart's geography and mileage are at variance at times. The route to the mines was described in *The Oregonian*, June 23, 1862, and in S. J. McCormick, *The Oregon and Washington Almanac, 1862*, pp. 35–36, who gives these distances: Walla Walla to Dry Creek, 9 miles; to Cap-a-lil Creek, 7; to Touchet, 7; to Touchet Crossing, 4; to Pataha, 15; up the Pataha, 12; to the Alpawa, 6; down the Alpawa to Snake River, 7; up to Buckley's Ferry, 2; up to the mouth of the Clearwater, 8; up the Clearwater on the north side to the Nez Percé Agency, 15; up to the Ferry on the Yak-to-in Creek, 5; up to the forks of the Clearwater, 20; to the foot of the mountain, 10; to the mines, 32. A total distance of 163 miles from Walla Walla to the Clearwater mines.

[13] Freelon Schnebley, known as "Stubbs," had a store at the Touchet River crossing. In 1862, he and a neighbor, Richard Learn, stole some government mules and tried to run them to Canada. They were overtaken at the mouth of the Okanogan and shot by their soldier pursuers. WPA., *Told By the Pioneers*, II (1938), 146–149. Reinhart details more of this in Chapter XII.

posite the Centre Market (I had traded him pie and cake for marbles more than once). He had a brother named Washington Brady. So I made myself known to him and he was awful glad to see me. We had not met for over 16 years, when we were boys—he was ten or eleven years older than I. He was on his way to the mines, and he told me he had been through the Mexican War and was in Col. Stevens' regiment that came through Mexico and Lower California to Monter[e]y, where they and Col. Fremont and Commodore Foot took Monterey and came through to Los Angeles and on to San Francisco, where the regiment was discharged in 1849.[14] And he was one of the earliest of the 1849ers in the gold discoveries in California. He said his brother Wash was at Vancouver, running a gambling saloon, and was married, but he would no doubt be up to Walla Walla, and maybe to the Nez Perce mines in a short time. I traveled with them a short time, and stayed overnight where they did on some creek between T[u]can[n]on and the head of Elpowa. The boys with Jim Brady were a hard set of roughs, mostly Irish, and on the gamble and drink, and one named Jim Burn, a shoulder-striker or prizefighter.

I had quite a talk of old times in New York. . . . [We spoke] of Mike Mitchell that used to sing so well at the public school Number 5, Mott St. He was a widow's son, and he took up with a circus troupe after leaving school, and after a while used to travel with some Negro minstrels of fame and notoriety as a song-and-dancer, and went to California. He won a gold belt at Sacramento, or Marysville, as the champion jig dancer of Pacific States in 1858.

A day or two before I started to cross the river at The Dalles, on our way to Fraser River, the company Mike Mitchell was with came to The Dalles and performed several nights. They had another number, one jig dancer with Mike Mitchell, his name was Johnny Ewing, a light and graceful dancer, and they danced their celebrated match-dances together, as rivals. Old Eph Watson, of whom I spoke at Roseburg, was with them. I had not seen Mike Mitchell for over 16 years.

After I left, I heard the troupe traveled to British Columbia, and through Oregon and the mines, and I heard the winter of 1860 they performed at Portland, Oregon, and Mike Mitchell got on some awful sprees

---

[14] This is an extremely garbled version of American military activity in California during the Mexican War. The regiment, which came overland from Fort Leavenworth through New Mexico, along the Old Spanish Trail into southern California, was commanded by General Stephen F. Kearny. Frémont had been in California since the winter of 1845. Commodore John D. Sloat, soon after succeeded by Commodore Robert F. Stockton, initially seized the port city of Monterey. Colonel Jonathan D. Stevenson brought a regiment, the First New York Volunteers, around the Horn to California, to give added support to the military activity of Kearny. The regiment reached San Francisco Bay during March, 1847. James Brady was a member of that regiment and was discharged in the fall of 1848 in San Francisco. Guy J. Giffen, *California Expedition*, p. 30.

and one night after leaving his crowd quite late at an engine house where a lot of them had been carousing, he left them alone to go home to his hotel after midnight, a very cold night. Next morning Mike Mitchell was found dead, frozen, lying in the gutter. Such was the death of as fine a boy when he went to school as I ever saw. I was awful sorry to hear of it—to die such a disgraceful death away off from all old friends and relations.

In the morning I started on. The boys were all on foot, could not keep up, and I traveled with some other horseback men, and found some splendid farms and ranches taken up, and still some vacant on Elpo[wa] and the Pat[a]h[a] creeks, and valley, and next day got to Snake River, where there was an old Indian Agency and an old Mission where a large piece of land was fenced, and some buildings inside, and a large orchard of apples, peaches, and pear trees growing—the largest I had seen in Oregon or on the Pacific Coast, but I do not think they were budded fruit, but all seedlings, and had been planted 15 or 20 years.[15] Six or eight miles brought us to a small government post and there the Nez Perce Indian Agency and some soldiers were stationed.[16]

From here it was but four or five miles to the Snake River Ferry, kept by a white man named Craig, who had married a Nez Perce Indian squaw and had been with them as a sort of chief for over 25 years.[17] On the other side of the river, where the ferry ran to, was the city of Lewiston. Old Craig had the charter of the ferry for six or eight miles below and above the Indians, and no one could put in a ferry in opposition to him. It was a big thing for him, from $50 to $200 per day during the gold excitement. I crossed over (and it cost about $1.50 for a horseman) to Lewiston, stopped a little while to look at the town. It was nearly all tents, and every other one sold whiskey.[18] Gambling saloons were in full

---

[15] The tiny hamlet of Silcott is situated here today. The Rev. Henry H. Spalding gave the orchard seeds to Chief Red Wolf in 1837. In 1861, Sam Smith established a trading post. Reinhart has apparently confused this site with Lapwai Mission which lies somewhat southeast of Lewiston.

[16] Reinhart's reference would be to Fort Lapwai, some eleven miles southeast of Lewiston. Victor, *Oregon and Washington*, pp. 120-121.

[17] William Craig was born in Green Brier County, Virginia, around 1799. In 1829, he was employed by William Sublette and launched his career on the fur frontier as a trapper-trader. During the successive decades, his travels criss-crossed the Rocky Mountain's eastern basin. As a member of Joseph R. Walker's party, he journeyed to California in 1833. He took up permanent residence in the vicinity of Lapwai Mission in 1840 and died there in October, 1869. Walla Walla *Statesman*, October 23, 1869; LeRoy R. Hafen, "Mountain Men—William Craig," *Colorado Magazine*, XI (1934), 171-176. Thomas J. Beal, a long time friend of Craig's, wrote his recollections (published in the Lewiston *Morning Tribune*, March 3, 1918), which recount many of Craig's adventures.

The ferry operated across the river where Clarkson is situated today.

[18] George Cole recalled Lewiston in 1861, writing:

> At first it was a city of tents, and its population often reached 7,000 or 8,000, but of so shifting and transient a nature that it is hard to strike an average. However, as

blast night and day. I kept on up the Clearwater River (Lewiston was at the mouth, where it ran into Snake River; the new diggings were on the head of the left-hand forks of Clearwater) until I got to the forks of Clearwater. A town had just been laid off, Slaterville.[19]

I stayed overnight, and next morning left the rivers and took out over some mountains, and next day came down on the creek called Orofino. We passed several ranches where they herd horses the miners have to send back, at $3 per month per head on grass, and some had taken up milk ranches to supply the miners with milk. We had to go quite a distance up the Orofino to the main town, Pierce City, close to where the late discoveries had been made, and a creek just below the town came into the Orofino, named Rhod[e]s Creek, after an old whiskey-head, a Virginian named Bill Rhod[e]s; he looked to me to be about half Negro.[20] The most mining was being done on Rhod[e]s Creek for the Orofino was still too high to work the bed of the river.

Orofino was a lively city full of miners and whiskey saloons; fighting and gambling was the main thing around town, and all the worst roughs in the country had congregated here.[21] I sent my horse out on a ranch

---

> time went on permanent buildings were gradually erected, among the first being the old Luna House, a famous hotel of those days, which supplied the members of the first and second territorial legislature with a home and later became the courthouse of Nez Perce county.

Talkington, "Story of the River," pp. 188–189.

[19] When the *Colonel Wright*, the first steamer to navigate the Columbia River above The Dalles, reached the Clearwater in the spring of 1861, it tied up one night at the mouth of Lapwai Creek. The next day, despite the effort to push ahead, the river proved unnavigable. As George Cole recalled: "Failing in this and not finding a suitable landing spot, we returned to one we had passed and landed the passengers and freight. Slater [one of the passengers] put up his tent and opened a store, which we called Slaterville." *Ibid.*, p. 188; *The Oregonian*, June 18, 20, 1861.

[20] Orofino had about sixty houses in June, 1861, while Pierce City, two miles below, had a population in the neighborhood of 2,000. *The Oregonian*, June 24, 1861.

Rhodes Creek emptied into the Orofino a mile and a half above Pierce City. Rhodes and his company of eleven miners were reported as producing three hundred ounces of gold per day. Another report merely advised that a miner could average between ten and thirty dollars a day; others were skeptical and charged that the strike was grossly exaggerated. *Ibid.*, May 20; June 1, 4, 5, 1862; Hubert H. Bancroft, *History of Washington, Idaho, and Montana*, p. 240, note.

[21] W[illiam] A. Goulder, *Reminiscences: Incidents in the Life of a Pioneer in Oregon*, p. 204, writes:

> By the first of June [1861], thousands of eager gold-hunters had pitched their tents along the level alluvial bottoms of the Oro Fino Creek, and along the streams and gulches that empty into it, far up into the heart of the Bitterroot Mountains. Into the fastnesses of these mountains went representatives of every civilized nationality on earth, with many whose native lands could hardly be thus classified. Oro Fino and Pierce City, two model mining towns with houses built of pine logs and roofed with "Shakes," had already taken up their positions within a mile and a half of each other on the banks of the same stream which had given a name to the new camp and to one of the new towns. The other town had been named in honor of Captain Pierce, the leader of the discovery party.

with some others I had come in with, and stayed in town a day or two to look around. I found wages were from $2.50 to $4 per day, but as yet scarce until the water got lower. I looked around to see about getting claims, but did not like the looks of the way the land and claims lay, and what I did like was already taken up with notices on them. I saw and found out lots of men were holding more claims than the law allowed, but a person would have to be posted to find out who and where they were.

So I looked around for a chance to whipsaw out lumber but when I saw the timber, a white fir, I did not much like the looks of it. I came across the Pollards and Thomas Wilson, who used to be mine and my borther's partner on Sucker Creek in the bowling alley there. He was here whipsawing out lumber, but did not like the kind of timber. I saw several others from Althouse and Sucker Creek, two half-brother Swedes named Johnson and Monson and one called Chuck....

A man came to me at Pierce City; he wanted a miner to work his interests in a mining claim he had down the Orofino River, where two brothers named Gold[st]en had each an interest and knew nothing of mining. The man's name was Clarence Gifford; he and the two Gold-[st]en boys had farms on a creek nearly eight miles from Walla Walla. Gifford had a partner in ranch and stock named Erk Davidson, a violinist. The Gold[st]en family lived on Gifford and Davidson's farm. Gifford had to go home to Walla Walla, for Davidson was not well, and wanted Gifford to come. So he hired me to work with the two Gold[st]en boys, one named Dan, the other Perry....

I found out I had seen this man Clarence Gifford before at The Dalles in 1858. He had brought a band of wild or half-broke horses, or cayuse and Nez Perce ponies to sell, and I had traded him my navy revolver that I had got of S. M. Charles at Kerbyville for one of the ponies (but it did not get half way to Frazer River). I spoke to him about it and he remembered the pistol; he had sold or traded it off soon after getting it from me, and he never knew that it was cracked in the barrel, so all was right.

Gifford came back in about three weeks and we found our claim did not pay much, and one of the Gold[st]en boys, Perry, was going home to Walla Walla, and I thought I might as well go too, and get to run a threshing machine in the valley, and I would be where I could make some improvements on my claim on Dry Creek, where I had laid up my foundation logs.

I got a pony and some money for my work, from Gifford, and Perry had a horse to ride, so one morning we saddled up and rode out....

---

For an over-all description of the mines, *The Oregonian*, August 20, 1862, presents a fair statement.

## Winter in Walla Walla

It took us some four days to get to Perry Gold[st]en's house. I went with him a little out of my way to Walla Walla, for I had some clothing and blankets on my packhorse to carry, and he owed me some money, which he was to pay when he got home. So I stayed overnight at his parents' house, and saw a sister of Dan's and Perry's. Erk Davidson was rather sweet on her, and Erk lived with them, or they lived on Erk's place and in his house.

Next morning I got to Walla Walla and stayed all night. Next day I went out to Dry Creek, where I had taken up my land claim, to see if it was still all right, or had been claimed while I was gone. I found it all right.

So I come back to Walla Walla, bought some provisions, a chopping ax, hand saw, a couple of augers, a hatchet, some nails, a shovel and a spade, and went back to the claim and commenced to cut down some trees, not too large, for I could not haul or handle them, so cut them about 6 to 8 inches through and 8 feet long and spliced them in the middle by putting posts in the ground on each side of the center of the building, which was 14 by 16½ outside. Where spliced, I joined the ends to the side post, having hewed it flat first. The corners I saddled, as all log cabins are put up. At the ends, one was a fireplace, and I had posts set in the ground 18 or 20 inches, to make the side jambs. At the front end I had the door posts set in the ground same as the sides and chimney-posts, so I had no large logs to handle, and they were very scarce anyhow. But the main thing was, I did not have to have anyone to help me, and I could do all the work as it suited me, at no expense—only my own time. My two horses I kept staked or hobbled while I worked.

I hewed off the outside and the inside a little as I carried up the logs, and pinned them to the posts, and carried up the chimney, back and sides all together. The timber I used was birch, alder, and some cottonwood. When I got to the top of the fireplace, I put in small logs to reach the 14 feet until I got to the commencement of the gable end, both above the door in front and the fireplace in the back. I did not leave but one opening for a shutter or window looking east towards the upper end of the claim and the wagon-crossing to Walla Walla, or towards The Dalles.

After I got up the side logs I laid ribs to lay on my roofing—straight poles—and then put on some thick bundles of rye grass close together, so that nothing could go through, then put on a dirt or clay roof, having two large side pieces at the lower side to hold the dirt on. My poles on the roof were mostly split willow, flat side down, to make it smooth and close or tight when the rye grass was put on. I was not over eight or ten days at it. I split out some wide pieces of willow and alder and made puncheon or heavy plank, dressed smooth with the ax, for the door, benches and table, and put up a bunk to sleep on, and I put on strong wooden hinges and a bolt that could be fastened by a secret string on the outside when

I went off, for a common padlock could be too easy picked.

My nearest neighbor was John Broughton on the Walla Walla River, and Anderson of the Bunt[on]s lived close, adjoining Broughton's about two or two and half miles from me.[22] On my creek—Dry Creek—James Holt lived about three and a half miles below. He was married to a half breed Nez Perce Indian girl; her father a white man named Charley Adams, had been an old sailor and mountaineer, and had lived with the Nez Perce Indian, with old Craig of Lewiston a good many years. During the gold discoveries at Jacksonville, Oregon, old Charley Adams had got to Jacksonville to prospect. Charley, my brother, got acquainted, [and] then I saw him a few times. He was a whiskey-head and would get drunk when he could get whiskey. He used to speak of the Nez Perce Indians, and his daughters, and old Craig, but at that time I did not dream of ever seeing his daughters. After Charley Adams got home from his Jacksonville trip, he kept with the Indians until he died, and left his two daughters with Craig, who adopted them both. They were Craig's wife's sister's children, and Charley Adams had been a widower some time.

James Holt was a runabout government teamster and was around old Craig a good deal, and at last married, or run away with the oldest Adams girl, named Mary. And he took up a claim two or three years before on Dry Creek below me. I used to know him when I worked, cooked and stopped at Caleb Grover's about one or two miles below James Holt's place. Jim Holt was from Iowa and he had a spot in his eye, and it confirms my opinion of all spotted-eyed men, for he, like all of them, would bear watching. More of him in the future.

I did not like to ask anyone to help me build my cabin, for James Holt was off teaming, and most men were busy, and I would have to pay them for the work that I could do best by myself. So after I got my cabin done, I could leave one horse, and my provisions and tools. I used to ride around to look for something to do. While at Walla Walla one day there was a gift enterprise to throw dice and you would get a blanket or some jewelry or money. It was the old Dianna or ten-dice game, a regular take-in. I was a little green and invested a few dollars, but instead of letting them keep the jewelry at their own figure or discount, I kept a few ladies' sets to trade on.

One day ... I went out to Charley Russell's, on Russell's Creek about four or five miles east, not far from the foot of the mountains. He had a fine place of four or five hundred acres of the finest land in the valley. It had been held by the government and reserved for a farm, but Charley Russell and Albert Mix—I believe Mix was either quartermaster at Fort Walla Walla, or wagon-master—and between them both they had some-

---

[22] The Buntons had settled on Whiskey Creek in 1859. The Broughton and Anderson places were near at hand as Reinhart indicates. WPA., *Told by The Pioneers*, II (1938), 138, 146.

way smuggled the land and all the cattle and mules to work on the lands. They hauled freight to Colville . . . and to other government posts, getting the contracts on private terms, and using the stock belonging to the government to fill their contracts with, and charging the use of its own teams to the government, and lots of property was condemned and bought in as useless by Russell and Mix: horses, mules, oxen, threshing machines, all kinds of implements, seeds, sacks, or harness, and the profits would be divided up by the officers.[23]

It was strange all these favorite contractors were invariably Rebels or Secessionists or Southern Sympathizers without any exception that I heard of. Albert Mix's father-in-law, Major Green, commander of the post [Walla Walla], Albert Mix, Charley Russell, Walter Davis, and several whose names I now have forgotten, were rank Rebels shortly after getting their contracts.[24] The commander of the post afterwards got command of the Pacific Department—I will think of the name and put it in here after a while.[25]

Charley Russell had in a big crop of barley, oats and wheat. The barley was already contracted to the government. He had a four-horse threshing machine he could not do much with. I fixed it up and he engaged me to run it for him. I made it do the best work it ever done, and with four horses would run through as much grain as the big machines did with eight horses and double our men in numbers. . . .

About in October, a large emigrant train under a United States escort, arrived at Walla Walla. It was under Col. or Capt. Crawford, and the escort, of fifty or one hundred dragoons. There had been several small trains of emigrants; some had stopped in Montana on the west Gallatin, some went to Virginia City, some in Grand[e] Rond[e] Valley, and this the balance, came on through to Walla Walla Valley.[26] There were

[23] Charles Russell was born in Boston, Massachusetts, September 18, 1828. He served in the army during the Mexican War and was discharged in New York in 1850. He then traveled to California and gained employment with the quartermaster for the Pacific Department. Accompanying Colonel Steptoe, he reached Walla Walla in September, 1856. Until 1859, he was in charge of transportation at the fort. In that year he was appointed a county commissioner. Gilbert, *Historic Sketches*, pp. 221-234, and Appendix, p. 34.

Russell's partner may have been W. A. Mix. It could not have been James D. Mix since the latter did not settle in the area until 1863. *Ibid.*, p. 261; Appendix, p. 27.

[24] Walter Davis was appointed a county commissioner, January 19, 1859. He owned a large farm on Dry Creek. Gilbert, *Historic Sketches*, pp. 222, 281. Reinhart is in error concerning Major Green. See following *note*.

[25] Colonel Justus Steinberger of the Washington Territorial Infantry succeeded Colonel Thomas R. Cornelius as commandant of Fort Walla Walla, July 28, 1861. He arrived in early May at Fort Vancouver with four companies of volunteers raised in California for the defense of Washington Territory against both Indians and any possible Confederate activity. He served at Walla Walla until 1865, when he was mustered out of the service. Evans, *Pacific Northwest*, II, 16; Aurora Hunt, *The Army of the Pacific*, pp. 228-231.

[26] Captain Medorem Crawford, commanding a company of eastern volunteers, left

some 80 or 100 mules and horses, and 75 or 80 yoke of cattle, wagons, and the whole outfit was advertised for sale in thirty days by the government. There had been an appropriation by Congress to purchase this outfit to haul supplies and escort all the trains and come on through with the last ones to Oregon or Washington Territory, to Fort Walla Walla. They had a surgeon with them, named Dr. Nimrod (I did not know him or get acquainted until 1871 at Chanute, Kansas).[27] The bids to ranch and grass the cattle and horses was let out, and Albert Mix and Charley Russell got the contract, and Charley Russell gave me a job herding the cattle, and someone else the horses and mules, but we were close together. There were some fine matched horses among them. They were sold at auction, one span Van Dyke and Whiteman got for $198—they were large duns or creams, silver mane and tail—they made the crack sleigh team that winter, and used to go out to a double sleigh at $10 an hour. All the horses, mules, cattle and wagons sold in teams and there was some money made by those who had money to invest. Mix, Russell, Davis, and some merchants and gamblers bought whole teams consisting of five to seven yoke of good cattle to a new strong wagon. The auctioneer, William (or Billy) Kenley, of Kerbyville, was part in the auction and got some five per cent. Another, named Good, both gamblers, got the auction contract by smuggling in with the officers who ran the thing.[28]

About this time I got a letter from my brother Charley; he had sold out at the Gordon place and had started to go to The Dalles but took down with the pleurisy of the lungs and he was laid up. . . . He was somewhere near Portland and had sent the letter some time before, but somehow it had been delayed and it had been over one month until I got it. He asked me to send him some money to come up to Walla Walla on, and just now I was out of work, my two horses dead, and winter coming on. I do not think I had over thirty or forty dollars left besides my bay horse. This must have been some time in December, 1861.

One day as I was in town, Charley, my brother, came to me. He had just got in from The Dalles where he had left Elizabeth and George, the boy. They had waited for me near Portland, but I had not come, nor had they heard from me, so they had sold out the team of oxen and the wagon

---

Omaha on June 16. Their purpose: to guard the northern trails for westbound emigrants. On June 28, the escort took under its protection the first emigrants at Fort Kearney. After an uneventful trip west, Crawford's detachment reached Walla Walla on October 14, 1862. Walla Walla *Statesman*, October 25, 1862; Medorem Crawford, "Journal of the expedition organized for the protection of emigrants to Oregon . . .," *U. S. Senate, Executive Documents*, 37 Congress, 3 Session, Document 17, 14 pp.

[27] Captain Crawford lists Dr. J. A. Chapman of New York as his command's surgeon. *Ibid.*, p. 9.

[28] The auction took place on October 21, 1862. "Six mule teams with wagon, at an average of $793; a number of mules at an average of $65; several horses at $55 each; rifles and revolvers at $20 and $21, the whole property bring $13,037." Gilbert, *Historic Sketches*, p. 282.

and what they could not bring had sold, and come on to The Dalles, and Charley thought they had better stay awhile, for Elizabeth was bout to be confined. So he left her at The Dalles, at a family's and came on to here. He came by land; at Wild Horse Creek he had left his horse, about give out, and got a chance to come with someone else.

He thought he might find a place to work at a hotel, so I went around and saw who I knew. He got the promise at the Walla Walla Hotel, run by Allbecker and Goldsten, a cousin of Dan and Perry Goldsten, whom I had mined with at Orofino at the Nez Perce mines. Their bookkeeper and manager, Waddingham, a New Yorker, and brother of Wilson Waddingham, a Wall Street broker; he thought he could give Charley a job as pastry cook and baker for the winter. We saw Jim Buckley, the sheriff; he was glad to see Charley, and that night Charley run across Wash Brady running a gambling table, and his brother Jim was walking on a wager 100 miles in 100 hours. We went in and saw Jim. Charley had seen neither Jim nor Wash Brady for 16 years. They were glad to see us.

We stayed around Walla Walla a few days, when Charley got a letter written ten or twelve days before, saying that Elizabeth had a little girl, and would come up the river to Wallula and from there by stage to Walla Walla, and sure enough, next evening while Charley had commenced at the Walla Walla Hotell, here come the stage with Elizabeth, George, and the baby, not two weeks old, and took us quite by surprise. I was working, too, as second cook at this same hotell, both together. Charley was to get from $75 to $100 per month the second month. I was getting $45 the first month. A young man named Charles Ebert was one of the partners in the hotel. He had bought out Goldsten. He was of German descent, from St. Louis. There were about six men waiters, some five or six men to run the kitchen; they fed from 100 to 200 men each meal; board per day $1.50 to $2, per week $10 to $15. No bedding or lodging. Single meals 75 cts.

Charley had to rent a small frame house, not plastered or ceiled, for $7 per month. Rents were high and houses scarce.

It was an open fall up to the 23rd of December, when it commenced snowing. I had not stopped on my place since about the middle of November, for I had no crop in and no team to do anything with, so I had concluded to get some work in Walla Walla, where I could make my living. I had one horse left, and when the snow commenced I had him at a feed stable for a day or two, and concluded I had better sell him than to let him eat his head off, so luckily got a chance to sell him to some miner for $57, and it helped to keep the family and get a few things for the house, for it turned out to keep on snowing until the snow was three feet deep and drifted some places eight or ten feet high.

It proved to be the hardest, coldest, severest winter yet experienced in Oregon or Washington Territory. The snow lay on so long and kept so

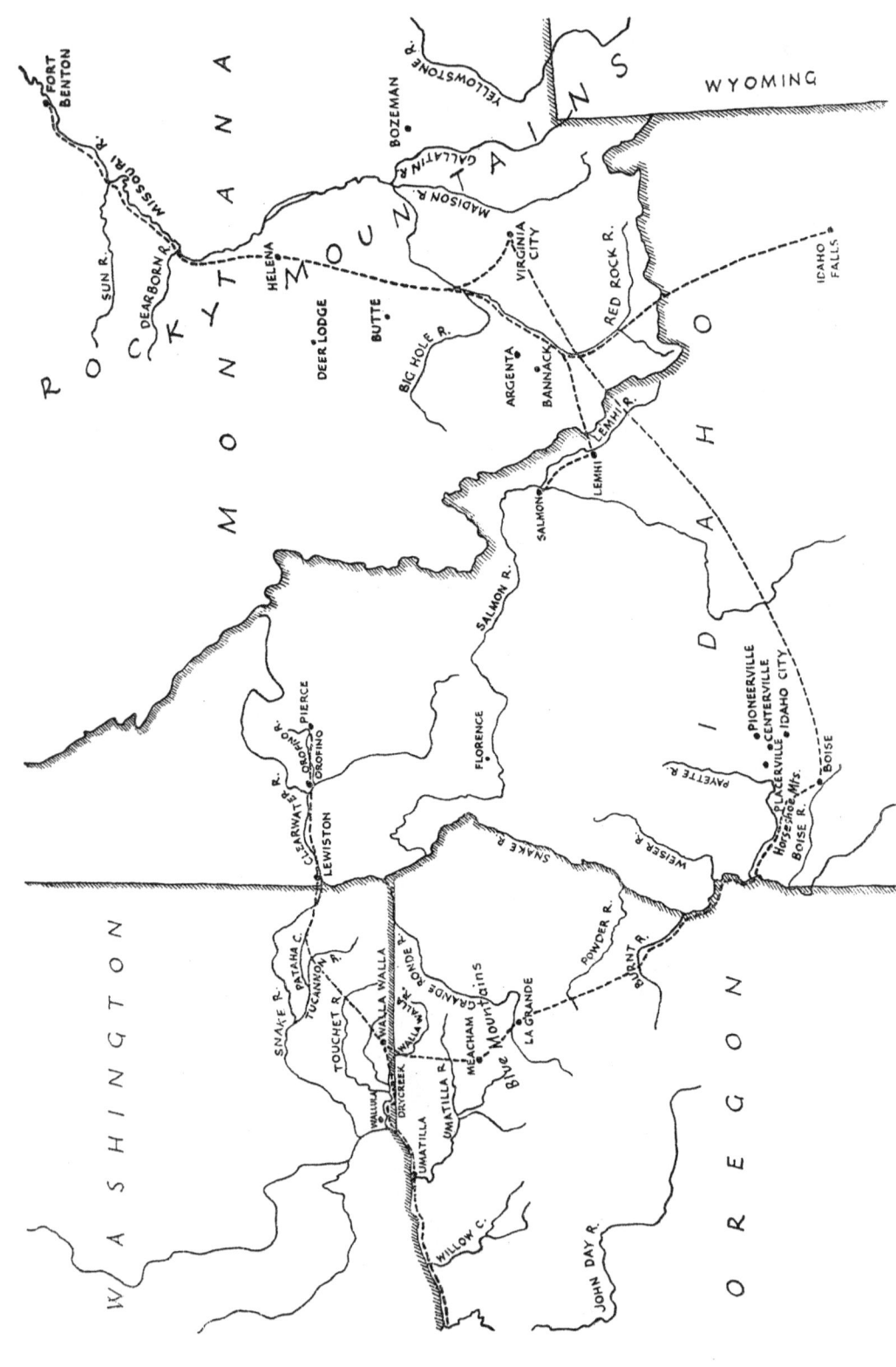

very cold, from 15 to 37 below zero. Most houses were flimsily put up and a person would nearly freeze by the fire, or stove, and fuel got scarce.... The weather [got] so cold they could not go out to chop it in the timber, and bought and borrowed from each other, expecting every day the weather would moderate and teams would bring in wood to sell, for all the creeks a short distance from the city were lined with heavy timber. But the cold hung on and the roads kept impassable, and no wood came in.

A stage loaded with passengers started on their way to The Dalles. The men were merchants and miners on their way down to Portland, and some to the Willamette Valley. They were cautioned not to start in the snowstorm and cold, but they were so anxious to get to their destination that they would not take advice, for they were big stout robust miners, most of them. There were nine until they took in a ferryman on the Lower Umatilla to go to The Dalles. I knew three or four of the men, one, Dick Bolton, or Cribbage Dick, from Althouse Creek and Prattsville and Gold Beach; one named Brown, a Jew merchant from Walla Walla, partner of Burnstein, from Yreka, California; old John Mulkey, a butcher from Salmon River, a large fleshy heavy man, uncle to Cy or Bill Mulkey, who were gamblers; Aulfrey, the ferryman, and some six or seven others. When they got to Willow Creek they all had to walk, and some proposed to stop at Richmond, so they walked to keep warm and be easier on the horses, and kept on to John Day River, where the stage driver would go no further. Some three or four stayed with the stage at the hotel overnight.

Next morning still cold, no change, the wind blowing, but still some would go on, and persuaded some others that would have stopped. But Dick Bolton, Mulkey, Brown, and Aulfrey would keep on, and said they could walk and run and keep going until they should get to De[s]ch[u]tes River. But not long after leaving John Day River the Jew, Brown, commenced to complain of getting tired, and he told the others he had a large sack of gold dust, and they offered to carry it and did, but he had got too much fagged out before he told them; he had been afraid to trust them with the facts. He soon commenced to drop back; the others urged him along for awhile, but at last he gave completely out, and said they had better leave him to himself and save themselves.

So there was no help for it. The others had all taken turns carrying Brown's gold dust, about 50 lbs, until they too were nearly give out. So they dug down in the snow and covered him with a blanket and put his sack of gold alongside of him, bid him goodbye, covered him up in snow and told him they would send help as soon as they got to a house. And they kept on.

Next the ferryman gave about out, and they worked to get him along as far as they could, and at last had to leave him behind, coming along slowly. The balance had to keep going to keep from freezing. The next to give out was Mulkey, the butcher, and he had to be left behind, and the others kept on, then one more dropped back, and they left him to come along slow. The others said they would hurry through to a house and send help right back. And on they went. And when all were about to come to a standstill and freeze to death, they came in sight of the Desch[u]tes House or Ferry, and motioned to some they saw at the House to come to them.

They went out and brought in the men and when told about the others, got up the help that could be got with team and provision to carry the men, and four or five persons went back towards the men that dropped behind. Some of the men who were able and one or two at the Ferry then struck out for a town called Celilo, between Desch[u]tes and The Dalles, and got help to go back with fresh horses and light wagon and some from The Dalles, where most all were known, there was got up a big crowd to go out, and well prepared, with blankets and stimulants. And they commenced picking up the men and brought all in to The Dalles. Brown was dead when found and he and his gold brought in. Aulfrey lost his two legs and died. Mulkey had one leg and one arm taken off, I think after, died. The four had feet and hands froze. All were some frozen hands, fingers or ears, and lost some of them.

The mercury was down to 36 or 37 below zero. It kept men more quiet, and they stayed closer to shelter and fires. The people of Walla Walla ran out of wood and burnt their fences, and rails sold for $60 per cord, and those that had no money took hand sleds at night and went out to the government fence of rails and slabs and posts and brought in every thing that would make heat. Crowds of miners wintering in Walla Walla ran out of money and went to the bakeries and hotels and made them give up bread, cheese, bologna and meat to keep from starving.

Many hotels had to close up, and the Walla Walla, where we were, done so. The proprietors could not give their meals away and pay rent, and wages to their waiters and cooks, and pay their beef bills, so Charley and I were both out of employment. We could go to the saloon and keep warm, but the family could not, and we had to rustle to get wood to burn to keep Elizabeth, George and the baby not over three or four months old from freezing. It was fearful. We had to go on the creeks close by and cut willows to burn, and raid on some other woodpile.

Old Caleb Grover's cattle were all fat that fall. There was quite an emigration to the valley, and lots of poor cattle and cows came in. The Fall was so open that Grover was very much dissatisfied, and wished we would have a hard spell of winter so that all the poor cattle that had come in so late in the fall would all die and he would keep his fat and they could stand the winter. And in the spring, cattle, cows and steers would bring a big price. He was a selfish beast and cared only for himself. Nobody put hay for stock or cattle in Walla Walla Valley, for the winters were mild so that cattle could live on the bunch grass all winter, and if a little snow did come, it usually went off with a south wind from the ocean up the Columbia River. It would take the snow off like it was warm weather and was called the Chinook Wind, or from the southern Indians [East Indies] on the Pacific Ocean.

So when the snow commenced falling on the 23rd of December, Grover was well pleased, only he was afraid it would not snow enough, or lay on long if it did. But when it came three feet deep and cold, it just tickled him to think how the poor emigrant cattle would die off, and his cattle all so nice and fat. But after it lay on a long while, and it was so cold he could not keep around his cattle to look at them every day, he got a little suspicious that it might last awhile. He could not keep the cattle together; it was too cold to go out with them to keep them together, and the snow drifting, and the cattle commenced to wander off for water. The creeks had frozen dry and the cattle looking for drink broke through the ice and snow and could not get out, and no one to help them out. Some wandered on for water to the mouth of the Walla Walla and the Columbia thirty miles away, and would go out on the ice to drink, break in and drown.

Then Grover got scared, no feed to feed them, too cold to see to them, and he commenced to look to the south for the Chinook Wind, but none came. Every morning and many times a day old Caleb Grover would step out of his cabin with his hat off, looking wistfully to the south for the Chinook wind, but none came. He had over 600 head in November, cows, calves, steers and all—he offered to sell the lot for $6000, but did not find a purchaser. He had at one time in Oregon been sunstruck and it had sort of affected his brain, but got over it; but this continuation of the snow and cold, and the anxiety for the loss of his stock nearly drove him to madness (which it did nearly two years after). But the snow was still there, and no Chinook Wind to take it off. At first all his calves died, then his old cows, then his yearlings, next his young cows and heifers and some of his two-year-olds, and by the time the snow did go off, in March or April, he had but 165 head of two, three and four-year-old steers left, and they looked quite poor for beef. But he was not the only man that lost cattle for all the emigrant cattle had died, every hoof. . . . Four-fifths of all the stock in Washington Territory (Northeast) and Northwestern Oregon died, and in the spring of 1862 was the highest water, and the hardest winter known up to that time.

# 10

## SETTLING DOWN ON DRY CREEK

(1862–1864)

There had been some rich new discoveries late in the Fall of 1861 on Salmon River, Idaho Territory. By going to Lewiston you kept on [southeast] and came down on Salmon River over 100 miles [below] the mouth where it empties into Snake River, some 50 or 60 miles [below] Lewiston, but hardly accessible from the mouth up, but by going over a mountain between the . . . [South] Fork of Clearwater [and the Snake] you come down on Warren's Diggings and a short distance Salmon River.[1] There had been a town laid off named Florence,[2] and the diggings were extremely rich, but provisions was scarce, and the snow and cold and high water kept them from being worked much. That winter it had the name of the roughest place in the West—gambling, drinking, fighting, shooting and killing was all the rage. All the miners from Warren's, Orofino, and Clearwater and the Nez Perce and all that part of Idaho came to the great Salmon River Mines,[3] and got in late and there was no provisions,

---

[1] James Warren is credited with the first discovery in the Salmon River mining district, August, 1861. The best route to the new field was from Walla Walla—

> The route follows the old emigrant trail across the Malheur river, and down it to the Snake river; across Snake river to the mouth of Payette; up Payette 75 miles, and thence across the mountains in an easterly direction to the diggings [350 miles]. The mines are in a basin forty miles long. . . .

*The Oregonian,* November 6, 26, 1862.

[2] Florence lies some fifty miles south of Craig Mountain and is situated in a small valley at the end of Meadow Creek. Here a large granite butte rises to a height of some 1,000 feet. The mining town, located on the butte, had a brief, but productive existence. The site lay about seventy miles northeast of Fort Boise. Rossiter W. Raymond, *Mines and Mining of the Rocky Moutains,* pp. 249–250; *The Oregonian,* November 4, 18, 1862.

[3] Reinhart is confusing the Idaho mining fields. The Nez Percé Mines were consid-

## Settling Down on Dry Creek

so they had to go where there was to get provisions or to winter.[4] Some stopped at Lewiston, but most of them came down to Walla Walla. Some went on to The Dalles and to Portland, and to Oregon and California to make arrangements to come back to Salmon River in the Spring. Those that had some money gambled at Walla Walla all winter, and the roughest lot of miners and gamblers I ever saw were at Walla Walla.

After the Walla Walla Hotel closed I was out of work for about a month. So was Charley, only when he got a chance to help get up a ball supper, and he got jobs of that kind.

Some time the fore part of March, two partners in a bakery and brewery and saloon wanted me to keep bar and saloon, as Henry Kroney one of the partners, the baker, was agoing to Salmon River as soon as the roads were open, to start a butcher shop at Florence, and Emil Myers, the other partner, the brewer, was to be in with Henry Kroney and some old butcher, a German from Corvallis, where Henry was from. Myers was to brew, I to keep bar and store or saloon, and they had a little German to do the baking of bread, cakes, pies, and cook for us three. I was to get $75 a month and board, and I and Myers, the proprietor, and the baker always slept in the store or shop. They done a good business and sold beer, liquors, cigars, bread, cake, pies, crackers, cheese, sardines, bologna, cover oysters, tobacco, browned coffee, tea, cider, and took in $150 per day or $1,000 a week, and we retailed about $70 worth of beer a day at 12½ cents per glass, whiskey, brandy, gin, wine, rum at 25 cents per glass or drink, and sold lots by the bottle.

In our saloon backroom we had three or four tables to play or drink beer, and we had a big soldiers' trade. Greenbacks went only from 65 to 75 cents to a dollar in coin or silver or gold dust. I was kept very busy from 6 in the morning to 10 o'clock at night, and many days would not step off the sidewalk to sweep off the walk. While at my meals the baker

---

ered to include the Orofino-Clearwater area; the Salmon River Mines included the Warren and Florence sites.

[4] William A. Goulder, reporting on the winter of 1861-1862, writes:

> Though the snow lay deep on all the roads and trails from December till April, there was but little interruption to the tide of travel that went on. All winter long men were daily leaving Oro Fino on snowshoes with heavy packs on their backs for the distant mining camp, Florence in the Salmon River country being the principal point that attracted them. The whole country above Lewiston lay buried under a thick mantle of snow with the exception of some narrow river bottoms favored by lower altitudes. Pack-trains could not move from their winter quarters; in the lower country the winter was one of unusual severity, the mercury falling far below zero much of the time. The merchants of Oro Fino and Pierce City had taken the precaution to lay in ample supplies of all articles needed in the mines, so that when a scarcity began to be felt in the other camps, especially at Florence, where the latest discoveries had been made during the preceding autumn, the camp of Oro Fino furnished the much-needed base of supplies, though strong men had to be carriers and take the place of pack-animals.

Reminiscences: *Incidents in The Life of a Pioneer in Oregon and Idaho*, pp. 219-220.

would keep store in my place, and it kept the boss, E. Myers, busy brewing beer so as to keep some on hand and not run out. There was but one more brewery in town—"Lager Beer Joe," by name Joe Hellmuth, who lived the second house south.

Henry and his partner butcher did not go right off; the snow lay on longer than usual, and so were all ready to start whenever the roads were favorable for getting in to Salmon River. Thousands were eager to start, and fitting out and buying horses to pack. Horses got high and in great demand, and times opened up lively. Miners from California and Oregone were already coming in so as to be into the diggings as soon as the Salmon Mountains could be crossed with animals.

There was a theatrical troupe stopping at Walla Walla (where Charles Pope, the actor, had built a theater, got up fine like an Eastern theater),[5] named the Robinson Family, consisting of old man and his wife, their daughter Susan, and a boy, Fred Robinson. They were a good troupe and were great favorites in Walla Walla.[6]

Susan had just got married to Charlie Gestler, a Wells Fargo Express employe and a resident of Walla Walla.[7] The firemen had a ball, and they had tendered a benefit to Susy Robinson, or Gestler, that night, and the firemen were in uniforms, and a great deal of drinking was going on all the day before.

The gamblers and roughs at Walla Walla and all the mining camps were Rebels or rebel sympathizers—some no doubt were afraid to say they were Union men, for the Rebels at that time were jubilant over the defeats of the Union Army on the Potomac, and these gamblers and desperadoes were the terror of the Union people, who dare not give their sentiments amongst the desperadoes and Rebels. When the soldiers from

---

[5] Charles Pope was born near Weimar, Germany, February 17, 1832. Later that year his family emigrated to America, and settled in Rochester, New York. When sixteen years old, he joined the touring company of Augustus A. Addams, the tragedian. In succeeding years he played in theaters from New York to New Orleans. In April, 1861, Pope married actress Virginia Howard, and in August they started on a combined honeymoon and professional tour to the West. After a prosperous engagement at Maguire's Opera House in San Francisco, they toured northern California, Oregon, and Washington, and other western states. July, 1864, found Pope returning east to continue his career. He retired in 1887 and died in New York, July 2, 1899. John S. Kendall, *The Golden Age of the New Orleans Theater*, pp. 458-466; Alice H. Ernst, *Trouping in the Oregon Country*, pp. 24-25.

Bernard Berelson and Howard F. Grant, "The Pioneer Theater in Washington," *Pacific Northwest Quarterly*, XXVIII (1937), 134, state that the first theater built in Walla Walla was the Gaiety Theater, 1879.

[6] The Robinson Family were great favorites in San Francisco. "The unequaled and celebrated Infant Prodigies, Miss Susan and Master William Robinson," their elder sister Clara, and father, Joseph B., composed the troop. George MacMinn, *The Theater of the Golden Era in California*, pp. 452-454.

[7] In a letter to the California Historical Society, December 2, 1957, Miss Beth Fachine, a great-granddaughter, states that the name was "Getzler." *La Petite* Susan died in Sacramento at the age of twenty-six.

## Settling Down on Dry Creek

the fort came to town and got to drinking, they were slurred as "Lincoln's Hirelings" to kill Southern people and drove them away from saloon bars and would not allow them to drink, and sometimes would make a soldier or volunteer drink to the health of Jeff Davis or Beauregard, and if he refused, kick him or knock him down. Most of the large gambling houses were kept by Rebs or they showed Rebel sympathies and some would not let the soldiers or volunteers in either to play or drink.

A company of California Volunteers had been sent up and were temporarily stopping at the garrison in their quarters.[8] (John McClancy was with them). When they left Red Bluffs, or Colus[a], California, there was a man called "Cherokee Bob," or Robert Talbot, a gambler, had got into some fight and shot or killed a man, and the volunteers helped to get Talbot away from town, and from being arrested, had smuggled him off, and when the company was sent to Walla Walla, brought Bob Talbot along and paid his way for him as he had nothing. But he was a rank Rebel, and as soon as he got to Walla Walla, took up with the Rebel gamblers there and was the worst to the boys who had saved his life at Coluso, Calif. and instead of being thankful to them, took every opportunity to insult and abuse them for being Union volunteers.

Now our sheriff, James Buckley, was a rank Rebel, so was George Porter, a deputy, and our city marshal and their assistants, all Rebels. And these same men belonged to the fire company, to give the complimentary benefit to Susy Robinson. Among the California volunteers was a seargant named John L[a]t[ze]nheiser, a fine young gentleman of modest and fine appearance, a Pennsylvanian of good family, well educated. He with his lieutenant came to our bakery, and when asked to drink, did not, but took a cigar. There were some others, but none so well behaved as John L[a]t[ze]nheiser. They said they were a-going to the Fireman's Complimentary at the theater. My boss, the baker, Henry, was a fireman, and floor manager. After the theater there was to be a Fireman's Ball. Our Henry was on good terms with all the officers and soldiers at the garrison, who were our best customers.

The theater was crowded with firemen and friends, their wives and children, and many gamblers and Rebels who were highly elated at some recent Rebel victory, and were drinking a good deal. There were some saloons a short distance from the theater and between acts they would go out to drink, and some had bottles full with them; some were very boisterous and noisy and were reproved by the city marshal Porter, and James Buckley the sheriff, who were drinking too much.

And as a lieutenant and L[a]t[ze]nheiser came in, George Porter, the city marshal, roughly called for them to be quiet. The lieutenant said

---

[8] One of four regiments, raised by Colonel Justus Steinburger in 1861, stationed at Fort Walla Walla. Hubert H. Bancroft, *History of Washington, Idaho, and Montana*, pp. 228–229.

he had made no noise, but Porter, in his bulldog way of speaking, told him if he did not set down, he would make him, and the Lieut. said he guessed not, when Porter came to them with a pistol in his hand and told him to "Hush up! You son of a ———!" and shoved the pistol against him. John D. L[a]t[ze]nheiser knocked the pistol out of Porter's hand, and Bob Talbot (Cherokee Bob), who sat by, pulled out a pistol and shot at the lieutenant, and the gamblers all drew their pistols. The lieutenant jumped back to the gallery where there were some of the volunteers. One gave him a revolver, and the gamblers and officers of the city had commenced to fire on the volunteers. Some few jumped down from the gallery and got where L[a]t[ze]nheiser was, and all were shooting through the theatre. The crowd of ladies and children ran out the back way, over the stage, and all was confusion.

Fifty to a hundred pistols were used, and after John D. L[a]t[ze]nheiser had emptied his revolver, he jumped on a bench and said he was unarmed. James Buckley, the sheriff, put the muzzle of his pistol to L[a]t[ze]nheiser's mouth, said, "Take that, you Union son of a ———!" And shot him dead. Another volunteer got shot several shots, and got out of the front door of the theater and fell dead. His name was Hubbard. One or two more were wounded, but not seriously. Of the citizens, George Porter, the city marshal, was shot in the thigh and knee, and the deputy marshal was shot in one thigh and arm. Both lay on the stage, covered with blood. Many others were hit, but only slightly. It was after ten o' clock, and it broke up the performance and the ball. Bob Talbot was one of the first to fire, and he fired on the very man that brought him a way from his trouble in California. He skipped out that night to Lewiston, and so did Sheriff James Buckley, for as soon as some of the volunteers heard who had done the shooting, and how, they arose as one man, eighty men in the company, took their arms, and were about to take two field-pieces of cannon, but the officers got them away, or they would have shelled the town and residence of the sheriff. So they, without officers, marched down and surrounded the sheriff's house, and they found he had skipped, looked the whole town over for him, and Cherokee Bob, and he too was gone, and all those who had any immediate hand in the shooting and killing of L[a]t[ze]nheiser and Hubbard, whose body they took back to the fort or garrison and buried it. Some dragoons followed on to catch Bob Talbot, but he got past Lapwa[i] Fort and Indian Agency before they got there, and they did not have permission to follow further.[9]

There were hot times at Walla Walla at that time. When they took

---

[9] Reinhart's report of this incident involving Cherokee Bob [Henry] Talbot is an accurate one. George W. Fuller, *A History of the Pacific Northwest*, pp. 281-282, presents a brief account. A more detailed version, which substantiates Reinhart's, is given in Nathaniel P. Langford, *Vigilante Days and Ways*, I, 106-110, 120-124. Talbot fled to Lewiston, later to Florence, where he met his death.

*Settling Down on Dry Creek* 207

the names for the draft, it was found that it would not be safe to take Union men away, for it required all to keep the Rebels from breaking out then, and if the forts had not been there and at Lapwa[i], Colville, Fort Simcoe, and The Dalles and Vancouver, I believe they would have molested the Union men, women and children, and took what spoils they could have taken, and struck for the South.

In April and May the rush for Salmon River was at its height. I saw old acquaintances from all parts of Oregon and California on their way to Salmon River. Some time in the last of May or the first of June I heard that some man with a family had jumped my claim on Dry Creek, and I saw someone and they told me that James Holt had encouraged the man to take it, that I had abandoned it and was living and making my home in another state, or territory, and he (Holt) would help him to plow and put in a crop. I went out one day and saw the man, and it was the man Stonebraker that was living at Marshall's [McKay?] when I came up the Umatilla the spring before. He had got Holt to plow him a few acres for a garden, and fenced a little, and was about to enclose a large field with a ditch-fence. He talked like he was going to stay, that I had left the claim, and Jim Holt told him I had said so. So I went back to Walla Walla to see about my getting back my claim. I stayed on at the city brewery and bakery until July. Old Henry Wilson, who had left Coffee Creek and then took Tom Kavanaugh's horse, and had unknowingly struck and discovered Gold Hill, came in where I was keeping store one day, on his way to Salmon River mines.[10] (A young man named John Robnette that owned a farm close to James P. Hogue's had married last fall the young lady I had been going with some little—her name was Catherine C. Brock, an only daughter of Vineyard C. Brock, a farmer quite well off, and he was on his way to see the mines, too. I had seen her two brothers a year before at Pierce City, Nez Perce mines, Idaho, and they had told me all the news.)

About this time my boss's partner, the baker, Henry, that had gone into the butchering business at Florence, failed there, and lost three or four thousand dollars, and Myers was afraid he would be holden for some liabilities of the firm. The other Corvallis butcher had nothing, so Myers bought Henry out at Walla Walla on the sly, and Henry went back to Corvallis to his bakery, where he had come from less than a year before.

There used to be a man named Old Tex—his right name was James Helm—he had a brother called Boon[e] Helm, at Florence, Salmon River mines. They were both hard cases and rebels. The brother, in a drunken row, got in a wrangle with Dutch Charley, a shoulder striker and gambler and Boon[e] Helm cut or shot Charley so bad that he died, and Helm got away by the help of his rebel friends, or the miners would have hung

[10] Reinhart's story of how gold was discovered at Gold Hill, Oregon, contradicts the accepted version. (See Chapter VIII, *note* 14, *ante*.)

him. Old Tex heard of it and went from Walla Walla to help him get clear. This same Boon[e] Helm was afterwards taken with a lot of road agents or highway robbers, and was hung at Virginia City the Fall after he killed Dutch Charley. Old Tex himself made some money at Idaho City and the Boise mines, came to Walla Walla, got drunk, and his horse threw him off and broke his neck short off, and he died, in 1863 or 1864.[11] (He is the only drunken man I ever heard of getting hurt or killed while drunk—it had always been thought that a drunken man could not be killed by an accident.)

Some time in July Stonebraker proposed to let me have half of my claim back, and to enlarge it and go to work together, and ditch the whole large field. So I had saved over $125 or $130, bought a lot of provisions and tools, and made preparations to go on the place and improve more of it and have more broke up. Charley was glad to take my place tending bar, and he and Myers got along well. We could all speak German and our trade was good for them.

Now before I go further, during the winter a young man whom Charley was acquainted with, named Samuel Hibbard, he lived on the adjoining place to Hardy Ellif[f], on the south side of the canyon [in Oregon], had been up to the Clearwater and Nez Perce mines, but had nothing left, so he lived with us and Charley that hard winter, and nothing to do. We used to go to the saloons to get warm, and at nights we did not use fire. He had a brother-in-law, Henry Armer, and he had taken up a land claim up the country, but the winter was too hard for them to stay, so came to Walla Walla and stayed with him awhile. He was acquainted with Armer on the Umpqua River, and a relation of Elizabeth's who had married a man named John Ladd. He had had a pack-train to pack to the mines, and he had his wife to come to Walla Walla. They had no children, and no house to be rented. Got in with Charley and they was company to each other, but Mrs. John Ladd was a dissatisfied woman, and made mischief and had Charley's wife nearly persuaded to leave him, and go back to Yoncolla Valley, their old home. Now this Sarah Ladd had been very sweet on a man I knew at Canyonville, keeping a saloon, named John Story, a gambler. He got acquainted with Mrs. John Ladd at Oakland, either before or after her marriage to Ladd, at some dances. He, Story, had kept gambling saloon at Oakland, and they had been corresponding while Ladd was here. He bought into a livery stable.

One evening John Story came up on the Wallula stage, stopped at the City Hotel [Walla Walla], and at evening found Mrs. Ladd and came to

---

[11] Thomas Dimsdale, *The Vigilantes of Montana*, pp. 167–168, details Boone Helm's capture, trial, and death. He observes: "Helm was the most hardened, cool and deliberate scoundrel of the whole [Plummer] band, and murder was a mere pastime to him." Langford, *Vigilante Days*, I, 156–175, summarizes Helm's earlier criminal career, and dates "Old Tex" Helm's death in 1865. Actually, Helm was hung January 14, 1864.

*Settling Down on Dry Creek* 209

see her. My sister-in-law saw Story, and had known him at Oakland and Roseburg. That evening Sarah Ladd took a walk with Story, and came in that night late. And early in the evening John Ladd went up to the stable to feed and when he came home his wife was out. He saw Story the night before at the hotel, but thought nothing of it until then. He went up town to the hotel and found Story and her. She said she was a-going home to her folks at Oakland, and Story was going, and she would be under his protection. Ladd was awfully taken aback, and tried to get her to stay, but she said she would not, and so Ladd left her, and she took the stage with Story, and went towards home. I do not think they ever got together again. Ladd run a pack-train next season to the mines and done well, and afterwards kept a public house and nice farm in Grand Ronde Valley, or Bu[r]nt River Valley, I forget which.[12]

Charley said his wife had been very much dissatisfied and Sarah had tried to get Elizabeth to go too. I heard John Story left her somewhere in Oregon to shift for herself. She was a self-willed devil, a minister's daughter and of no account. I had heard of her from Dave Ransom. She was a cousin of the Ladd's at Portland, bankers, and someway the Ladds were cousins of Dave Ransom's.

So the fore part of July, 1862, I went out to my old claim. I boarded with Mr. and Mrs. Stonebraker. I furnished the provisions and Mrs. Stonebraker done the cooking. We commenced to ditch at the lower end of the field, close by the cabin, and worked up toward where the road crossed the creek above. The ditch was four feet wide on top, sloping three and a half feet deep to about twelve or thirteen inches wide on the bottom. Grass sod was cut regular, so as to build up from the inside of the ditch up about 3½ feet high, making 7½ to 8 feet to the top of the sodded ditch, and a post was set in every eight feet, and a rail or board nailed to the posts made it so high that no stock could jump in from the outside, and few would attempt it from the inside. The grass sod would grow together and make a durable wall, it was hog-tight, too, and in wet places would answer for a drain as well as fence, and we could dig from three to four rods per day each. And it made a cheap, durable fence and drain.

It came quite hard to me to go right to hard work in the heat of the summer, for I had done no outside work all fall and winter up to July, and the hot sun on the side hill soon tanned me as dark as an Indian, and it made me mighty warm and to perspire quite freely, more so than if I had been at hard work, for I had to keep up my amount of ditching, so as to do my share with Stonebraker, who was used to the work. In our new

---

[12] Reinhart refers to Grand Round Valley on several occasions, and the reference is confusing. Grand Ronde Valley lies in Polk and Yamhill counties while Grande Ronde Valley is situated in Union County. Popular reference to these two different valleys is universally misspelled. Burnt River Valley is in Baker County—all of the three are in eastern Oregon.

location of the way of taking the two claims, we claimed one mile long and a half a mile wide to a half section, or 320 acres, and by so doing took in the road at the crossing and some good farming land on the other side of the creek below the cabin or house. So we cut a quarter of a mile of ditch on the side going up the creek, then put in posts and spiked on three or four rails or poles on the cross-fence, then went down along the sidehill on the south side with the ditch to where we commenced at the beginning on the other side. Part of the way we did not have to ditch, for the hill was perpendicular and no stock would ever dare to go up or down.

Stonebraker was a sort of carpenter or a wheelwright and could stock plows and do wagon work, which he did when in demand. Toward fall he asked me to buy him out and I would have it all to myself again, and said he never would have jumped it from me, only by the advice of Jim Holt, our half-breed family neighbor. He said if I bought him out he would go about two miles up the same creek we were on and take up a claim at a cottonwood grove and some good garden and farming land, and I could help him put up a log cabin.

So in Nov. I bought him out and helped him put up the cabin. He had another man to help us, named John Harper, from Pine Creek, who had been in the mines and had come home to winter. When Stonebraker got his cabin done he moved in, and I batched it again. . . . I put in the winter cutting poles, rails, and ditching and clearing off some ground for breaking up, cut up willows and thorns and grubbed out the roots.

Now I will go back to the spring of 186[2]: After the winter was over, Grover had some 165 steers left, 3 and 4 year-olds had stood it when all the balance had died. Right above the ranch and cabin where Grover lived, a man named Samuel Johnson had a stock ranch and farm adjoining Caleb Grover's. He had a partner named Fairchilds, some relation of Gov. Fairchilds of Wisconsin.[13] They had a pack train of mules and horses, and had a wagon train of cattle to haul freight to the Colville and other mines, and at Salmon River, at the city of Florence, they had a large general store in with two more partners named Calvin and Ward, who ran the pack train and had some mining claims, and they carried on the Saml. Johnson ranch and farm. They bought one half of the 165 head of steers from Caleb Grover for $6500. In the fall, before the winter set in, Grover had offered to take $6000 for the whole 600 head of steers, cows and calves, and after having lost 435 head, got double what he had asked for all, but so many had died that what was left brought high prices. Samuel Johnson and Caleb Grover drove the 165 head to British Columbia and got a big price. They realized some ten thousand dollars clear of expense.[14]

---

[13] Lucius Fairchild served as governor of Wisconsin, 1866–1872.

[14] A lively cattle trade developed between Oregon, Washington, and British Columbia

## Settling Down on Dry Creek 211

During the winter of 1862 and 1863 the Salmon River mines were so overrun with prospectors for more mines, and Warrens, Elk and Clearwater were discovered in Idaho Territory,[15] and in Montana the Bannack and Alder Gulch and Virginia City,[16] and many others in Oregon. The John Day mines, Canyon City, Powder River, and Auburn were struck,[17] and a great rush took place to all these places.[18]

I will here relate of two young men who used to be on Bowling and Sucker Creek; one named Jack Desmond was a partner in the rich claim where the big seven pound, two-ounce piece was taken out in 1856, where Jimmy Hope, a tailor from Yreka, Calif. and James Dooling were partners. Dooling got into a drunken row with a man working for them, named Ed or Ned Boyles, and Boyles cut and shot Dooling so he died. Boyles was under suspicion of having killed a man in California, but here got clear. When Jack Desmond left Bowling Creek with Joe Labenee, a boy only 18 or 20 years old, they went to the Auburn mines not far from Grand Round Valley and were working or prospecting in the mines.[19] One day they were both in a saloon playing cards, and Joe Labenee was playing single-handed with a Spainard who was drinking and quarreling with anybody he could. He tried to cheat the boy, Joe Labenee, and they had some words together. When Jack Desmond went to try and quiet the

---

in the early sixties. In 1861, Victoria imported 7,081 head at a value of $313,797. By 1866, the trade was valued at $348,292. William J. Trimble, *The Mining Advance into the Inland Empire*, p. 107.

[15] In 1862, placer mines were found at Elk Horn and Placerville in the Boise, or Grimes Basin (not to be confused with Boise City), sixty miles southeast of Pierce City. A rush to the Owyhee district also matured and brought into existence Ruby City, Boonesville, and another Oro Fino. Raymond, *Mines and Mining*, pp. 235-252. As a result of the heavy growth in population and the political situation in Washington, D. C., the Territory of Idaho was established on March 4, 1863.

[16] The gold regions of western Montana were discovered in July, 1862, and were located in the Jefferson Basin and Gallatin Valley. The former was at one time a rendezvous site for fur trappers in the upper Missouri Basin and was commonly called the Bannack district. Grasshopper, Stinking Water, and Alder Gulch Creeks were the sites for rich placer and quartz diggings. The seventeen mile length of Alder Gulch reputedly had 10,000 miners working its steep banks in 1863. In this area four "cities" were spawned: Virginia, Nevada, Central, and Summit. Last Chance Gulch, which became Helena, sprang up in Gallatin Valley around gold-bearing quartz veins, along with other famous, but short-lived gulch camps. *Ibid.*, pp. 269-317.

[17] These mines were located in Oregon. Gold was discovered on Griffin Creek, in Baker County (Auburn), in October, 1861. This provoked a rush into eastern Oregon. Miners by the score poured into the Blue Mountains. Auburn was founded June 13, 1862. The John Day River flows through much of eastern Oregon. Canyon City, Grant County, is situated in a canyon some two miles south of John Day River, while the Powder River flows through Baker County. Trimble, *The Mining Advance*, pp. 62-84, treats in summary the various gold fields discussed in *notes* 15 and 16 as above presented.

[18] It is impossible to arrive at accurate figures on the number of miners who were lured to the inland empire gold rush. The best approximations vary from 50,000 to 75,000 for the period 1861-1863.

[19] Auburn, Baker County, is situated quite south of Grande Ronde Valley.

trouble, the Spainard drew his knife and cut Labenee, then Jack Desmond, both so bad they dropped dead, and before the bystanders knew what had been done they died. The Spainard flourished his knife, and ran out of the saloon for his cabin.

The miners were awe-struck at the double murder before their very eyes, right in their midst, they could not realize that two men were dying or dead, and both were well-liked, harmless and unoffending, and they had not made any attmept to harm the Spainard, merely remonstrated about the cheating, and his friends saw the whole game and hardly anything had been said by either to provoke the attack. So the miners rushed after the Spainard to arrest him, but he would not be taken, and used his knife on everything, and his pistol, too. He got into his cabin and threatened to kill everyone if they came near him. So the miners fired his cabin, and he had to come out, and they threw a rope noose around his neck and dragged him through the street, filled his body full of bullets, hung him up and strangeled him and shot him to death and threw him onto his burning cabin and let his body burn up and scattered his ashes over the ground and street. He had some Spanish friends, but they skipped out until the excitement was over. I did not see this, but got it from an eyewitness who knew both Jack Desmond and Joe Labenee.

In the spring I got some little garden and corn put in and kept my place in repair and commenced to get out some logs to put me up a house at the head of my claim close to the wagon-road crossing from The Dalles to Walla Walla, so that I could keep people from camping and burning up my rails and posts to cook and warm by, and leave my fence so that stock would get in and eat up my crop while I was gone to town. Charley was still at Myers Brewery and Bakery.

One day a man we used to know in the Indian Creek mines, California, named Jake McClain came by. He had left the mines and gone down to the Willamette Valley and married a widow who had a lot of cattle and other property. He was on his way to the Idaho mines with a bunch of fat cattle to sell. He had along a pair of large gray work-horses which he worked to a wagon in harness, and concluded to sell the wagon, harness and horses, and pack some ponies or mules, as he could not take his wagon and team into the mountains, and Charley bought the outfit cheap, part down and the balance on time.

Charley concluded to cut and put up some hay, some for himself and some to sell in the city of Walla Walla, and I went to find to a place where we could get grass. I found a good place about ten miles off on the old Hudson Bay Claim Meadows. The Hudson Bay's claim run out in 1859, and there were 100 hundreds' of acres of meadow land down on the Walla Walla River where my Dry Creek and Pine Creek formed a large bottom or meadows.[20] Charley hired a man to help mow and get up the hay.

[20] The Hudson's Bay Company charter expired in 1858, but not its land titles. In-

*Settling Down on Dry Creek* 213

We made us a camp on the ground on a creek, and the water ran over and through hundreds of dead cattle that lay in the creek, as many as ten in a heap, who had all died the hard winter before when so much stock died. We had been using the water some time before we saw all the dead carcasses, and we soon hunted up some other place for water to drink and cook with.

We hauled a lot of hay to town, but the roads were bad and some very hard hills to pull over. We crossed the Walla Walla River near the old Whitman Mission, where in 1855 or 1856 the Indians massacred the whole family of Dr. Whitman; the Presbyterians and Methodists of Boston, Massachusetts, had sent them out and established a mission, and the Hudson Bay Company's priests had got jealous of the Methodists and persuaded the Indians to kill everyone connected with the Whitman Mission.

It happened this way: The Indians were having the chills and fever, and Dr. Whitman gave them medicines to cure them, but through their mode of bathing every morning in cold river water, one of their Chiefs died, and the priests told them that the quinine Dr. Whitman had given them was poison and caused their deaths. They, being excited, went and killed every man, woman and child of the mission, and the government sent in troops and they caught the murderers and hung some and punished others, but the main cause, the priests, made their escape right after the massacre.[21] A minister and family were now occupying the mission, by the name of Dr. Eals, one of the earliest missionaries in Oregon. He used to live around Astoria and Portland before coming here.[22] It was about 7 miles from my claim or ranch.

After we got done haying Charley put his team to hauling lumber and timber to put him up a house and barn on some lots adjoining Mr. Kroney and Emil Myers, who had built a brewery in Idaho City which Kroney and Myers kept while Charley run the Walla Walla Brewery and Bakery, while Myers was gone.

After Charley got his hauling all done, he sold the harness and wagon

---

stead, Americans simply became squatters. Ralph R. Martig, "Hudson's Bay Company Claims, 1846–69," *Oregon Historical Quarterly*, XXXVI (1935), 60–70.

[21] Marcus Whitman and his wife came as medical missionaries to Oregon in 1836, along with Henry H. Spalding and wife. The Whitmans built a mission, the Waiilatpu Mission, near the mouth of Mill Creek, a few miles east of Fort Walla Walla. They were slain by a band of Cayuse Indians, along with eleven others, November 29, 1847. Reinhart's version of the cause is incorrect. The real provocation was not the priests, but the culmination of the fear and suspicion which had long been brewing among the Indians against Dr. Whitman. Clifford M. Drury, *Marcus Whitman, M.D.: Pioneer and Martyr*, pp. 390–411.

[22] The Reverend Cushing Eells came to Oregon as a missionary in 1838. He spent his early years in northeastern Washington; later transferred his activities to the Willamette Valley. In 1859, he founded Whitman Seminary, which later became Whitman College (1883) in Walla Walla. He died February 16, 1893. Myron Eells, *Father Eells*, p. 33, *et seq.*; *The Oregonian*, November 23, 1877.

and traded off his grey mare for a sorrel trotting horse to work in a delivery wagon single, and he let me have the grey horse left. I traded him off for a large yoke of work oxen, and the same man let me have another yoke of oxen to work with the others for so many cords of 4-foot cordwood, to be delivered in town (hard wood—birch and alder). I hired a heavy wagon at 50 cents per day of Jack Monroe, a freighter, and I found two German men who had been up at Salmon River mining and done nothing to make it pay; they used to come to Myers Saloon, Bakery and Brewery. I engaged them to cut me 50 or 100 cords of hard wood. I furnished them provisions and they done the cooking.

I took them out to our Dry Creek, in the cabin, where Old Caleb Grover had kept his cattle that he had taken to British Columbia. I leased the cabin a-while, and the wood to be cut was just below Grover's claim on Government or vacant land. I made a trip every day to Walla Walla with the two yoke of cattle. I could haul about 1½ cords and I got from 7 to 8 dollars a load, and I got back to the creek in time to load, but it kept me busy. Some of the road was bad. I delivered my contract, and bought the wagon from Jack Monroe for $45.

After a while the men got a little money ahead, thought the wood too hard to cut and split, but they had cut about 100 cords, and they quit and I paid them off. I still had wood to haul, for when they chopped they could chop from 3 to 5 cords per day, and I could haul but one load a day. So I kept on until I finished hauling the wood.

When I got through I went to plowing at home and hauling logs from the mountain (pine and fir) for my house at the road. I put me up a stable of posts, crotches and rails, and covered with hay and straw, top and sides, which made it warmer than any house, and I had room for my 4 head of oxen and a cow and pony.

Once in a while I took a load of wood to town to sell, or to Charley. One day he said he had bought a large 6-mule government wagon. It had been smuggled away from Fort Walla Walla by some teamsters and sold to a blacksmith on the creek below town. Charley had bought it and put it in his barn and covered it with hay. He told me to take it out to my place and use it, and if I could, sell it or trade it for cows or horses, or sell mine and I could keep his. So one night I took his wagon, and kept around the garrison on the Mill or Yellowhawk Creek, and got it home all right, and I got a chance to sell mine to the Lewellen Brothers on Pine Creek, and I used Charley's until Fall, when he sold it to a cattleman on our creek, way below where we cut the hay....

The Spring of 1863 or 1864 there was a rush for a new discovery of gold mines in the Boise Basin, Idaho,[23] and all the miners who had wintered at Walla Walla started out to cross the Blue Mountains on the

---

[23] The rush to Boise Basin started in 1862 and to the Owyhee district in 1863. *The Oregonian*, November 8, 13, 1862; May 14; September 19, 1863; August 10, 1864.

## Settling Down on Dry Creek

Meachim or Umatilla Reserve Road, where the Emigrant Wagon Road comes in off the plains.[24] The travel had commenced and every few hours parties passed by my creek at the wagon road....

About November some three miners who had been up in the Salmon River mines came down to Walla Walla to winter, calculating to go to the new-discovered Boise mines in Idaho. They came to my cabin and wanted me to let them winter their mules and a pony at my place. They proposed to furnish coffee, tea, sugar, and meat (beef or bacon) and I was to furnish flour, cornmeal and potatoes, onions, cabbage and corn (to make hominy of). They would do the cooking in my cabin, and eat together, and they would sleep in my new stable that I had put up for my cattle, when it was not too stormy for my cattle to run outside. I agreed to the offer, for I would not then be so lonesome, and I had plenty of flour and cornmeal and the other things I was to furnish. The three were, the youngest about 19 or 20 years old, named John C. Beckney, a good cook, and he was from Pennsylvania; the next, Jim Amberson, about 35 years old, owned a pair of mules and had hauled logs in the mines to put up houses or cabins. The third man's name was Charley Sumners—he was from Mass. and had been a sailor; he was not over 25, and the best behaved or raised of the three. He had nothing, and the other two were paying his way, and after they got to Boise Basin, he was to work and pay them back....

The two men, Beckney and Amberson, were hard cases, and Charley, the third man, acted like he did not go much on John C. Beckney or Jim Amberson, or did not trust them fully. They both went to town often and drank considerable and would bring bottles home with them with liquor, and give some to Charley and me. Beckney told me he was well acquainted with the Volunteers at Fort Walla Walla, but he kept clear of them. I have an idea that he was a deserter from some post and was afraid of being known by some of them. He was the hardest boy of his age I ever saw, and he had plenty of cheek and appeared to be overbearing and domineering. I will speak of him in several places in this volume.

Along in April or the fore part of May, as soon as the snow on the Blue Mountains was so that they could cross, they packed up two mules and one pony and kept one pony for a saddle horse so that they could take turns in riding, and they all three started for the Boise River Mines in Idaho. So I was alone again, but I had plenty of work to do to put in my crop....

So I went to breaking up raw land with my two yoke of cattle. It was

[24] The Meacham toll road, called the "Blue Mountain Wagon Road Consolidated," was a former part of the Emigrant Trail through northeastern Oregon—connecting Umatilla and Walla Walla with Boise and Idaho City. It was owned by Harvey J. and Alfred B. Meacham (later U. S. Indian agent). Lewis A. McArthur, *Oregon Geographic Names*, p. 397; Irene Paden, *The Wake of the Prairie Schooner*, pp. 242 (map), 347–352.

hard work plowing up the bunches of rye grass and willow and thorn patches, tough enough for six yoke to pull through. I had to drive and manage the plow and be careful not to break the plow. When I got all my work done, I turned the cattle up the creek to rest and get fat on the good grass, and every three or four days would go to where they were to see how they were doing.

One time I had been in Walla Walla a few days, and I went to see how the cattle were getting along but I could not find them. Hunted all day up the two forks of Dry Creek and in the hills, but could not find them. So next day I got on the track of them—they had gone towards Walla Walla River, crossed over, and then I lost the track again. I made inquiries, and at last a man told me some one on the Tum Lum River had taken up some work cattle that had broke into his field and destroyed some corn. So I went to see and found they were my cattle, and the man (Biglow by name, next to the Walter Davis place) claimed $7 or $8 damage, and I found my big stag, the one I got from Beckney and Amberson, was a breachy rascal, and would break down or jump fences, and he let my other three into the field, and I had to pay $5 to Biglow, and took my cattle home. But whenever they got the chance, the stag would lead them off into troubles.

So I told Charley and he said he knew some men that wanted to buy some work cattle, and he saw the men, John Ding and his partner. They saw my cattle and would buy, but had not the money to pay right down. But they had a lot of barley to sell to Myers the brewer, and to Joe Hellmuth (Lager Beer Joe) and I could get the $200 when they sold the barley, and Charley keeping the store for Myers, he would see that I got the money. So they paid me some $25 down and were to pay $225 more and they took the four head of cattle and hauled freight from Wallula to Walla Walla.

A short time before my money was due they claimed that one of the oxen was lame and wanted me to make a reduction of $50 on the balance due. They had used him four or six weeks and had said nothing about his lameness until then. So I would make no reduction and they would not pay, so I attached some of their property for the debt, and I had to give security for the cost. I got Mr. Em[i]l Myers to go my security. The trial would come off before Judge John Kelly, a Mason. The day the trial came I told Mr. Myers; he told me to see Judge Kelly and tell him he wished to see him before the trial. (Myers was a Mason and my brother Charley had made application to join the same lodge.) So when Judge Kelly came, Myers told about the case and how he was my security, and that my brother was about to join their lodge, and he wanted Judge Kelly to do his best for us (or me). We had a round of lager beer and the Judge said all right, and some time shortly after, during the forenoon, the Judge got into a little poker game in Myers' saloon and borrowed a $5 note from

Charley, and in the afternoon when the trial was called the opposing council tried to dissolve the attachment, but the judge overruled the motion. They then tried to get time for a new filing of an answer, but their council delayed some way, and the Judge ruled in my favor for the judgment and costs. I had paid my lawyer in gold (his name was Lassater) in advance, and when John Ding and partner came to pay my judgment and cost they paid in legal tender (or greenbacked me), by which I lost over $100, for they were only 60 or 62½¢ per dollar, but it was legal tender and I could not help myself, but I was to have had my pay in gold.

That Fall I got an emigrant named John Davis, just acrost the plains, from Iowa, to harvest and help me thresh out my grain, do some work and plow for me. He had not been long married and had no children (she was but 14 years old when married). They lived in my cabin, and I got into a job driving a 6-yoke of cattle team to haul lumber from the old Mullen Mill (then called Lincoln & Brush), from the mountains some 28 or 30 miles to the yard in Walla Walla. I got $80 a month and had my provision furnished, but I had to cook and camp out most of the time, and in rough weather it was quite hard and disagreeable. The team belonged to Phillip Scheible, a milkman running the City Dairy. I liked the work and drove some three or four months.

During the winter John Davis moved to a place on Mill Creek, and my brother sold out his house and lots in town to Charley Johnson, the Postmaster, who had married Taylor's daughter (that kept the City Hotel). Charley had quit keeping saloon for Myers and wanted to go to farming, and wanted to buy a half interest in my place (farm and claim). I sold it to him and he paid me some down, and I was to wait for the balance. He moved out early in the Spring. He had a span of brown mares and a couple of colts, two or three cows and chickens and hogs, wagon, harness, and farming implements. He moved down the new log building that I had put up at the crossing of the road, and not quite completed, to where I had put up my cabin at my old location, as it was the best place, and he put up the new house again and added to it some side rooms and porch and made it one and a half storys high. He got some lumber and built stables and barn, chicken houses and granary. . . .

# 11

## HAULING TO THE BOISE BASIN

(1864–1866)

About the first of April I took sick with a sort of lung fever, and had to lay up a week or ten days. I stopped in a small house of Phillip Scheible's and took medicines and got some better.

When Scheible got two loads of freight to haul to Idaho City (Boise Basin or Mines), about 300 miles from Walla Walla, he had another team of five yoke of oxen, together with the one I drove, of six yoke. When he got the loading he found he could take some more, and he bought some for himself (groceries and flour). He asked me if I was able to drive my team, or would I have to stay? I told him I thought I would go and drive my team, and I got me two or three boxes of liver pills to take along, and we loaded up one day. I had to put the loading in good shape. We loaded in 11,000 pounds, 5,000 for the 5-yoke team and 6,000 for the 6-yoke team which I drove. When we got all loaded up I was about give out, I was so tired I could scarcely move, and we would have to cross the Blue Mountains in deep snow at the Umatilla Reserve crossing, called the Meachim or Emigrant Road.[1] Several men advised me not to go, that I could not stand the snow and exposure, and drive cattle with my lungs at that time, that it would kill me before I could get acrost the mountains, for the roads were awful rough at that time of year, and the snow soft and melting.

But in the morning I felt better, and so we struck out with our two teams. A young Englishman drove the 5-yoke team, and Phillip, the owner, was along on horseback to help us along. I done the cooking and George, the Englishman, the getting up the cattle, and Phillip on his horse

---

[1] *The Oregonian*, June 23, 24, 1863, describes the traffic over this route.

to look out for bad places, and best roads, and at night good camping places. We got to Wild Horse Creek the first day out, and to the Umatilla River in two more. We forded the river near the Indian Reservation. We had to double at the crossing, for it was high and rocky and steep banks, and eleven yoke had all they could do to take my heavy wagon over, and out of the river banks.

On the fourth day we commenced to go up the Blue Mountains. We fell in with some other teams that were crossing, and we all had a fearfull time, such rough roads and hard pulls. Sometimes we would hitch on 15 or 20 yoke of big oxen to one wagon. We had to sleep on the ground or snow under the wagons for shelter, for the insides of the wagons were full of goods to the tops of the bows, and the snow and rain kept us continually wet, and our blankets were soaking wet, but we would dry some to put next to our bodies by the campfires at night after we stopped. I used to cook a large camp kettle full of beans and bacon every night, have hot bread, coffee or tea, sometimes fresh meat, and as sick as I was when starting I had a good appetite all the time on such food, and I took three of my liver pills every night, and with all the exposure and loud hallowing at the cattle, I was improving in health.

Our trip was as bad as could be, and we had our wagons upset in the snow many times, and had to load and unload often, but it came to an end at last, and we got to the town of La Grande, in the head of Grand[e] Ronde Valley, in about 12 or 14 days from Walla Walla. We laid by part of a day to repair some breaks on the wagons and chains, and we dried our blankets, changed clothes, and bought some such things as were needed for our eating.

This is one of the most beautiful valleys in Oregon, but was already well settled and good improvements made. La Grande was a lively little place of about 700 or 800 inhabitants. Next day we started on and had to ferry the Grand[e] [Ronde] River again; we had already crossed it once or twice in the mountains, once ferrying. It took us several days to get through the valley to Pyle's Cove and Canyon, where we had some heavy pulls and high hills to double over.[2]

We passed a fine farm belonging to James Hendershott, who used to be our sheriff in Kerbyville. He was located in Grand[e] Ronde Valley at a point next to a valley called the Cove, where there were three Brothers Boicy, whom I used to know at Indian Creek, Calif. and Waldo (Sailor Diggings) in Josephine County, Oregon. They ran a pack train to the Boise mines. Hendershott kept a public house and farmed some. His brother Sidney Hendershott had been our Representative at Kerbyville.[3]

[2] Pyle Canyon, named for James M. Pyle, is situated south of Union.
[3] The Hendershott brothers were born in Illinois. They came west to Oregon from Iowa in 1853 as gold seekers, settling in Josephine County. Samuel B. Hendershott was a delegate to the 1857 state constitutional convention. After moving to Union County, James served as state senator in 1868 and 1870.

We next got to Powder River Valley, another fine body of land, and had to ferry the river. Next we struck Burnt River Valley and passed several hotels; Straw Ranch was one, then the Express Ranch, then the California, then the Eagle; all these hotels done a good business off of the travel going to the Boise or Idaho mines. Miller's was the last on Burnt River. I knew him at Walla Walla; he kept the stage station here and made a good thing; meals at $1, and stabling and feed very high.

Next day we got to Snake River and went up a few miles to Old's Ferry, who done the best business on the river, and he had all he could do to take all the teams and pack trains over, and $400 to $500 a day was not an extra thing for him to take in. His ferry sold for $5000 for one third interest, the first year of the rush to the Idaho mines. He, Rubin Olds, had the only charter for a crossing on the road. I was well acquainted with Rube Olds on Althouse Creek, Oregon.

We crossed our teams over Snake River and kept on up to the We[i]ser River; here we had to ferry that river, and it took time and money too. We had to swim all the cattle, but one yoke to each wagon, and some cattle do not like to take to the water, and we always lost the day in ferrying and in many places we had to wait, for there were some other teams ahead, and all have to take their turns. Sometimes teams had to lay over three or four days on that account, or the river [was] too high to ferry with safety.

We were several days from We[i]ser to the Payette River, and had to lay by a couple of days to cross. Then we took up the Payette River Valley, for three or four days. Ranches, hotels and farms on both sides, and the roads lined with wagons and Pack Trains and large mule and ox trains, all loaded or empty, going or coming to the great Boise mines.

When we struck the Horseshoe Mountains, and then the Boicy (Boise) Mountains, we had our hardest pulls, but the roads were clear of snow, and not so muddy as we had in the Blue Mountains.

We found some old acquaintances on the Payette above Horseshoe Bend; one named John Flanagan, and one Bill Dunn, had farms and doing well. They were from Indian Creek, California. In about a half mile after we got over the Horseshoe Mountains, we came down on a creek and a ranch and hotel kept by Shaffer (who was with the volunteers at Gold Beach, and at the ball when the Tootootenay Indians killed their guard, above Prattsville, in the massacre of 1855, when they killed Ben Wright and the 32 settlers and burnt up my Pacific Ranch).[4] He was doing well here.

It was but 14 miles to Placerville, but all uphill. Some six miles from Placerville there was a hotel kept by a German from Walla Walla; his name was Herzog; he was a barber and dentist when at Walla Walla

[4] See Chapter V, *note 2, ante.*

where I knew him. He was doing well. All these hotels and ranches kept liquors and meals, and had no license to pay and sold their drinks for 25 cts, and meals for $1—everything very high.

We got to Placerville and found quite a large mining town and very lively. Found many I knew, and Kroney, my partner from Victoria to Kerbyville, was there mining and butchering. Lager Beer Joe (Joe Hellmuth) of Walla Walla had a brewery and saloon and doing well.

We camped close to town on the creek and I went uptown and I was around the saloons seeing them gamble, and I had a few half dollars in change, and just to pass the time away I bet a little at Spanish Monte. I did not bet over one dollar at a time and when I had won about 18 or 20 dollars went back to camp.

Next day we passed through Centerville, another nice town, and found that Wm. Tincture and Ed Hutch[in]son kept a saloon, the Magnolia.[5] Ed and Bill were both violinists and used to play for us at Indian Creek, and I saw Ed at Victoria [British Columbia] last, on his way back to Oregon and California. He was glad to see me. A stream called Hogem passed close to town, and there was a town called Pioneer City some 9 or 10 miles above Centerville.[6] We had to go through toll roads from Shaffer Canyon to Placerville, then again to Centerville, then again to Idaho City, and every toll road charged from $3 to $5 per wagon.

We at last got to Idaho City, to where our goods were to be delivered, and we laid there a day or two until Phillip Scheible could sell out his goods. Idaho City was larger than either Placerville or Centerville, and full of saloons and gambling houses, dance houses of evil repute, and hurdy-gurdys of fair repute. The mines were rich and money plenty and the miners gave it a chance. Wages were high, and the workingmen spent their money drinking, gambling and dancing.

After we unloaded, Phillip Scheible did not get his money right down and had to wait, and he told us we might take the teams and start for home, and he would come on and over-take us, as he had his horse to keep with him. I found another old miner from Sucker Creek, named Leeman; he was keeping a large wholesale liquor store and saloon, and he got some of the loading we had. His partner was named Sallaro (French or Italian). Robert Leeman was glad to see me. He was from

---

[5] Reinhart is possibly mistaken. Vardis Fisher, *Idaho: A Guide in Word and Picture*, p. 385, states: "[Placerville] was built around a square which, strangely enough, was called the Plaza, and its chief saloon, the building of which still stands, went under the fragrant name of Magnolia." Most of the mining towns mentioned by Reinhart are now ghost towns, at least those spared by forest fires.

[6] Pioneer City, sometimes called Fort Hayes, much later Pioneerville, was the first mining camp on Grimes' Creek, "but owing to the selfishness of the original discoverers, it received from those who arrived subsequently the euphonious appellation of Hog'em." Hubert H. Bancroft, *History of Washington, Idaho, and Montana*, p. 261. All three mining towns mentioned are within a radius of less than ten miles and lie, triangularly, a short distance northwest of Idaho City.

Philadelphia, Pennsylvania.

So we left Idaho City and started back to Walla Walla. We had a good time on the road. We got butter and buttermilk at one dollar per gallon, and we feasted on the milk and butter at one dollar a pound. We got back to Walla Walla, having been gone nearly three months. We turned the cattle out to grass and rest up until Scheible should return, expecting him every day.

I had nothing to do, so had to lay around until I should hear or see about Phillip Scheible. He had let or rented out his cows and dairy, and his farm, to Havard Brothers. I was around the saloons a good deal and saw lots of gambling going on. I tried to get some hauling to do for the merchants from Wallula, but the farmers' teams were at leisure, and they could pay them in goods, and turn in old accounts, so I did not get any loading. I did not know if it would suit Phillip Scheible for me to engage a load for the teams on a long trip, so let them rest until he should come home. William Woosley, whom I knew at Wolf Creek, had been to the Salmon Mines and stopped in Walla Walla a while and went down to the Willamette Valley to stay a while; shortly after I heard he took sick and died.

While I was at Myers Brewery, keeping saloon, Abel George passed through Walla Walla on his way to the Salmon River Mines; he looked bad, and very much dissipated. I heard that the Spring of 1862 he and his family were living on a farm in the Willamette Valley and the River raised very high, so that he took his two children in a boat to a tree, and tied them till he went back for his wife, and while he was gone the water raised about the tree and drowned both, one a girl, the other a boy. (I had seen them both when the trial of Abel George for the murder of Hough McCaslin was going on at Kerbyville.) When Abel George and his wife got to the tree and found both dead, his wife became frantic and I heard she only lived a short time after. (Here was retribution for his misdeeds, and the innocent had to suffer for it.) He went up to Salmon River to the mines, and had some difficulty with a man named Moses Milner from Corvallis, and Mose Milner shot Abel George dead, and he got clear by Abel George being in the wrong. So that was the end of him who at one time was a good and peaceable man, but let liquor get away with him—a sad warning to many.

While I am here I will relate of another man I knew on Sucker and Bowling Creek, a miner and gambler named Crossroad Jack, or by right Jack Elliot from Boston. He went up to the Boise mines and one day, while drunk, shot a Jew storekeeper about some clothes he was about to buy. It was considered a cold-blooded murder—he shot him acrost the street with a shotgun, and the miners just took Jack Elliot and hung him up until he was dead. He used to be a sailor, and while drunk was overbearing and brutal.

I came very near forgetting to relate the balance of the life of Cherokee Bob, or Robert Talbot. After he made his escape from Walla Walla he went to Lewiston, a short distance above the Indian Agency and military post at Lapwa[i], to where the soldiers had given up the chase after him, and he gambled around Lewiston and ran a saloon and boarding house called the Lunar House with some woman of doubtful reputation. During the following winter he went up to Florence on Salmon River to a Grand Masonic Ball. He brought the woman I spoke of as a partner, but the committee of arrangements would not let him take his partner in. So he threatened to kill the doorkeeper and the floor manager, Jakey Williams, who used to be at Walla Walla, and Bob Talbot had another desperado with him named Red Face Bill or Poker Bill (I do not remember now), and they both were determined to take the woman in to the ballroom and dance, in spite of the whole of them. So Jakey Williams and some of the men armed themselfs and when Bob Talbot and Red Face Bill came to force their way in with cocked revolver, they all commenced to shoot, and both Bob Talbot and Bill were shot to death on the spot. Some others were hurt, but none bad. Jake Williams gave himself up to the authorities and was cleared (Talbot and Bill had commenced the shooting). So John B. Latzenheiser and Hubbard were avenged.[7]

When Phillip Scheible came back from Boise Basin he was not in good humor; he had got to drinking and had spent most all his money, and after he got back he was abusive to his wife (who had shortly before been sent home cured from a lunatic asylum, where Phillip, her husband, had placed her), and he would get drunk, and while drunk would vent his abuse on her until she could not live with him with any safety, and she stopped with Mr. Havard, who had the milk dairy and farm rented.

I asked him to settle up with me, as I wanted to go into something else for myself, and he said I would have to wait for my wages, something over $300. I told him he would have to get it for me. But he would not, and he was very much in debt, so . . . I commenced suit for my wages, and after a while I had to take legal tender for the whole amount, which should have been in gold. I lost over $150 in expense discount and fees, so I had not much over $150 left, for I sold my greenbacks for about 60 cents on the dollar. . . .

Some time the middle of July Charley said he believed if I would take some 10 or 12 pigs of his, and buy some more so as to make 30, a wagon-load, and take them to the Boise Basin, and sell them one or two as I went along that they would sell well. So I looked around and found enough to

---

[7] The account given here is accurate. Bancroft, *Washington, Idaho, and Montana*, p. 452, *note*. Although Bancroft spells the last name Talbotte, most accounts agree that the name was Henry Talbot. Nathaniel P. Langford, *Vigilante Days and Ways*, I, 151–154. Talbot's partner was William Willowby. The ejector was Jacob D. Williams; Cynthia, the mistress.

make up the 30; they would weigh from 14 to 25 pounds apiece; I paid $2.50 apiece. I took Charley's light wagon, put on a sort of a double deck pen with small troughs running the length of the coop, put in one cross partition below and one above, took his brown mares (his work was done), and loaded in some wheat and shorts or middlings for the pigs and a few sacks of oats and barley for the horses, and my provisions and blankets, in all about 20 hundred pounds, but it would be getting lighter every day, the horse feed and pigs, too.

So I started July 23, 1864, and Charley helped me over the worst part of the mountains on the new Thomas Road.[8] He put on the third horse in bad steep places and when I got to the top, he had his saddle along and he got on and went back home. Whenever I got to a creek I would pour in some water for the pigs to drink, then over them, and wash out and cool them off. I drove 30 miles a day and made the fast driving on good roads.

When I got to Snake River I met the emigrants coming in from acrost the plains, and they were awful well pleased to see my nice pigs and some girls said they wished I would give them one for a pet. They asked what they were worth; I told them $30 apiece for the choice. They were astonished and thought they should not be over a dollar or two anyway, that back in Indiana they could get them for 50 cents apiece.

At Farewell Bend on Snake River I sold one average pig for $20, about one dollar a pound, and I next sold two for $40, or about the same as the other. On the Payette River I sold two for $50, choice of the lot in pairs of one of each sex. At Thompson's on Payette Valley I sold the next pair at $45. At Barnett's, near Horseshoe Bend I sold one pair at $42.50, and the balance I took into the Boise Basin, to Placerville, and was about to retail them in pairs when a German young man working for a butcher offered to buy the whole lot, 21 head, if I would put them down reasonable. So I offered them to him at $20 apiece in gold coin, and he made me an offer of $18.50 in gold coin, and I sold them to him. But when he came to try and get the gold coin, the men wanted to blow his dust too much, and would only let him have $16 per ounce for his dust when it generally passed for $18 per ounce. He came and told me he would have to let them have his gold dust at the $16, and if I would take it at that he would rather I got it (for they would have thoroughly cleaned and blowed it for him, so he would have lost if I had not taken it). So I took it at $16 per ounce and I got 24½ ounces of nice Placerville (Grimes Creek) gold, and I got $18 per ounce for it at Walla Walla, which gave me $48 more. I had done uncommon well on the pigs. Some of the 21 would not have weighed 13 pounds, or nearly $1.50 per pound. You may think that a large price, but

---

[8] The Thomas Road was built by George F. Thomas and his partner, J. S. Ruckle, to accommodate their stage line which ran from Walla Walla over the Blue Mountains. Bancroft states that the road did not open until April, 1865. *Washington, Idaho, and Montana*, p. 423, *note*.

the same man, Fritz, afterwards told me he made more money off of them pigs than he ever made on that amount of money, $388, for he had nothing to pay for their feed, for they lived on the offal of the slaughterhouse, and he had barrels at the hotels and got their swill for nothing only his trouble of hauling it to his pigs. He raised them to weigh from 200 to 300 pounds apiece before he butchered them, and then they were worth $60 per hundred pounds, which brought him $150 to $180 apiece in less than six or eight months.

As soon as I sold out I put a notice on my wagon "Passengers Wanted for Walla Walla," and I got four passengers at from $22.50 to $25 apiece and they were to board themselfs, and none had but little baggage or bundles of blankets and few clothes. So I took off my coop and washed out my wagon, and next morning started back for home.

My passengers were going home, two to the East, Boston and New York; the first named Gray, and the last Pat Graham—he was going right to Brooklyn or Hoboken, where his wife and family were. He used to be a gardener for Robert L. Stevens, who built the Iron Ram at New York or Brooklyn.[9] He had made several thousand dollars in a short time on Grimes Creek near Placerville, and sold out. Mr. Gray had been his partner or neighbor, and had done well, too. He was quite an old man. The other two; one stayed at Walla Walla. We had a pleasant trip to my ranch, and we stayed over night with Charley and next morning I took my passengers to Walla Walla, where the two took the stage for Wallula.

I had made my round trip in less than 22 days, and after deducting the price of my pigs and all my expense, had cleared $550, or $25 per day going and coming, and the team was not over-drove.

I stayed in Walla Walla overnight and next morning bought two ponies to work in harness and on a light wagon, and took some provisions and something for Charley's two children, and went out to our place. Being as I had done so well, we prepared to buy up pigs and sows so as to take two wagonloads, and start as soon as we could get ready. I bought another light wagon, harness, and made inquiry for hogs. We could not find any small pigs, so concluded to take some from 60 to 80 pounds, and make up the two loads with some three large sows for breeding. I bought another pony, a roan, for $55, and we got our loading all ready and were about to start when Charley Kraft, a stable keeper in Walla Walla, persuaded Charley to buy a big grey horse. I did not like him, but Charley thought he would work, and would pay for himself on the trip. We only had to pay $65 for him, and Charley could leave one of his mares at home, or one of the ponies....

[9] Robert L. Stevens was a distinguished engineer, naval architect, and inventor. He died in 1856. His efforts in designing an ironclad ship led to the launching of the *Merrimack* and *Moniter*, famed for their naval encounter during the Civil War. James P. Baxter, *The Introduction of the Ironclad Warship*, pp. 224, 228.

[We] loaded up and started, so as to get through to Placerville and sell out and get out before the cold or bad weather would set in for the winter. We drove quite fast . . . and we were heavy loaded. At the Express Ranch on Burnt River there was an election, and my brother was anxious to vote for McClellan for President, and he talked me into voting the same way, for I had voted for Abraham Lincoln in 1860 in the Willamette Valley. Several teamsters were away from home and thought they could vote at any polls, no matter where, but when we offered to vote, the judges at last refused to let us vote, and I was mighty glad afterward. I reflected that I did not get in a vote for Little McClellan, or I would have been forever sorry for it. . . .

By hard driving we got into Placerville. The butchers heard my brother say he was in a hurry to get home, so concluded to hold off and let on that they did not want to buy, and at last I found a man named Moore who made me an offer of one dollar a pound for all the smaller ones, but did not want the larger ones at all. Now there were about 700 pounds of the small ones, and the three larger ones would have brought 65 cents a pound, which would have come to over $1,000 for the lot. I was in favor of taking the one dollar a pound but Charley would not sell unless all could sell together, and the butchers told Moore not to buy them at what he had offered, and so he would not take them at a dollar a pound, and Charley got discouraged and was homesick, and Moore and the butchers knew it.

So I told Charley to get ready to start for home, and I went to a butcher and hired the privilege of keeping my pigs and sows in his slaughterhouse yard for a short time, until I could go to Centerville or Hogem (Pioneer City) to sell my hogs to some of them. I unloaded Charley's wagon and got him ready to start home at night. He was to leave early in the morning; I was to stay and sell all together or retail them out as best I could do. So when the butchers found they could not scare me into selling at their price, they came and made an offer, but I told them if they wanted to buy they would have to do so that night before my brother left, or I would not take the amount I offered to take, and when Moore found out the butchers were after them, he come and made me the offer for the small ones, and I took it, and got $700 for them. We sold the three large ones for $400, got our money, and I got ready to leave when Charley did.

I took my wagon apart and put it in Charley's, and he took the two mares, and I the grey and sorrel, took a set of harness, and we traveled together eight or ten miles. I had some $400 with me and Charley had the balance. I was to go by a cross trail to Boise City, about 30 miles off, where the emigrants come in off the plains, and see if I could buy a good new wagon cheap, and then drive on and overhaul Charley with his team and two wagons on the Payette River, some 60 miles ahead, on our road to Walla Walla. I had along a new self-cocking English navy size revolver; there had been a case sent by Wells Fargo & Co. express to Walla Walla by some English firm, and the consignee failed to get them and they were sold at auction. I bought one for $3.50 or $4 and I kept it loaded with me. My ride was awful lonesome and long and I was afraid I would not find Boise City. I got so tired riding on a blanket and no saddle. I got in in the night and could see the lights for seven or eight miles—I thought I would never get there. I was sore and had to ride slow, and it must have been eight or nine o'clock at night when I got to the city. I went to a feed stable and put up my horses, and went to the Overland Hotel, the best in town. I knew the clerk at Walla Walla. I got supper and went to bed soon after. In the morning I went to the stable, and found the gray horse quite sick. He had been sort of droopy for a few days, and he was quite sore from his day's ride, and he had no appetite for his feed. I did not like that, for how was I to haul a wagon with him and his mate and catch up to Charley, who had 45 or 50 miles the start of me? I looked around town for a wagon to buy, and found a man who had been in but a few days off the Plains, with a large span of black horses, about 15½ or 16 hands high, five and six-year-olds, quite poor, and a good wagon and harness, bows, cover, and all. He was on the sell and asked $400 for the outfit. He did not want to sell the wagon separately, and I could not find a wagon to sell that would suit me. I liked the team, but it was poor, and I had a drive of over 300 miles to make, and more than likely I would be caught in the snow before I could get across the Blue Mountains, it being late in December. And my sick horse could not pull a wagon. So at last I traded the man my English self-cocking revolver for $65 and paid him $335 in gold, and got ready to leave that same evening.

I found several old friends in Boise City: James Buckley, our used-to-be-sheriff at Walla Walla, was keeping a saloon, and one Samuel Driggs from Corvallis was keeping a livery stable with James Agnew, whom I knew at Walla Walla, and Cyrus Jacobs, brother of Richard Jacobs, was running a large grocery and general store.

When I started, a young man who had no money wanted to go with me 80 or 100 miles, so I took him in for company, tied my two horses behind our wagon and drove out to the Junction House on the Boise River. My gray got to running at the nose, so I knew he had a distemper, and did not travel free. At the Junction House they kept a store and whiskey, and either they were Vigilantes or Horse Thieves, one or the other, and I was suspicious of the outfit. They played cards and drank whiskey and spoke in strange ways of hunting and keeping stock or herds at the Great Falls of Snake River, but I got my supper and horses kept all right, and I paid my bill and started early to make the best time to catch up to Brother Charley as soon as I could, and on the third day got to the ranch on the Payette called the Hay Press Ranch, and I found Charley had been waiting for me to come up with him. He was surprised to see me with a team and wagon and would not believe I had bought it, and thought it belonged to the man with me, and he did not like it that I had bought and paid $400 for it. . . .

We had some cold rains and sleet and we kept the gray blanketed, and when we got to Burnt River he was nearly give out. We could not stop for it was snowing in the mountains; we had to cross to get home. One night we stopped in Powder River Valley and the old gray lay down and died.

We pushed on as fast as possible and got to the foot of the mountains and found the snow falling fast and deep. We stayed overnight at old Lazarus Wright's who used to keep store, postoffice, hotel and mill at Myrtle Creek, Douglas County, Oregon, not far from Burnetts where we kept hotel and postoffice in 1859. He had got broke up and come to Grand Ronde Valley and taken up a new claim. His boys and girls were running around in the snow barefooted playing at daylight when we men were glad to be near a good log fire in the fire place. . . .

At last we got home, the third day of January (1865) having had a long hard trip. Our loading sold for $1,100; our expense and cost, $400; lost one horse, died, valued $125; got back clear on this trip, $575. Bought the horses and wagon and harness at $360 at Boise City, and lost two pigs; so that on the two hog trips we cleared $1,125.00.

I got two light wagonloads of flour to freight to Boise City, to Cyrus Jacobs, 4,000 pounds at $9 per hundred pounds. I got a man named Pilcher to drive one team. My expense was $60, Pilcher's wages $40; cleared on the trip with the four horses, $259.34. (Third trip to Idaho, May 26, 1865.)

(Fourth trip to Idaho, September 9, 1865). I bought up 26 pigs for $85, and 232 chickens for $98, took two light wagons and four horses, and left the black team to work to farm. Sold the pigs for $325, or $12.50 apiece. The 232 chickens brought $417, many of which brought $2.50 apiece, and I got one passenger back as far as Straw Ranch on Burnt River, making $751 for the loading and $354 for expense, left $396 clear on the trip.

I took one span horses and a light wagon, some 15 pigs, and 1,000 pounds of barley; some of the barley I sold on the road for feed, at 10 cents per pound, and the balance of the barley and the pigs I took into the Boise Basin, passed through Placerville and went to Hogem Creek, Pioneer City, and sold my barley to a brewer keeping a saloon, and he took the pigs, too. I cleared about $300 on the trip.

While in Pioneer City I came acrost John C. Beckney, who was keeping a gambling saloon and hurdy gurdy dance house at Eldorado. He had got up a lottery scheme for some houses and his saloon and a lot of jewelry, the drawing to take place at Idaho City, where some more of the prizes were. He was glad to see me and wanted me to buy some of the tickets for his lottery, the drawing to take place on Christmas Day. I did not take any. I asked him about the two other boys; he said they were mining somewhere. So I got two passengers to Walla Walla and got home all right, and I had a pleasant and profitable trip.

When I got back from my trip with the pigs and chickens in September, I was in favor of taking the black horses and two others I had bought at Walla Walla and make up three wagons and teams and get some freight to the Boise Mines or to Owyhee, and make it the last trip, and then let our horses rest at home during the coming winter. Freight was always higher late in the Fall and merchants laid in their winter's supplies as late as they could, but would always have to pay a higher price. But my brother had been farming all summer and he had got through, and the big black horses were in splendid order; he was anxious to make the trip, for I had been having a good time and had good luck, and we had eight horses to work. So he proposed for me to let our horses rest until the last days of October, take three wagons and the eight horses and go to Umatilla Landing[10] and load with high freight, make the one trip, and get back in 30 or 40 days in time to escape the storms and be home for the holidays with plenty of money to spend, and after I first did not like to take the chances, I agreed to do so.

Now I will go back to something that happened during the Spring, which I did not relate before: Some time in the Spring of 1865 the Wells Fargo & Company stage was robbed by some masked highwaymen in the Burnt River Canyon, some 8 or 10 miles above the Straw Ranch. From the Powder River side of the canyon it was but one or two miles from the stage station. The coach had on a 4-horse team when the driver was hailed by the masked robbers, one of which had the near lead horse by the bit. The stage driver, Little Jack, whipped up the horses and undertook to run the robber down. The lead horse plunged forward and came near knocking the robber down, but he turned his revolver and shot the horse in the flank, and the horse fell over dead. The robbers covered the driver with their pistols and he had to keep quiet, and two of the robbers went into the coach and robbed the passengers, one a Jew merchant of $8,000 in gold, and several with small amounts, then left the driver and passengers to hitch up two horses and go back to the last station, where they had just before changed horses, and got a fourth horse for the one the robbers had killed, and went on. There was great excitement about the robbery, and Wells Fargo & Co. offered large rewards for the robbers, but nothing was heard or seen for a long time.

Some time in July some gamblers came down from the mines, some from Boise, some from Owyhee, where some new rich discoveries had been made and several cities built up, Ruby City and some others. There

---

[10] An important Columbia River depot for river traffic for the Boise Mines, situated at the mouth of the Umatilla where it joins the Columbia. It was laid out as a town site in 1863. *The Oregonian,* May 16, 1863.

had been rich diggings on the South fork of the Boise and other camps. One among the gamblers was named James LaMar; he claimed to be a son of General LaMar of Missouri or Arkansas.[11] He was a rough looking desperado, and drank considerable, but somehow the gamblers did not go with him much, and kept clear of him. He made several winnings at short cards, poker, and beat some monte bank, and he had lots of money in his pockets.

One afternoon some gambler asked him to take a walk with him; he had some plan of winning some money and he wanted to talk to Jim LaMar alone. They started east past Myers' Brewery and walked out to a creek, and the young man asked LaMar to go acrost the creek with him, but somehow LaMar began to suspect something wrong, and said he did not care to cross on a fence rail, and turned to go back towards the road toward Walla Walla. As he turned several shots were fired from the bushes and LaMar fell, and there were some ten or twelve shots fired with shotguns and revolvers, and he was left for dead.

Some men who saw the firing at a distance came in to town, and the City Marshal and a lot of men went out to where the men had heard the shooting, and found Jim LaMar dead and a rope laying on the ground with a noose on it, and several boot and shoe tracks in the brush and on the edge of the creek, where they had jumped over from bank to bank. The body was brought in and taken to the police court and a coroner's jury was called to make examination and take evidence. I was one of the jurors.

The body was searched and twenty or twenty-three $20 gold pieces were found in his overcoat outside pockets loose, and a revolver and knife in his belt. He was shot in a dozen places, in the head, breast, back, side and one arm, and in the temple. Several witnesses were called. They saw the man LaMar and another walk out East, and Myers and some of his brewers said they saw several parties of two and three together with shotguns, as if they were going hunting, pass by the brewery, and were seen near the creek where the dead man was found. None would testify as to who the different parties were, but no doubt they knew but were afraid to tell.

It was privately known that Jack the stage driver had come to Walla Walla for the Company and he happened to see LaMar and he said he was the robber who had caught the lead horse by the bit, and when he struck the horses to run over him it had knocked his mask off and he saw that was the man. The stage agent, old George Thomas, and the Vigilante Committee done the work.[12] They expected to get LaMar acrost the creek out of sight, throw the rope over his head and try and make

---

[11] I have been unable to identify "General LaMar."

[12] George F. Thomas, with his partner, J. S. Ruckle, operated the stages between the Boise Mines and Walla Walla. (See *note* 8, *ante*.)

*Hauling to the Boise Basin*　　　　　　　　　　　　　　　　　　　231

him confess and tell on the others in the robbery with him, but when he turned back and would not go acrost the creek they were afraid he would get away, so shot him to death right there, and no one touched the body until the Marshal and posses from town got to him, and all his effects were still on his person. The Vigilantes were in great force around Walla Walla, and I expect while I lived there, in four years there were 10 or 11 men hung and shot by them, and it was a good thing, or it would not have been very safe for honest people or their property.

About 30 days after the death of Jim LaMar, the Sheriff of the County sold a mare, saddle, bridle, spurs and saddle blanket as his effects at a public sale. The mare was an iron grey, well and stout built, and I bought the whole outfit for $68.50. The auctioneer said the mare had been rode from 75 to 80 miles per day, and she was very tough. I bought a mare to match her, the same color and size, from a tinsmith in Walla Walla for $73, and they made a fine team.

So about the 26th of October (1865) we took our eight horses and three wagons and went to Wallula to see if we could get some freight. We could not, so we kept down the Columbia River to Umatilla Landing. It was awful bad road, but we saved over 60 miles by not going around by Walla Walla. We got to Umatilla, quite a little business place, and inquired for loading. We found some for the Owyhee, Ruby City, and some for Idaho City, Boise Basin. I had never been in Owyhee, so thought it best to load for Idaho City. Leeman and Sallaro, liquor dealers, had one lot of liquors in barrels and kegs and cases, and some bales of clothing for a Jew firm named Emmanuel, and some more liquors for Gans Brothers, all of Idaho City, at 15 cents per pound, in gold bars, which was better than gold dust but not as good as gold coin.

So we got our three wagons loaded at Powell's and Company, and we took 6,800 pounds at 15 cents per pound in gold bars. We got our provision and feed for our horses and started November 3rd. From Umatilla Landing we had 12 miles of heavy sand to pull through to commence on, and it took us all one day to go over it. The second night we stopped on the Umatilla River, close to James Daniels' place, and he came out to our camp to see us. I had not seen him since I saw him at Fort Hope or Langley on Fraser River in 1858, when he started for South America. He took a fancy to my grey mares, one of which was the LaMar mare, and told us he had a large fine pair of bay horses about 16 hands high and in good order, but he would give us a trade for our two mares, as he wanted to raise horses and so the mares would suit him best. We were very heavy loaded, and his two large horses would answer our purpose better than the mares, so we made a trade....

We kept having [so many] delays on account of bad roads [and trouble with the wagons] that we did not get to the Horseshoe Bend on the Payette River until the 9th of December, when we should have been home again by that time. So, the morning of the 9th of December we had to commence and double up our teams and take our wagons up the Horseshoe or Shaffer Hill; it took us until after dark to get the three to the top and commence to go down the hill. We had on rough locks and chains to go down the rough rocky steep road, and dark as could be. In one place the chain on the rough lock broke and the wheel horse, Black Jim, held back so hard that he broke his breast chain and he fell under the front wheel, which held the wagon. Charley got his foot under the wheel, too, somehow, and called for me to find a rock to chock the wheel with, and I had a hard time to find one in the dark, frozen ground, but at last found one and propped up the wheel and got Charley's foot out. We unhitched the horses from the wagon and took all the horses and some blankets out of the wagons, and left them stand until morning when we could see, for we had no lantern with us, and we went on to Shaffer's Hotel and put up our horses, unharnessed and fed them, and got some supper and soon after made down our beds.

It looked like storming and when we got up in the morning found that it had snowed about a foot deep, and still snowing. We fed and got breakfast and took two horses to get down the wagon. When we got up to where we had left the wagon with the broken lock chain, we found that if we had gone ten feet further out of the road the wagon and horses would have been thrown down a steep rocky point over 100 yards below, and would have killed the horses and broke up the wagon and destroyed the loading, all liquors in barrels, and champagne wine in baskets and boxes, worth four or five thousand dollars. But the brake had caught the shoe as it came around, and the horse no doubt could see the precipice and threw himself back with the force to break the breast chain, and he falling under the wheel and axle tree, stopped the force of the wagon. We had to get more of our horses and pull the wagon back into the road, and ever since I have been against traveling with teams at night. By hard work we got all our wagons to the Shaffer Hotel, and it kept on snowing.

The mail driver, named Strickland, brought us a letter from Leeman & Sallaro, who owned the most of our freight. He expected that the letter would meet us some 75 or 80 miles back. It said if we wished we might take our loading to Boise City and leave it, and they would allow us 12 cents per pound. We were sorry that we were not back 80 miles and we could have gone into Boise City without crossing any mountains or snow, but now to go back was nearly impossible.

The Shaffer Hill we had just crossed was steepest on this side, and it was as much as a team wanted to do to pull up an empty wagon. So we had to do the best we could. It was but 38 or 40 miles from Shaffer's to Idaho City, but it was all mountains and deep snow, so we concluded to let my brother take one of the wagons and four horses and start for home, for it was uncertain when I could get my goods all delivered at Idaho City, and we had to do that or we would not get our freight money, and we had contracted some $700 debts for expense on our way already, which had to be paid as we went back, for our feed and stabling and meals on the road.

So in the morning Mr. Green, our driver, and myself helped Charley to take the two light wagons and four horses over to Shaffer Mountain. He was to leave my black horse, Jack, at the Hay Press Ranch until I got back to there, and I would use him to go home with. He was too free, and worked too hard and got weak and needed rest. Charley was to take both the light wagons as far as Powder River Valley, where we had left part of our broke-down wagon, and leave one for me to bring along when I came. So I give him what money I had to spare to pay his way back home, for when we left there we had expected to be back in five or six weeks, and had, in a manner, left them without wood for the winter, and there was to be hay and grain bought for the horses and cattle. Some emigrants had come in and occupied my little cabin, and the women were company for Elizabeth, Charley's wife.

After Charley left I loaded up a light load to four horses and the one wagon, and we had stored the balance of our loading in Shaffer's storeroom. I and Mr. Green started with the team to Placerville, about 18 miles off. We had the creek to cross 9 or 10 times, and the crossings were icy. Our horses were shod, but not sharp enough to stick, and we had awful times, their slipping so much.

The first day we only got to Herzog Hotel, or Mountain House, and we had to stay overnight.[13] Next day we got into Placerville and had to get our four horses fresh or sharp shod, and we had to stay overnight in Placerville. We put up at a livery stable and had to pay $12 for the four horses, just for hay and stabling. We had our own grain to feed. We had to give $2 for our suppers, sleep in the barn, and $2 for our two breakfasts. We left our loading in the stable at Placerville, and got back to Shaffer's.

The snow was from twenty inches to four feet deep. We loaded up again and started with the second load to Placerville. In many places one of us had to walk ahead and break the snow in the road, so the horses could walk in it. The creeks commenced to freeze over, but not enough to hold up the horses, and we had hard times to make the horses break the ice—it would cut their feet.

We got into Placerville with our second load, and we stayed again at the stable, and left our loading and in the morning got back to Herzog's, and hired a pair of bob sleighs, hewed out of crooked pine or fir roots, not

---

[13] A. B. Meacham, later of Modoc War fame, and his brother, Harvey, had settled at Lee's Encampment on the Blue Mountains. They built a toll road and "erected what was known as the Mountain House. . . ." Bancroft, *Washington, Idaho, and Montana*, p. 424.

any shoes on, but we could do no better. I was to pay $7 per day for the use of them. We got back to Shaffer's and took a larger load with the sleighs, but it was still bad hauling and hard on the horses, from daylight to dark, and every meal we eat cost one dollar and every night we were at Placerville we paid $12 for the four horses to hay.

We at last got all our loading from Shaffer's to the Placerville stable, then we started one load for Idaho City, and the farther we went the deeper the snow. Mr. Green and I both in some high places would have to leave our teams and break a trail for the horses and go that far with the team, and go again. So on the first day we got but two or three miles past Centerville, and the night came on us dark and cold, and still ten or twelve miles to Idaho City. No hotel or stopping place near. By good luck we found an empty miner's cabin that had been a blacksmith shop. We tore down the forge and the sleeping bunk and built up a fire in the fireplace, and got our horses to stand in one side, and we had hay along, so the horses fared well. It snowed most of the night. We cooked us some supper and made up our beds in front of the fireplace. We burnt up the jambs of the old forge and we considered ourselves lucky to get stabling and lodging so convenient.

In the morning at daylight we hitched up, and we had to repeat the day before, only worse. The snow in some places was up to our arm pits and it took us all day to make the ten or twelve miles to Idaho City. We took our goods to their owners, and then took our team to Dryden's Livery Barn, and asked how much he would charge for the four horses to hay over night. He said four dollars each horse, in gold. I told them they were higher than Placerville, but they said hay was harder to get and further to haul so I had to pay $16 for the four horses one night to hay, in gold, and if I had wanted oats or barley I could have got it at 25 cents per pound. After feeding our horses we went to a restaurant, and we had good suppers at $1 apiece.

I got some of the freight money from the Gans brothers, who kept a saloon and brewery, so that I could pay some of our current expenses while hauling. I had some for Emmanuel, and we found one of the bales had got chafed by rubbing against the wagon bows and it had cut through some coats linings and I had to pay $9.50 damage on them, or buy the coats, so paid the damage. I had some little along for Leeman and Sallaro. They were surprised that I got in at all, and expected that I would have to store the goods and take my teams home to Walla Walla, as nearly every freighter had done who got catched when I did, and never got their freight money, but had to pay the damage for keeping the owner out of their goods all winter and no supplies.

They were glad to hear that I would bring everything in all right and they treated Mr. Green and I to their best brandy and wine. We were wet every day from morning to night, walking through snow and water and

## Hauling to the Boise Basin

our clothes would be froze on us and our boots too would not get dry until night by the fire. We could not ride for the cold, and it was that much more to the poor horses, but they had splendid appetite for their hay; every night they would eat from 25 to 40 lbs. apiece, and kept eating all night. We had to make four or five loads to get all hauled to Idaho City, but after the first night at Dryden's stable we made out to stay at Placerville or Centerville, where we only had to pay the $12 a night for hay.

On Christmas Day we lay over at Placerville, and we always had our meals at the hotel, kept by a man named Hart. They got up a Christmas Dinner, and I and Mr. Green took some and we had to pay $2.50 apiece, or $5 for us two, but we had everything nice.

When we got our loading delivered at Idaho City and got our receipts of the amount we had taken for them, I found we had two barrels of liquor too much, and it did not belong to either Leeman & Co. or the Gans Brothers, but belonged to someone who had shipped it to Powell and Company at Umatilla Landing, and through their carelessness had got it in our wagon and they had not checked it off right. It had caused our four-horse team to be overloaded some 800 lbs., used up the horses, and caused the two breakdowns of the wheels, and had actually kept us out so as to get into the snow storm, and $700 would not have paid the indirect damage done by it. I sold the one barrel of Cologne Spirits, and one of High Wines or alcohol at Placerville for $448 in all; some I got $7 per gal. for. So after I had sold out and got my money we made preparations to start for home. Leeman and Sallaro were very fair to me. While doubling up our wagons at the Shaffer Mountain, someone had went to our big wagon while we were away and broke open a case of wine and had taken out two bottles and nailed it up again, but Bob Leeman would not charge me for it, for it was not our fault, but we should have had to pay for it if they had wanted pay. They give me a present of a large buckskin gold purse, to put in the gold bars and gold dust, and give me a bottle of wine on the road. My loading came to one thousand and twenty dollars, and the two barrels $448 more, in all one thousand, four hundred and sixty eight dollars. We bought each some new boots and socks and underclothes in Idaho City. I found John C. Beckney and a man named Leby[?] keeping a hurdy-gurdy dance house in Idaho City (Beckney married one of the girls but they did not get along together).

We left for home and got to Placerville, and paid a hay bill of $35.55 and went to Frank Rivers' Hotel, past Placerville, stayed overnight, and paid $11.25. Next day got to Herzog's. . . . And so on, until I got to the We[i]ser River House where Saml. Hawkins kept a hotel and ferry. Here I heard that Charley had to lay over, for he could not get over Snake River, and he had stayed some four weeks, and cooked for a ball they had Christmas, and when he left, gave them an order on me for some $70 for hay and feed, board and ferrying.

When I got to Baker City, Powder River Valley, I stopped overnight and heard Charley, my brother, had left there the morning before. When the next stage came along I sent him word by the stage driver to wait for me one day at Uniontown, Grand[e] Ronde Valley, and when I got there I expected he had stopped.[14] But he had not—he did not want to wait, he was in such a hurry to get home. We drove the faster to catch up before he should get to the Blue Mountain, so that we could cross together, but when I got to Henry Trumble's, four miles from Summerville,[15] I found that Charley had got there two days before, and made him a sled and put his wagon on it and had left that morning early, and I got there in the evening. I had to make me a sleigh next day, and while making my sleigh it commenced to snow, and snowed all night, and in the morning I tried to travel and got a mile or two, and my horses kept getting down, so I had to turn back to Henry Trumble's, and I had to lay there five weeks, the snow too deep to go. . . .

I got tired of laying still and concluded to try and go on and cross the mountains, and got about seven or eight miles and found it impossible to go further, the horses would strike through every step and plunge, and would have soon killed themselfs, so I had to turn back to where a man kept a sort of ranch and hotel, name John Harper, some six miles from Trumble's.

One day a large government mule train of ten or twelve wagons and six to eight mules to each wagon concluded to try and cross the mountains. They had sleighs under some of their wagons and had a lot of men and teamsters ahead with shovels, and dug out the snow some. [Capt.] John Creighton was the wagon-master. He had been caught during the winter on Snake River and Boise Valley and he was on his way to Umatilla Landing or Walla Walla. I was awful glad to see them go over, and I got ready and followed right after, and by hard work and awful wet and cold, we got across, and in three days more, were home.

Now if my brother had stopped at Henry Trumble's and made a sleigh for my wagon, I would have got there, and we both could have gone over all right, but my one day's stopping to make the sleigh made a difference of 35 days time, and over $150 worth of expense in feed, hay and our board bill, besides Mr. Green's wages three months at $30 per month, my wages three months $150, Charley's wages two months $100, expense paid out ten hundred and eleven dollars. In all, one thousand, three hundred, and eighty-six dollars. And the loading and two barrels, deducting the cost of it, $1243, making the loss on the trip $143.

[14] Baker City is simply called Baker today. Uniontown is Union and was established in 1862.
[15] Summerville is in Union County and was founded by William Patten in 1857.

# 12

## BADMEN, GUNMEN, AND VIGILANTES

Two emigrant families . . . stopped in the little cabin until they built them each a house up the creek four miles, and when they got it done they moved up. One of the girls of Mr. Kinyen, (who had been a widower, and married a young wife) by his first wife, named Cynthia Kinyen, lived with us to help Elizabeth while we were away on the second trip to the Boise mines, Placerville, with the large hogs. She stayed with us after her folks moved up to their cabin, [because] she and her stepmother did not like each other. She was about 16 years old, tall, fair and good-looking, and I paid considerable attention to her when at home, and I come to the conclusion to ask her to become my wife.

So in the spring of 1865 she was going home to stay, and I took her up to their place a-horseback. On the way I took the opportunity to pop the all-important question to her. She was rather uncertain if she wished to change her mode of life, and only objected to having to live with my sister-in-law. Elizabeth had rather treated her too much like a hired girl, she was afraid she would continue to do so if I took her home as my wife. I told her she would be as much mistress of the house and place as Elizabeth, and I could put up a new house and we could have our own home, as half of everything on the place belonged to me, but she seemed not to have the courage to do so, and was afraid she and Elizabeth could not get along together well, and I did not insist too strong against her wishes. So I remained single. It was the first offer I had ever made to a lady or girl.

Cynthia got married that summer to a young man whose mother was a widow and she had rented her farm to Mr. Kinyen, and they were living on the place when she married him. He was a light-complexioned young man about 27 years old, rather fast, and he was arrested a few months after his marriage for horse-stealing. His mother spent a great deal of money to defend him, but he was convicted and sentenced to five years in the penitentiary. And so they were parted. The stealing horses had been done the summer before.

I heard some talk from a man that knew the family Kinyens on the Plains, and talked rather unfavorable of both Mrs. Kinyen and of Cynthia, and I became satisfied that I had done well not to marry her.

The discovery of Virginia City, Montana, and Alder Gulch, and Bannack City, caused a rush from Portland, Oregon, and all the northern part of Oregon, and a merchant from California and Portland named M[ag]ruder took up a large stock of goods, and he with some other partner had a large pack train of mules, and they had done well, sold out their goods, and Mr. M[ag]ruder came down with the pack train from Montana, and he had a large amount of gold dust with him to take to Portland, where his family were stopping. He had taken in several men to help him with the pack train, and was bringing a few passengers. I will give the names of some. I do not know who was the partner, but I knew some of the men working for him.

I spoke before of Roma[i]n, who kept the book and stationery store at Yreka, California, where William Cochran clerked in his store, and how he put me in mind of John Musgat, my brother-in-law in Wisconsin. He was one of the party; one named Lowry; one Howard, and a lame man named Page, from Josephine County, Oregon. I had seen him there frequently. The party got to Lewiston, and some inquiry for M[ag]ruder being made, the men said they had been ahead and that M[ag]ruder would overtake them before they got to Lewiston, but the train did not stop but came down to Walla Walla and left the train at some ranch, and Howard, Romain, and Page skipped down the river, first to Portland, and from there to San Francisco, California; Page to his home near K[e]rbyville, Oregon.

A few days after the train passed Lewiston some more merchants and business men of Montana came on, after expecting to catch up with Mr. M[ag]ruder to keep his company to Portland. They inquired if M[ag]ruder had passed, and some men who knew M[ag]ruder said he had not, but the train had passed and reported M[ag]ruder still behind to catch up. The men now here said they saw M[ag]ruder start when he left Montana, and knew that they had not passed him, for they had heard of his being so far ahead, and with his train, some two or three days before they got to Lewiston.

So they suspected foul play, and a party went back to track and find where M[ag]ruder was last seen, and trace him as far as they could, and the sheriff and a detective started after Howard, Page and Romain. The party that went back found their camp, and at last found signs, and found where M[ag]ruder had been murdered and part burned and buried, and they brought the body back to Lewiston. I think there were two boys found and some more men, and the men who had come down from Montana started down to Walla Walla and on to Portland, saw the detectives and found where all three of the murderers had gone.

Mrs. M[ag]ruder and family came up as soon as they heard Mr.M[ag]ruder was missing, and when at Walla Walla they heard his body had been found and went right up to Lewiston. Page was found and brought back and, after a little scaring, and some lawyer's advice, turned State's evidence on Howard and Romain, and it seems to me some other man. They were all brought back from San Francisco and convicted and sentenced to the penitentiary for I do not know how long, but I believe the most guilty was Page, and he got off the easiest.[1]

A few days after a highway robber killed a man named Scott, I think he was a stockman. The man who killed him, Dave English, I had known for a desperate gambler and he always had a bad reputation. . . . Dave had got to Wallula and was in a saloon gambling when Jim Buckley, the sheriff, had a man to go in and invite Dave to take a drink at the bar. Dave English did not know that there was anybody after him; he walked up to the counter with the man and took his glass in hand to drink, when the Sheriff stepped in with a double barreled shotgun cocked and told him to throw up his hands, and another man with the sheriff done the same, and Dave English had to obey. They put handcuffs on him and put him on the stage and brought him to Walla Walla for trial. I do not remember now what was done with him, but I think he was taken to Lewiston and hung by the Vigilantes as soon as he got there, and if I am

[1] Reinhart's narrative of the murder of Magruder is garbled. Floyd Magruder, a prominent citizen of Lewiston, was returning from Virginia City, located in the Idaho panhandle, after a profitable trade season. He realized some $14,000 in profit. To assist in driving his mules back to Lewiston via Elk City, he employed James P. Romaine, Christopher Lowery, Daniel Howard, and William Page. The employment was a fatal one since the first three mentioned planned to rob Magruder of his trade receipts. Romaine, sometimes called Dr. or Doc, was a gambler; Lowery was a blacksmith, who had worked on the Mullan wagon road construction; Page, an old trapper. Howard was "a goodlooking, brave young man, of kindly temper, but reckless morals."

On the return trip Horace and Robert Chalmer (two brothers from Missouri), Charles Allen, and William Phillips from the Willamette, joined the party headed for Lewiston.

As they camped one night six miles above the Clearwater's crossing, Lowery slew Magruder; Howard and Romaine murdered the two Missouri brothers and, soon after, Allen and Phillips. Page seems to have had no part in the blood-letting. The bodies were wrapped in a tent and thrown over a precipice; all the animals except eight horses, were taken into a canyon and shot.

The murder band were able to reach San Francisco, but were arrested on telegraphic requisition. They were brought back to Lewiston on December 7th to stand trial. This was the first case in law before Idaho courts and was tried at a special term. On January 26, 1864, the jury rendered a guilty verdict and Judge Samuel C. Parks pronounced the sentence of death by hanging. On March 4, surrounded by a detachment of U. S. infantrymen from Fort Lapwai, the gallows claimed their victims. Page, who turned state's evidence, was murdered in the summer of 1867 by Albert Igo. Hubert H. Bancroft, *History of Washington, Idaho, and Montana*, pp. 452–455; Bancroft, *Popular Tribunals*, I, 655–663; Nathaniel P. Langford, *Vigilante Days and Ways*, II, 97–144.

not mistaken Howard and Romain were served the same way, for the murders had been committed in Idaho, and most of the trials there ended in hanging.[2]

The winter of 1863 there had been quite a lot of roughs and desperadoes stopped in Walla Walla and got up to Montana and formed a band of robbers, many of which I knew, and I will give some of their names. The leader of them, Henry Plummer, captain; George Ives, first lieutenant, was from Racine, Wisconsin, a widow woman's son of good family (he was government herder the hard winter and stole over 150 mules and horses and run them off and reported them dead, but he and a man named Stubbs run them to British Columbia); John Cooper; John Skinner (Cooper's uncle); Boon[e] Helm; and 15 or more whose names I do not now remember, were all hung by the Vigilantes of Montana that winter. All caught at highway robbery. Some of their names I give whom I did not know were Bill Graves, Haz[e] Lyons, John Turner, Jack Gallecher, Buck Stinson, "Little Red" Steve Marshall, George Waggoner, Clubfoot George, ——— Parish, George Shears, Alex Carter, Robert Zakeria, some at Virginia City, some at Deer Lodge, at Bannack and Alder Gulch. It was the means of driving away a good many obnoxious desperadoes.[3]

The winter of 1865 George Porter, who was City Marshal when the riot at the theatre took place, where Latzenheiser and Hu[b]b[a]r[d] were killed, had another quarrel with a hotel keeper in Walla Walla named Hartman, a man with a fine family. George Porter was paying attention to some girl that was working for Hartman, and Hartman and Porter had some words, and Porter shot Hartman. He died a few minutes after. Now George Porter had lots of friends, and the gamblers, saloon men and Masons all took sides with Porter, or he would have been hung by the crowd. But Hartman was nearly a stranger and had not lived long in town, and Porter had men that would swear to most anything, and he was cleared. He had proved that he had not commenced the shooting, but most of the people knew better. When he was cleared many of the citizens were highly indignant, but the roughs were in the majority in Walla Walla and it was not safe to speak one's mind too freely. Then murders and shootings were everyday occurrences.

While I am at this part I will relate some other tragedy, but to do so

[2] David English, William Peoples, and Nelson Scott, a notorious trio, robbed Joseph and Jerry Berry, packers, of one hundred ounces of gold dust between Lewiston and Florence. They were arrested, taken from the sheriff and hanged by a company of expressmen and others. Langford, *Vigilante Days*, I, 134–140. *The Oregonian*, November 11, 1862, reports their burial at Florence on November 5.

[3] Henry Plummer was hung at Bannack, January 10, 1864. The Plummer Gang's notorious deeds and end are described concisely in Bancroft, *Popular Tribunals*, I, 684–688; fully in Thomas J. Dimsdale, *The Vigilantes of Montana, passim*. A more recent treatment is given in Wayne Gard, *Frontier Justice*, pp. 168–184.

Reputedly, William Bunton was second lieutenant. Reinhart's names are incorrect in most cases. Corrections are: Cyrus Skinner, William Graves, Jack Gallagher, Steven

will go back several years, with the principal in this action in the Indian War of 1854–1855. There were volunteers raised in Yreka, California, one company commanded by Capt. William Terry. While the company lay at Crescent City, California, Capt. Bill Terry had a quarrel with a man named Smith, and Terry put his revolver to Smith's face and cow-hided him, and Smith dare not resist or he would have shot him.

A short time after, Capt. Terry had a quarrel with two other gamblers, one named Ferdinand [J.] Patterson, and one named Hough Tate, and they had a shooting scrape at Yreka and Tate got the best of the other two, wounding both but slightly. It was in the dark, and each would shoot at the other at the flash of the pistols, so they could see each other. After they had got well, one day Ferd Patterson and Capt. Bill Terry had a shooting scrape, and Terry put his Colt's dragoon pistol to Patterson's forehead, pulled the trigger, and Ferd Patterson fell over. Captain Terry got on a horse and rode about 120 miles before he stopped; his horse gave out. He saw a stage driver from Yreka and asked what was going on at Yreka, and if they were after him for killing Ferd Patterson, but the driver told him Patterson was walking around town next morning, but very little hurt, the ball having glanced on his forehead, ran over the skull between the scalp and skull, and had but stunned him a while. William Terry did not go back, but went to The Dalles and kept a liquor store and saloon.[4]

Ferd Patterson got in with Pres Standifer and one or two more and in 1856 came to Sucker Creek, and they located some bank claims adjoining ours, and worked. But Ferd Patterson gambled and traveled with some loose woman with bunking games. Preston Standifer had been a well-to-do merchant at Yreka, California, and his brother Jeff Standifer was a wild loafer—had to leave California—went to the Sandwich Islands and got into some quarrel, killed a man of some notoriety, and was kept for trial for murder. Preston sold out his store and some mines he had there, and got all the money he could raise and went to the Sandwich Island and employed council to defend Jeff, his brother, and by using his money freely, got him clear. But it about broke up Pres Standifer, and he came to Sucker Creek in 1858. The two Standifers were on Fraser River when I was, and Jeff was smuggling liquors from Victoria up Fraser River to Fort Yale and Hills Bar. The English authorities got after him and he left and went to The Dalles in 1858 or 1859.

Ferd Patterson was at Waldo (Sailor Diggings), not far from Kerby-ville. He was living with some fancy woman, and another gambler named Fraser River George had a wife too in Waldo. Patterson used to get drunk and abuse his woman, and she took refuge with Fraser River

---

Marshland, George Wagner, George Lane, Frank Parrish, Aleck Carter, Robert Zackery. There was no John Turner.

[4] This report of affairs is accurate.

George's woman. Patterson came to George's house and abused him for keeping his woman, and he took out his pistol to go into the house to take her out. George told him to keep out, but Patterson commenced to shoot at George, who was in an entry of the house in the dark. Patterson was outside, shooting in. George could see Patterson, who could not see George in the dark, so George hit Patterson at every shot, until Patterson fell. He, Patterson, had emptied his revolver, but had not hurt George, only broke one arm. Patterson was shot several times, in the side, leg, and body, and was supposed to be mortally wounded, and looked for to die any minute. It was supposed to be impossible for him to live; three or four of his ribs came out in pieces, and his side was all shot to pieces, but to the surprise of all, he got over it. I did not hear from him for quite a while, until the winter of 1862, when he stopped at a hotel in Portland during the war.

Captain Staples of the Steamer Columbia of the mail route between Portland and San Francisco was with a party having a spree; they were all Union men, and were drinking at the bar at the hotel, when Patterson passed through the room. Captain Staples knew that Patterson was a Jeff Davis man and a strong secessionist, and he said to his companions, "Let us make Patterson drink the health of Abraham Lincoln!" They all gathered around Patterson and asked him to pledge to Old Abe, but Patterson said No, he would not, but if they said so, he would drink to Jeff Davis, and he went to go upstairs to his room. Captain Staples called, "Boys, come and we will bring him down and make him drink to Abe Lincoln!" They all started upstairs and Ferd Patterson stopped at the top of the stairs and told them not to come up after him, but they were all merry and did not mind him, and all started to go upstairs, Captain Staples leading.

As he came up, Patterson drew his revolver and shot Captain Staples and he fell back into the arms of his companions. He was mortally wounded. They took him to his room and he died soon after. The crowd tried to get ahold of Patterson, but he gave himself up to the authorities, and he was taken in charge and he had an examination before a justice, and he was discharged, for he acted in self defense. There was fearful excitement, but the Rebels were quite strong, and the Sheriff and City authorities upheld Ferd Patterson, and he stayed right there to spite them.

He got on a spree some time after, and he abused his woman and took her by the hair of the head and with a knife threatened to scalp her when some people interfered and he was arrested and brought before a police justice, and the witness who appeared convicted him and he had to pay a high fine and costs and put under bond to keep the peace.[5]

---

[5] Patterson was acquitted in Portland of Staples' death, 1861, when it was shown

Among the witnesses against him was a private night watchman of the hotel, and he had been one of the witnesses before when Patterson killed Captain Staples, who was liked by everyone in Portland that was not a Rebel. This night watchman's name was Thomas Donnahough and he was a very trusty man. After the trial Ferd Patterson saw Donnahough and he told him he would settle with him some day. During the Boise Mines excitement Donnahough had left Portland; he was afraid of Ferd Patterson and his friends, and he had a recommend to the merchants of Idaho City, and they employed him as nightwatchman there.

In 1864 Ferd Patterson came to Walla Walla and stayed a short time, and his friends, and Captain Terry, who was keeping a saloon at Walla Walla, his friends, fixed up the old grudge between the two. Both were strong Rebels, and when they met the first time Ferd Patterson made Captain Terry get on his knees and beg his pardon for the part he had played when he (Terry) had shot Patterson in the forehead (Captain Terry did not have Bullhead Smith to deal with when he put the pistol to his head and horsewhip[ped] ... him at Crescent City). Ferd Patterson got to be overbearing and would bully his fellow gamblers and they would not resist his actions, some for fear, others because he was one of them, and all the gamblers without any exceptions were Rebels in Walla Walla and all the mining towns of Idaho especially.

After a little Ferd went to Idaho City and gambled and done as he pleased in all the towns in the Boise Basin and Owyhee. One evening in Idaho City he came across Donnahough, the night watchman, and told him his time would soon come for that settlement, and Donnahough told the merchants that his life was not safe where Ferd Patterson was, and he took his discharge and a recommend from the merchants at Idaho City and went to Walla Walla, and he was engaged right away there as night watchman.

Ferd Patterson kept spreeing, and made himself very obnoxious to the law-abiding people of Idaho City, and many complaints had been made to have him prosecuted. Some had talked to Captain Pinkham, the then Sheriff of Idaho County, and some had reported to Patterson what Sheriff Pinkham should have said, and exaggerated the sayings on both sides, so that the gamblers and Rebels, who hated Pinkham, made Ferd Patterson their champion, while the law-abiding men and Unionists thought Pinkham their champion. Between the friends of both a certain rivalry was worked up until both parties happened to meet at the Hot Spring House, a summer retreat near Idaho City.

Ferd Patterson was better prepared and was going to bring things to a focus, but at the same time was going to have the drop or the advantage

---

they had had a quarrel. Reputedly he did not intentionally scalp his mistress. Bancroft, *Washington, Idaho, and Montana*, p. 459, *note*; Langford, *Vigilante Days*, I, 192.

when it did commence. Pinkham came to the Springs and did not know he was there, but Patterson knew Pinkham was coming, and was prepared, and had his friends posted to be witnesses and to help bring on the action. Pinkham was an ultra Union man and the Rebels wanted to get rid of him. He was a power with the Unionists, so when the Sheriff got to the Springs, something was said to bring on the thing, and Pinkham not being posted was unguarded, and when it came to shooting Ferd Patterson got in his work before Pinkham got his pistol fairly out of his belt. He was so badly hit that he died in a few hours.

Patterson's friends got the marshal to keep Patterson in a strong building and had a large guard of his friends, all well-armed, to keep the crowd from lynching him. The gamblers and Rebels stood guard day and night for several days and did not dare take him out for a hearing, and one night his friends smuggled him out of jail and got him in a coach and started with him. The crowd and Vigilantes thought they had taken him toward Salt Lake City, and started off a party after them, but instead the friends of Ferd Patterson had taken him toward Walla Walla, and had coach full of armed men with Winchester rifles to guard him.[6]

The Vigilantes had no idea that they would take him towards Oregon, but they did not care—any way to get away from Idaho—and after they struck Oregon they were not in much danger, as he could not be arrested and taken back without a requisition from one governor to the other, and when they crossed the Blue Mountains on the Thomas Toll Road, they soon got into Washington Territory to Walla Walla and knew there they were safe until the civil law arrested him.

So he lay around Walla Walla gambling and bullying the gamblers and his friends. He drank so much he became reckless and overbearing to his best friends. I have seen Patterson go into a saloon and go up to a game where a lot of men were playing faro and take some of his friend's bet off a card and go away and if the man would make any remarks [Patterson] would abuse him or throw part of it back to him and threaten to act so and so to him. Most of his friends through fear would say nothing, and Ferd would go to some other game or go out to spend the money. He was altogether kept up by his friends.

A week or two after he got to Walla Walla, one night he was passing some of the stores and he saw Donnahough, the night watchman, on his beat, and he went up to him and told him the time to settle would soon come. Next day Donnahough went to the principal merchants and leading men of the town and told them of Ferd Patterson's threats and said he left Portland on that account, then Idaho City, and now he would have to

---

[6] Reinhart relates the July, 1865, Patterson-Pinkham affair correctly except for two points: Pinkham was ex-sheriff, and Patterson was brought to trial, but acquitted. Langford, *Vigilante Days*, I, 182–209.

leave or be killed, and asked if Patterson's threats were to be relied upon. Every law and order man in town said that a threat from Patterson was considered dangerous, and if a man disregarded his warning, he would be himself to blame. Tom Donnahough was very downhearted; he had a family to support, and had lost two or three good places by the fear of the threats. Through some friends' advice, he came to the conclusion that Ferd Patterson would either kill him, or he would have to kill Patterson, and his mind became so discouraged that he was nearly out of his mind, and he brooded over the subject for several days.

One morning Ferd Patterson came out of the Bank Exchange Saloon and went into the barber shop next door and he took off his coat and vest to be shaved and got into the chair and the barber commenced to lather his face, when Thomas Donnahough came in with a cocked revolver in his hand, and said "Ferd Patterson, you or I must die!", and he commenced to fire on Patterson. The first shot struck Patterson in the mouth and the next in the breast. Ferd jumped out of the chair and ran towards Donnahough, who kept shooting into the head and body of Patterson until he got on the sidewalk and tried to go into the Bank Exchange where he had left his revolver, but he could not get in; he fell on the sidewalk, and Donnahough had shot every load out of his pistol. He went to a Justice of the Peace or Police Judge and gave himself up. Ferd Patterson lived but a few minutes and never spoke, and died.[7]

There was a great excitement in Walla Walla, and Ferd Patterson's friends were wild and wanted to lynch Donnahough, but the law and order men and all citizens that were not Rebels were on Donnahough's side, and were bound to have him have a fair trial. Preparations were made to bury Patterson; he had quite a funeral of his kind and sympathizers. In a few days the case came on for trial. I was chosen on the jury, but got off, as I was a non-resident. I lived in Oregon, but was teaming to town from Wallula. Donnahough got clear, but his friends advised him to go away, and made up some means for his family. He was afterwards arrested in San Francisco on the charge of murder by the friends of Patterson there, and came near being convicted, for Patterson's friends were bound to convict or kill or assassinate him. Next they got him arrested as a deserter from the Union Army, and gave him no peace, and he was hunted by the Rebel friends of Patterson until he enlisted in the United States Navy. During the war he got a good bounty for his family in advance to keep them till he could send them his wages. That is the last I heard of him.

A man named Stubbs used to keep a ranch on the Touchet River on the

---

[7] This is a fair account. Thomas Donovan (or Donahue) was Patterson's slayer. He was brought to trial, but the jury split. Later, he was rearrested in San Francisco, brought back for a second trial, but finally released. Reinhart adds much to the history of this killing. *Ibid.*, I, 209; Bancroft, *Washington, Idaho, and Montana*, p. 460, *note*.

road to Lewiston. I spoke of him as being a hard character, as I was going up to the Nez Perce Mines. In the Spring of 1864 he stole a lot of Government mules and run them off towards British Columbia; he and George Ives, the herder, had done so once before and got away safe. (George Ives was the lieutenant of the band of robbers headed by Henry Plummer, and they were all hung in Montana.) Captain John Creighton, a wagon master at the Post of Walla Walla, took one or two men and followed the tracks of Stubbs and the mules. While the other men were looking some other direction, Capt. John Creighton came on Stubbs with the mules hid away in a mountain valley on the British side. Stubbs knew Creighton and pulled out his pistol to shoot Creighton but John was quick. [He] shot Stubbs dead with his shotgun [and] got all the mules and drove them back to Walla Walla. They buried Stubbs first.[8] There must have been over 120 mules and horses in the lot, all stolen property. . . .

[8] A true version. See Chapter IX, *note* 13, *ante.*

# 13

## TEAMSTER IN MONTANA

(1866–1867)

Now after I got back from our last unprofitable trip and let our horses rest a few days, and done some hauling, I and my brother went to Walla Walla and got some things wanted for the house, provisions and some clothing, and we had some bills to pay and some settlements to make, for my brother was a great hand to run bills at stores, and it had discommoded me and cramped me more than once to settle bills that had been brought to me unexpectedly, when I could least spare the money.... When we got home I told my brother that I wanted to go to hauling freight from Wallula to Walla Walla, about 30 miles, and to let my hauling go on to the bills we were owing to the men wanting the goods hauled, and that I could pay part of the bills in that way. I told him to let me take the black team, Jim and Jack, the harness, and the big bay team, Prince and John, and harness and wagon, and I would pay $180 of the indebtedness [which amounted to almost $300], and he should take the brown mares, Dolly and Nelly, and the two others, the sorrel and the bay, and the harness, for the four, and the two smallest wagons, and he could keep all the stock, tools, and the whole ranch, farm or claims, and he was to pay the same amount, $180, on the debts, and we would dissolve partnership. When we went into teaming it was agreed that he should carry on the farm and I [was] to run the teaming or freighting. His interfering in two or three instances had caused a great loss to us; first, by the second load of hogs, then by the last load of freight from Umatilla Landing November 3rd, by which we made a difference to what I would have done, by $1,500 or $2,000.

So he took me up at my offer and I got my four horses and harnesses put in good fix and put bows and cover to my biggest wagon, and went to Walla Walla and commenced to haul freight to Harris and Marks, and Dick Jacobs, and paid them all up, and to Stine the blacksmith, and I put up at Krafts' stable and paid him part, and kept my team there. . . .

During the winter of 1865 and 1866 a rich discovery had been made in Montana, known as the Blackfoot Mines, between Helena and Deer Lodge, and quite a rush started to go in the Fall and Winter of 1865, but it was too late to prospect and provision awful high and scarce.[1] In the Spring of 1866 a big rush took place from all the surrounding country and cities in Idaho, Oregon, California, and Washington Territory. I tried to sell my four horses, harness and wagon for $700 but could not find a purchaser just then, and I wanted to go to the Blackfoot excitement and take up a rich claim if I could.

One day I came acrost two young men, brothers, who had come up from San Francisco on their way to Blackfoot. They were both German shoemakers, and had a stock of leather and a lot of readymade boots and shoes with them, and all the tools for themselves to work at their trade. Their name was Woolf. They had bought some six or eight cans of alcohol, to make up the lot to about 20 hundred pounds, and they were looking for a conveyance to take them and their stock or freight to the Blackfoot Mines by way of Virginia City and Helena, Montana, and if it suited them at either place they might stop.[2] I had been recommended to them by some of the Germans in Walla Walla, and I told them I would take their whole amount, twenty hundred pounds, at 35 cents per pound, or $350 per thousand pounds in gold coin, and go by way of Boise City, Idaho, on to near Fort Hall, and strike the road from Salt Lake City, Utah, to Virginia City, Montana.[3]

---

[1] J[ohn] L. Campbell, *Idaho and Montana: New Gold Regions. The Emigrant's Guide*, p. 26, recorded:

Produce at Bannock was high when we left, which was in October, 1863. Flour was $25 per hundred; Bacon, 30 cents per lb; Hams 60 cents; fresh Stakes, 15 to 25 cents; Potatoes, per lb, 25 cents; Cabbage, per lb, 60 cents; Coffee, 80 cents; Sugar, 60 cents; fresh Butter, $1.25; Hay, 10 cents per lb, or $30 per ton; Lumber, $150 per thousand. Wages ruled at $5 per day for miners and common laborers, and $6 to $8 for mechanics. Female labor ranged from $10 to $15 per week. Washing, from $3 to $6 by the dozen.

As for the richness of the mines, it has been estimated that between 1862–1869, Montana's gold production totaled $94,000,000. Oscar O. Winther, *The Great Northwest: A History*, p. 226.

[2] The Walla Walla *Statesman*, April 13, 1866, noted the impact of the Montana strike:

In the history of mining excitements, we doubt whether there ever has been a rush to equal that now going on to Montana. From every point of the compass, they drift by hundreds and thousands, and the cry is, "still they come." . . . In addition to the usual convenances, men of enterprise have placed passenger trains on the route between Walla Walla and Blackfoot, and these trains go out early, with full passenger lists. Fare, with provisions furnished, $80.

[3] Virginia City lay 400 miles from Salt Lake and 200 miles from Fort Benton. Rein-

They inquired and found my charges reasonable and customary, so engaged me to take them and their goods and blankets and provision, and they would buy ponies for themselfs to ride, and to cook, camp, and travel together. We made preparations to start as soon as we could well get acrost the Blue Mountains with my team. My four horses were in good condition and as fine a four-horse team as could be found in that part of the country, but my wagon was not quite heavy enough to take all the freight and the feed and other baggage I would have to take. So I sold the wagon and bought me a larger one, and heavier, and I had to put on a good brake, and a good spring seat, bows and top, feed sacks and hobbles, and have the horses all fresh sharp shod, and it cost me about $300 to fit out for the start. Another young man [Louis Heinige], a hotel cook and baker too, from California (Sacramento), paid me 25 cents more to carry his provision and blankets and he got himself a pony to ride. He was a German, too. So my loading and passengers came to over $735 (about 2,000 pounds of freight), or more than I had offered to take for the four horses and the other wagon, and if after I got to the Blackfoot Mines it did not suit I could leave, for I would have my team complete, and some money left. So I done well by not selling when I offered to, for I would have spent $200 or $300 to have gone to Blackfoot, and prospected around until I would no doubt have spent the balance of what the team would have come to, and I would have been in a manner broke.

So when we were all ready to start (I had drawn about $200 from the Woolf Brothers to fit out with) I paid Kraft, the stable keeper $65 (part of it was our company bill); Fred Stine, the blacksmith, $76.50, it being part company and all the other bills I owed; Jacobs, $23 company; Dick Richards, $31.50 Charley's; and Lynch the blacksmith $20 for Charley. I valued my four horses, harness, wagon and outfit complete at $1,100 cash.

I left Walla Walla some time in the last of April, 1866. I drove out to Charley's and bid them all goodbye. He had at that time three children; George, the oldest, was not over five or six years old; Bertha, the next, three or four years old; and Charley, the youngest, about two years old....[4]

We had a hard trip acrost the mountains, but we had along company, another wagon and some passengers that I had forgotten to mention, a German Jew jeweler named Rosenthal who kept a jewelry store in Walla Walla. He had a wife but no children, and he decided to go out of business in Walla Walla and go to Blackfoot or somewhere in Montana. He had a brother-in-law, a young man, brother to his wife, who had come

---

hart was receiving a handsome freight rate—the average at this time was twenty cents per pound on goods transported via the Mullan Road to Helena. Alton B. Oviatt, "Steamboat Traffic on the Upper Missouri River, 1859–1869," *Pacific Northwest Quarterly*, XL (1949), 97–98.

[4] Reinhart did not see his brother and family again. Charles died in October, 1888.

from San Francisco, California, and was on his way to Montana and the Blackfoot Mines. He had along a lot of jewelry, too, and he was to travel with his brother-in-law. His name was Joseph Biernbaum. They were from New York City. Mr. Rosenthal engaged a teamster in Walla Walla named Brown. He had a good four-mule team, and was not quite as heavy loaded in goods as I was, but Mrs. Rosenthal rode in the wagon all the time, so did Mr. Rosenthal and the two Browns (a brother being along). Joe Biernbaum had a pony to ride, and once in a while he would change off with Mr. Rosenthal.

We all traveled together and camped out in most of the places. We had to go slow at the start, as the roads were muddy and bad and the streams high, and when we got to Boise City we had to lay by a few days; the Boise River was quite high, and it did not hurt our stock to rest a few days. Joe Biernbaum traded for another pony, so that both of them could ride.

A few days out from Boise City we heard that Dave Updike, the sheriff of Boise County, had got hung on the South Fork of Boise River by the Vigilantes, and it made some of his associates look out for fear they would be made to suffer for him and like him. Dave Updike was hung by the Vigilantes for being complicated [implicated?] in horse stealing, in May, 1866.[5]

We had to travel a great deal by ourselves, and not much roads and no bridges on any of the streams and we had to go it blind, and drive into bad places and take the chances. Sometimes we would get into some awful bad places to get out of, but we could double in many places and go through. . . . We stood guard every night for fear our stock might be stolen, either by Indians or white horse thieves, who were supposed to be quite numerous and venturesome. So when we got to the main road from Salt Lake City to Virginia City we were glad, for we would have more company, and roads that had been worked, but we would have to pay toll, and we had been paying a great deal of ferryage and lots of toll. In coming through Oregon and Idaho the roads were lined with trains of mules and oxen wagons with freight and provisions for the Montana mines.

We crossed the Rocky Mountains and come down on Red Rock Creek, which is the headwaters of the Jefferson River, one of the main branches of the Missouri River. We passed down a canyon where Big Hole River runs into the Jefferson or Deer Lodge (I forget which); then we got to a tollgate through the canyon and had our toll to pay, some $2.50 or $3 per wagon.[6] Next day we crossed Black Tail Deer Creek with a ferry,

---

[5] David C. Updyke, ex-sheriff of Ada County, and Jacob Dixon, formerly of Shasta County, California, were hanged as road agents in mid-April, 1866, on the road to South Boise. Hubert H. Bancroft, *History of Washington, Idaho, and Montana*, pp. 461–463; Nathaniel P. Langford, *Vigilante Days and Ways*, II, 340–353.

[6] The Big Hole empties into the Jefferson River.

and in a few days we got to the ferry on Stinkwater and crossed, paying ferrying.

We kept up towards Virginia City, and camped until the two Woolf Brothers, Louis Heinige and Biernbaum [went] up to Virginia City to look at the place and see if they would like to go into business up there. We let our teams rest that day, and they got back at night. They did not like the city, but Woolf had been offered a fair price for his sole leather and some French calfskins, but would not sell. So next morning we left Stinkwater, one of the streams that empties into the Jefferson, and were again on our road to Helena.

We passed down the Jefferson River, then crossed some 50 or 60 miles from Helena. We had several toll roads to go over, and got to Pipestem Creek, and had some high creeks to cross. We got within two or three miles of Helena, in Prickly Pear Valley, and camped close to the road, and the boys went into town that evening. When they came out they said it was a lively place, but a hard one. Just before they got to town, on a creek where there was a slaughter-house and a large pine tree close by, they saw a man hanging, that had been hung the night before by the Vigilantes, for murder, and had been left hanging.

That night we turned our horses out and I hobbled one or two of them. Next morning Brown found his mules by themselfs, and mine not in sight. I hunted all around but could not find my four big American horses, and the other boys said they had been stolen. We hunted all day, but with no success. Mr. and Mrs. Rosenthal had gone to town, with Joe Biernbaum, and Brown had unloaded his wagon and had gone back on a mining creek where his brother had stopped with some friends as we came by when we came in. One of us had to stay in camp to keep anyone from stealing or robbing camp of the stock of boots and shoes and leather, and one or two of us hunted the horses, but after hunting for four or five days we got a team to haul the goods to town, and unloaded, and the boys paid me off the balance they owed me, some over $300 (I had drawn some on the road)....

There were herders that herded stock around Helena, and every night some two or three herders would take out the night herd and bring them back in the morning by daylight, and no doubt some of the herders had drove my four horses away so as to keep them out of their night herd, and they had rambled farther off themselves. Prickly Pear Valley was a large wide valley, and Ten Mile Creek, beyond the town of Helena, was considerable fenced up on the town side, and I did not go acrosst the creek to look. I was getting awfully discouraged and was afraid someone had stolen them or they had struck back to Walla Walla.

One day I went to a stable in Helena and inquired for the night herder, and saw him, a little Mexican I had seen at Walla Walla; his name was Spanish Jim. I think he was part Indian. I asked him if he had seen or heard anyone speak of seeing four large American work horses. He said he had not. I told him if he would hunt for them and find them and bring them in I would give him $20. He said he would see if he could get off to hunt for them, and in a short time returned and said [he would] if I would make it $25 (as he was wanting to buy a set of harness and would have to pay the $25). I told him all right, and he told me to call that afternoon, and he would go right out then. But I saw him after dinner and he had not as yet been out.

In the evening at 4 or 5 o'clock I went to the stable to inquire if he had gone out or come in, and there were my four horses in the lot already. He said he got them acrosst Ten Mile Creek, and I now believe he drove them there himself to get the reward if I did not find them myself. I did not like the Mexican's actions about it, but I paid him the $25 and took my horses, took my wagon up Grizzly Gulch, back of town, and in the night put them in the night herd, and I looked about town.

Helena was quite a nice mining town of about 1,000 or 1,500 population, built in a hollow and bar on Last Chance Creek. Just above town Grizzly Gulch came in, and Last Chance was quite rich.[7] The Woolf Brothers rented a part of a shoemaker's shop and put in their stock and went to work. Louis Heinige went to cooking at the St. Louis Restaurant and Hotel.

At the stable where I had got my horses I got acquainted with a teamster named Thomas Fruid, a Scotchman. He was lately from the States, having drove out a mule team acrosst the plains, but he used to live at Jacksonville, Oregon, and worked in a livery stable kept by John Drum and James [C]lugage, who with Skinner first discovered Rich Gulch. He knew most everyone I knew at Jacksonville and Yreka. He had gone to his home in New York, where he had been a coachman and hostler for Woolsey, the sugar refiner, and relatives of the Aspinwalls and Howlin's and Jeromes and the Vanderbilts and Lorrilard Brothers, the great tobacconists of New York. He had lived with several of the rich families of New York as a hostler and coachman, and had lived at Throgs Neck and Morrisina, at several of their out of town houses.

After he got home he stayed a short time, but soon got tired of the dull place, and fitted out a mule team, with some other men, and crossed the plains to Boise in 1865, and bought out his partners, and took a load of freight from Boise City to Bannack, and had come on to Helena, Montana, and he got a load of passengers to take to Fort Benton, on the Missouri River, about 160 miles from Helena.[8]

---

[7] Last Chance Gulch (later named Helena) was struck on July 14, 1864 and was 140 miles from Fort Benton and 125 miles from Virginia City.

[8] Fort Benton of the American Fur Company (formerly Fort Lewis) was built in 1850 by Major Alexander Culbertson. It is below the Great Falls of the Missouri at the confluence of that river and the Teton River. It was the last navigable port on the

In them days all the freight and provision and all kinds of goods were shipped up the river from St. Louis. The steamboats started with the rise of water in the Missouri River, and it usually took them from two and a half to five months to go up to Fort Benton, 3,100 miles above, and it depended on the stage of the water in the river if they made more than one trip during the Spring, Summer and Fall.[9] The steamboats were mostly light draft and carried spars, one on each side of the boat, to lift them up to pass over low shallow water on the sand bars. When they run aground they put two spars ahead like two crutches and had chains from the tops of the spars to the capstain and would raise the front of the boat, and have lines to the shore to some tree to pull by the windlass or capstain, and when they got the front raised [they would] start the engine and paddle wheels agoing and pull on the shore ropes and the boat would go ahead, sometimes clear off the sandbar the first time; if not, [they would] keep repeating the operation until they got over, and so on, if they got to move. The lower the water, the more sparring they would have to do. Sometimes the river would keep high, and steamers of more or larger capacity would go, and make two or three trips.[10]

There was a great rush to the mines right after the War, and the route from St. Louis up the Missouri to Fort Benton was a favorite route. The first boats would get to Fort Benton about the last of May or the first of June, and many miners who had been in the mines some time and wished to return East, started to go down to Fort Benton and return with the steamers as soon as they could unload their passengers and freight, and it only took the steamer from 30 to 50 days to go down the river to St. Louis, as they were light-loaded and ran with the current. It was a pleasant trip to go down, but fearful hard sometimes to come up. They would have to stop so much on the way up to take on cordwood to burn, some of which they could buy already chopped; if not, all hands would have to cut it and carry it on the steamer, and many men got a chance to work their way up

---

upper Missouri and as such became the river capital of traffic for goods being shipped into the Montana mines.

[9] The steamers left St. Louis or Sioux City in late March or April, in order to take advantage of the spring flood tide. They arrived after a 2,300-mile passage up river in late May, June, and early July, staying only long enough to obtain downriver cargo and passengers. In 1864 only eight steamers made the run. But with the rush to the gold fields, one thousand passengers were carried in 1865 along with six thousands tons of merchandise and twenty quartz mills. The trade "touched high-water mark in 1867," when seventy steamers tied up at Fort Benton during the shipping season. Hiram M. Chittenden, *History of Early Steamboat Navigation on the Missouri River*, II, 273–276.

[10] Added dangers came from the hostile Indians beyond Fort Randall. The flood tide would hardly have ever permitted more than one round trip. The records show that in 1866 two steamers made successful second round trips, but these were exceptions. Anonymous, "Steamboat Arrivals at Fort Benton, Montana, and Vicinity," *Contributions to the Historical Society of Montana*, I (1876), 317–325. (Hereafter cited *Montana Contributions*.)

on the steamer and save the price of the passage, which was quite high—from $75 to $300—for cabin fare the highest.[11]

So Thomas Fruid took down a load of passengers, and told me to get up a load and come down too, and I could bring back passengers from Fort Benton for Helena, and passenger freight was better than goods, for you got more pay, and in bad places they would or could get out and walk till you got to good road again. So I tried after Fruid left, and in two or three days got up a load of eight passengers and their blankets, baggage, and provisions for four or five days. I took the eight at $75, or about $9.40 apiece. I had a good team, for my horses had got rested up, and I went along nice.

We passed through Prickly Pear Valley, and went up a creek on the other side where there was considerable mining, and we went up one of the spurs of the Rocky Mountains, and about 17 miles from Helena struck the Mullan Road (a military road from Walla Walla, about 500 miles), on to Benton, some 140 more.[12] We went over the Medicine Rock, a high mountain and hard to cross loaded, but a firm of merchants of Helena, named King and Gillette, were cutting and constructing a wagon road, and it would be in traveling order in a few days. They charged $9 for a four-horse loaded wagon to go through, about 18 or 20 miles, but it cost them heavy to put in a lot of bridges and cut roads in the sides of the mountains and hills, and cut the heavy timbers. I expect the King & Gillette toll road was a paying enterprise.[13]

The Prickly Pear Canyon run some 20 odd miles, then you left the Prickly Pear, and the next stream, the Dearborn River, had to be ferried, and cost some three dollars. After that we had no large streams until we got to Sun River Ferry, where we met a lot of teams loaded with passengers from the first steamer up from St. Louis. Thomas Fruid had a load, and told me I would have no trouble to get passengers back, as some were there left, and more steamers were expected every day. We ferried over and paid $2.50, and in two more days got to Fort Benton, an old American Fur Company post, and a Hudson Bay post in ruins.[14] There were about

[11] The average fare was $150.00. Approximately one-third of those operating woodchopping stations above Fort Randall, between 1867–1869, were massacred by the Indians. Merrill G. Burlingame and K. Ross Toole, *A History of Montana*, I, 140–141; Chittenden, *Early Navigation*, II, 277, *et seq*.

[12] The Mullan Road which was constructed between Fort Walla Walla and Fort Benton, the head of Missouri River navigation in Montana. The road ran 624 miles and was built under the command of Lieutenant John Mullan, who launched construction in 1858. It was completed in 1862. W. Turrentine Jackson, *Wagon Roads West*, pp. 261–278, presents a concise history of the road and its use by gold seekers.

[13] In 1866, W. C. Gillette and his partner King purchased the Little Prickly Pear Toll Road and then began construction of "an excellent road, estimated to cost some forty to fifty thousand dollars." Within two years their investment was returned. Merrill G. Burlingame, *The Montana Frontier*, p. 146.

[14] Not to be confused with Fort Benton established at the confluence of the Big Horn and Yellowstone rivers in 1821 by Joshua Pilcher.

300 or 400 inhabitants, merchants, saloonkeepers, laborers, rangers, miners, and herders, and quite a lot of Indians, called Bloods and Piegans, and there was an Indian Agent living in town.[15]

I unloaded my load, and they all got off next day on the downgoing steamer. I engaged eight passengers back to Helena, at about $13.25 apiece, or $105 for the eight, and started back. I made the round trip in about ten days, and I got $180 for the trip; my expenses were not high, but the toll was fearful. When I got back I put my horses in the herd, and looked around for passengers.

A man who had just come through from California had a light new thorough brace 7-passenger coach, the nicest one I ever saw, [which] had just been made at Columbus, California; it weighed just 620 pounds, had a high seat for the driver, and good stage brake, and a baggage or trunk rack on behind, with a leather cover to buckle and fasten the baggage. I took a fancy to it when I first saw it. On Ten-Mile Creek, some four or five miles from Helena, was a hot spring bathing house and hotel, a race track close by, a summer resort for the town folks to ride to and pass the day bathing or gambling, or other amusements. It was kept by a German and his wife, and there was a hack running to it regular and charging high prices. I had been asked to put on a spring wagon or hack and run in opposition to the one now owned by Dr. Robinson (who a year after was assassinated by the Mormons at Salt Lake City for meddling with their wives, after being warned not to do so, several times).[16]

When I saw the little coach, I came to the conclusion to buy it, and run to Hot Springs Hotel, and I bought it for one-half its cost, $200 (it cost $400 at the factory), and I got a tin sign painted "Helena Accommodation Line." Dr. Robinson did not like me running opposition to him, and said if I did he would prosecute me, for he claimed to have the sole right and privilege from the former owner of the hotel, but I inquired and found all I had to do was to take out a United States license or common carrier's license for the Territory of Montana. I was directed to a Mr. T. C. Jones, who was the Collector of Internal Revenue for Helena, and I paid him $10 for the one year, and I could run as many wagons and teams as I wished by making my returns when necessary....[17]

[15] An Indian agency for the Blackfeet was located at Fort Benton in 1856. Gad E. Upson was appointed agent in 1863 and died in March, 1866. H. D. Upham, Upson's clerk, acted as deputy agent until the arrival of George B. Wright, who served until November, 1868.

[16] Dr. J. King Robinson was assassinated around midnight October 22, 1866, in Salt Lake City. Neither the motives nor murderers were ever discovered. Andrew L. Neff, *History of Utah, 1847 to 1869*, pp. 716–717, note.

[17] "Mr. T. C. Jones . . . has lived here at Chanute, Kansas, for the past sixteen years; he came three or four months before me [1873]." *Reinhart Manuscript*. Jones was probably an assistant collector since Walter W. Johnson, "List of Officers of the Territory of Montana to 1876," *Montana Contributions*, I, (1876), 326–333, does not mention him.

But when I got my license and got my team and wagon ready, a party of five men just from Virginia City, wanted to go right on to Benton, to go with a steamer expecting to leave there a certain day, and they gave me $25 apiece, or $125 for the five, and to go in the little coach. So I left my big wagon and one span of horses on a ranch and took my passengers down to Benton in four days. They got there in good time. We had a splendid time; they had everything good along to eat, and bought brandy, whiskey, and wine to drink; they were all well-fixed. I got four passengers back to Helena at about $50, and got back in five days, making nine days out and $173 for the trip....

When I got back I put the horses on the ranch and bought two small mules for $145, got the big wagon repaired some, and an extra spring seat for to carry the passengers with. When I got all fixed up I took the big wagon with one span of horses and the mules to it, and the little coach with two horses to it, and I had five passengers engaged for the coach: Charles Rickmoth, a saloonkeeper and miner (he had big diggings on Drummond or Montana Bar), at $25; William Coleman, merchant of Helena, $25; Ezra Millard, banker, of Helena and Omaha, Nebraska, and he was moving his bank, or gold dust, to Omaha, $29; Mr. Smith of New York, president of the New York Gulch Quartz Mining Company, with specimens, $29; and Mrs. Rosenthal, the jeweler's wife, who had come with us from Walla Walla. She was on her way to St. Louis to some of her folks, and then home to her parents in New York City, on a visit, with her baggage, $43.80. For the little coach I got for the five passengers $151.80, and for the big wagon load about $120, some nine passengers.

... Millard the banker paid one man's fare, and Smith, the president of the mining company, paid fare for a man named Swan, to guard their money, which was in boxes and Smith called quartz specimens, and Millard and Smith had three valises in the back of the coach under their seats. One was as much as I could lift, and when I lifted it to put it in, one of the handles pulled off. We all got along very well, and all were well armed with rifles and revolvers. Smith was a good shot, but he lost several bottles of wine shooting at the little prairie dogs with his rifle, for even if you hit them they fall right back into their ground holes. He had a splendid new Winchester rifle.

The last fifty miles before we got to the Seven Mile Spring, from Benton, we drove in the night, for there was no water from Sun River to the spring, and we got to the spring just before daylight. We all felt chilly and cold and we stopped at the edge of the ravine, where the spring is some 80 or 100 yards below. Some of the boys, first out of the wagons, went down the hill, near the spring, and built up a fire of grass and light wood and willow brush, while I was unhitching to take our horses down to the spring to water them, and we would then get breakfast. The fire was inviting to Ez Millard and Mrs. Rosenthal and Smith, and everyone.

The other teamster driving my team went down to the fire, so when we two brought down our six head of horses and mules, no one was left with the wagon or coach, and Mr. Smith happened to notice we were all down there. He said to Millard that the wagons were not guarded, and Swan and Smith went up to stay with them until relieved after breakfast. Millard said when we got to Benton that a rich haul could have been made on them wagons; no doubt there was $200,000 in the coach and wagon in gold dust, for Ezra Millard had been buying up gold dust with his banking capital, and what Smith had for his company. I have often thought, if I had been so minded and could have had a confederate to help me, that we could have taken the one valise from under the back seat of the coach, buried or hid it, and it would never have been missed until we got to Fort Benton, and no one would have known when it was taken out between Helena and Fort Benton. I believe there was more treasure went over the road from Helena, Virginia City, Diamond and New York Gulch, and Montana Bar, than any other place in the West, and I never heard of a robbery after the hanging by the Vigilantes in 1863 or 1864, and if that lesson had not taken place it would have been unsafe for a man to have had any money with him.[18]

We all got to Fort Benton in a few hours, and all parties made preparations to leave by the steamer soon to leave, on that day.

I took back to Helena one passenger and some freight, in all about $30 for expenses of trip. I got back to Helena and put my horses in pasture for a few days. While in Helena I bought at auction at Travis & Brothers stable a span of black mules and harness for $225, making eight head (four horses and four mules) and the two wagons (one iron axle California wagon I had bought for $65), and the little coach. After a while I got a load of passengers for the two big wagons, and left the coach on the ranch....

We got to Benton all right and on our way back brought some freight to the ferry on Sun River, and some to the Springs and to Dearborn River, and got into Helena to the ranch in time to hitch two horses onto my light coach and went up town and carried passengers to the prize fight between Con Oram and Hough O'Niel[1] (my old friend at the Three Sisters, Gold Beach and Port Orford, Waldo and Indian Creek). I charged 50 cents apiece from town, a few hundred yards, to where the ring was inclosed, and I kept so busy that the fight was over before I knew it. Con Oram won the fight by Hough O'Niel[1] striking a foul. Many said it was not square, but I think it was a sell-out, as most of the miners from

---

[18] Reinhart is correct about the value of the treasure—in one year alone, $1,250,000 was transported in one shipment on board the *Luella*. However, there were several coach robberies, one in 1865 in the amount of $60,000. John X. Beidler soon put matters right. Chittenden, *Early Navigation*, II, 276; Langford, *Vigilante Days*, II, 423–427.

the West—California, Idaho, and the Pacific slope—backed O'Niel[1], and the most money was to be made out of them.

I had the honor of carrying up the victor, Con Oram, and his second, Jim Connor, and a lot of roughs, about eight of them, and got beat or bilked out of their fare. But I got a load at night to some town six or eight miles above Helena, and some of them come back to Helena with me to spend the night a-fighting the tiger or at faro, to get back even what they had lost on O'Niel[1]'s fight....

In a day or two I bought a saddle horse and bridle, a pinto mare, for $35, of the Travis Brothers at auction, making nine head. I done some little freighting and let my horses rest, and one day in September (1866) sold my coach for $170 in cash, and one pinto horse valued at $75. It was too light to carry many passengers. October 16 sold the first pinto mare for $45 (I had paid $35 for her). So I had nine head still left. I sold a watch chain and four harnesses, or two sets, at $65, and one day some party wanted to go to Boise Basin, and wanted to buy my four mules, and I sold them for $280 in gold coin, and I took a load of freight to Blackfoot City and to the New York Gulch....[19]

Some time during the Fall of 1866 I bought another pinto or spotted horse to match mine, for $65, and a bay 3-year old filly half American, making me up to seven head, and at an auction one day bought a short cut-off two-barrel shotgun claimed to be one that George Ives used to own and had when hung at Alder Gulch or Virginia City. (I kept it until I sold out and came East in 1869.) George Ives was a lieutenant of Henry Plummer's, of the Robbers of Montana, all hung in [late] 1863 [and early] 1864.

Winter was coming on and I commenced to look around for a place to winter my horses, and Thomas Fruid, who had eight head of mules and horses, said we could find some good place on the East Gallatin River, where there was plenty of hay and grain, and the winters were generally mild, as it was about 150 miles southeast of Helena. So we both laid in our provisions and some clothing in Helena, and [set out] with another man who was going out to that part of the country to keep some stock, a Scotchman named Jim.

In two days we crossed the main Missouri River and kept down to where we struck the main Gallatin River (one of the main branches of the Missouri River) and we had to ferry quite a rapid stream. We kept up the main Gallatin to Gallatin City, and made some inquiries and went up the river to Morse Brothers Store, a large mercantile establishment, who owned a pack train, and run it to the different mines in Montana. One of the Morse Brothers was at home, and he told us of a good place in a swamp where there was a log cabin and part of an old stable we could get cheap to live in till Spring. It was in the timber and well sheltered,

---

[19] Situated in Deer Lodge County.

and farms all around where we could get hay, beef, buttermilk and such things, and groceries, provisions and clothing we could get at his store more reasonable than any store around and a better assortment.

So we went over and looked at the place, seven or eight miles acrosst the East Gallatin River. We forded the river in the Fall with our teams, and found the cabin, and we found we could fix it up, and build up our stable. We bought hay and straw, stacked some, and got out crotches and put up a shed against the strawstack, put on poles and covered it with brush first, then straw and hay, and poles to weight it down. We put in poles or chunks of split wood into the cracks of the old cabin, and built up the fireplace, made two little bunks, one on each side of the fireplace, for us to sleep single. We hung up a wagon sheet and a heavy blanket for a door. The cabin had no floor in it. We hauled up a lot of trees, the whole length, for firewood, then we had to buy a lot of wheat for our horses and mules. We got a large camp kettle, to hold eight or ten gallons, to boil the wheat in, and we got some oats, but they were scarce. (Wheat three dollars a bushel; it was the cheapest and best feed for our horses.) I went to Morse's store, and I had a lot of gold coin $20 pieces and Morse allowed me $23 each for them. I bought a pair of California mission goods pants for $16; one hat $4.50, and one pitchfork $4, and lots of other things.

It took more to winter us than we expected, and the winter the coldest I have ever experienced (1866–67) and worse, if anything, than the winter of 1861 and 1862 at Walla Walla, at least we felt it more (but our cabin was poor, with no floor). I am glad it did not last as long. We got butter at a close neighbor's, not a half mile off, and we got milk, butter and some vegetables. At James Martin's, a few miles off, we got meat, grain or flour, and bacon. The Martin boys had a large stock ranch and farm. . . . I sold them my 3-year-old bay filly in the Spring for $55, for I had three span without her, and she helped to pay my way during the last of winter and commencing of spring.

I got a little pointer, or setter, dog of my neighbors, a young pup, and got milk of them to raise him. I called him Tobias, or Toby. He was company for us two awhile; we had books to read, and talked of old times when I used to live in New York City, Brooklyn, and Newark, New Jersey, and Thomas Fruid had been back the year before and could tell how great a change had taken place since I left.

Some time during the Fall of 1866 or the Spring of 1867, a flouring mill owner from Deer Lodge Valley wished to make some contract with some parties beyond the Yellowstone River on the old Bozeman route through the Rocky Mountains, and he took the old mountaineer Bozeman with him as guide and traveling companion, and they passed through Gallatin Valley.[20] They were on horseback and led a packhorse with them con-

---

[20] John M. Bozeman was hardly an "old mountaineer." He was born in Georgia in January, 1837, and had ventured west in search of fortune, arriving at Cripple Creek,

taining their blankets, provision, and some changes of clothes, and they camped out. The Indians on that route, the Crows and Blackfeet and Sioux, had opposed the opening of an emigrant route through what they called their hunting grounds, but the Government took no notice of their opposition, but furnished the emigrants a soldier guard to protect their train, and the Indians had threatened the old guide, because he was the founder of that route.

Now when the flour contractor with old Bozeman got acrosst the Yellowstone River, some 20 to 25 miles from Bozeman City, West Gallatin Valley, Bozeman saw some strange Indian signs, and he told the miller that he did not like the signs, as they were Crow Indians and would trouble him if they could, and he felt uneasy about it. One day while they were stopping for noon and had their horses staked out and had got done cooking dinner, some Crow Indians came towards camp. Bozeman took his rifle, and told the other man to keep camp and not let the Indians come into camp, as he did not like the looks of them. Bozeman went out to get the three head of horses and was bringing them in, when the Indians shot Bozeman several shots, and he fell. The other man got on one of the horses and came back to a trading post on Yellowstone, and told of the shooting of Bozeman, and a party of men went out, some from Bozeman City, and brought in the body of Bozeman; he was dead, but he had wounded or killed some of the Indians, and a party from Bozeman City went out and tracked the Indians. They belonged some distance away, but were determined to kill the old guide for cutting through and opening the emigrant route, as they had often told him they would kill him for it.[21]

In the Spring of 1867, the governor, Thomas Francis Meagher, called for volunteers to fight the Indians on Yellowstone and Cow and Powder Rivers (Crows and Blackfeet and I think, Cheyennes, who had committed depredations on the settlers and miners and stockmen of Eastern Montana in about the middle of February, 1867).[22]

---

Colorado, in 1861 to do some placer mining. In June, 1862, he joined the rush to Virginia City's environs in Montana. The route Reinhart refers to, blazed in 1862–1863 by Bozeman and his partner, John M. Jacobs, was a route to the gold fields from the Oregon Trail. Simply: the route left the Oregon Trail at Fort Laramie, then went westward on the North Platte, northwestward, skirting the base of the Big Horn Mountains to the upper Yellowstone Valley. Bozeman Trail cut through Sioux Indian country and helped to provoke the hostility that erupted in 1867–1869. Merrill G. Burlingame, "John M. Bozeman, Montana Trailmaker," *Mississippi Valley Historical Review*, XXVII (1941), 541–568.

[21] Bozeman was slain by a small band of Indians on the banks of the Yellowstone, April 20, 1867. His companion was Thomas W. Coover. Grace R. Hebard and E. A. Brininstool, *The Bozeman Trail*, I, 221; II, 257, *note*.

[22] Montana Territory was created May 26, 1864. Thomas J. Meagher was territorial secretary, and at this time was acting governor since Sidney Egerton had resigned and the new appointee was still east. Meagher issued a call for volunteers in early summer, 1867. By July, three hundred mounted men from Helena and two hundred from Deer Lodge were assembled. Burlingame, *Montana Frontier*, pp. 122–123.

Four men came to our cabin and wanted to stay with us a few weeks till we got ready to go to Helena with our teams. They had been soldiers in the War and had been in the Crawford Emigrant Escort, they said, but after they had stayed with us a few days and we had got familiar with them, they told us more of their history.[23] One gave his name as Jack Farrel of New York City; one C. Miller of Baltimore, Maryland; one from Philadelphia (I forget his name); one from Indiana—we called him Long Jim—he was the youngest. After they had been with us a few weeks we found that they had been sent out to escort some emigrants, and they all four were regular enlisted soldiers and their company was stationed at Fort Phil Kearney, on the Bozeman Route. The Indians were very bad around Fort Phil Kearney and Fort Fetterman, and had killed a party of teamsters hauling hay to the fort.[24] Captain Brown went out to punish the Indians, but one small party had gone out a few days before to Powder River to escort the emigrant trains, and the four boys with us, with the consent of the sergeant or lieutenant in command of the small party, had deserted and come on to the East Gallatin Valley and hid. They had taken their guns, pistols, horses, saddles and bridles with them and had sold the horses and traded off their clothes and guns to miners and stayed out of sight till Spring and—lucky for them—an awful event took place after they deserted.

The Lieutenant went back to either Fort Phil Kearney or Fort Fetterman, and reported the deserters, but about that time the Indians run in some soldiers and came close to the fort, defying the Captain to come out and fight them. Captain Brown, of the company to which the four deserters belonged, and the lieutenant who had let them go (privately), all were of Brown's company, 96 strong. Captain Brown was a brave man and he took up their challenge, went out with some 80 or 85 men, and the Indians retreated to some ridges and gulches a mile or two from the fort, and the Indians let on that they were afraid to fight Captain Brown and his men, and kept retreating, and Captain Brown followed on and got into a fearful ambush of Indians who were concealed in the hollows and rocks, some 700 or 800 strong. [They] cut off Captain Brown, and butchered every man of the company, and nearly in sight of the fort.[25]

The commander of the fort could not go out to help Captain Brown, for the Indians had surrounded the fort and dared them all out to fight,

---

[23] See Chapter IX, *note 26, ante.*

[24] Fort Phil Kearney (not to be confused with Fort Kearney, Nebraska) was established in July, 1866, and was situated between the Big and Little Piney creeks on the east side of the Big Horn Mountains (Wyoming). Fort Fetterman was erected July 19, 1867, at the mouth of La Perle Creek, where it empties into the North Platte. It lay about eighty miles northwest of Fort Laramie in Wyoming.

[25] Reinhart's reference is to the "Fetterman Disaster," December 21, 1866, near Fort Phil Kearney: Captain William J. Fetterman's command was massacred, among them Captain Frederick H. Brown. The toll was eighty-one men.

but kept out of shooting distance from the fort. Some scout got out of the fort and went to another fort not far off, and the commander of the other fort sent reinforcements and they went out and gathered up the bones and bodies left of Captain Brown's soldiers. All had been scalped and maltreated. So they buried the remains, and an order came from Washington to abandon the fort, and take every movable article away to the nearest fort, and it was afterward destroyed.[26]

Now it was known all over the United States that Captain Brown's company had all got killed by the Indians, and it was supposed by the relations of the four boys who had deserted that they had been killed with the balance of their company, and it was lucky for the boys they had saved their lives by deserting, [as] they were not reported deserters on the soldiers' list in Washington and [no one] would . . . know of the desertion. They had all enlisted under false names. They were a hard set, and they told how, through the whole War, they had got the bounties given by cities, states, and United States, and would desert every chance they got, spend their money, and keep jumping bounties as long as the War lasted, and then when they got broke, enlisted again in the regular army, knowing they were to be sent to the forts in Montana, where they wanted to go, and when they got there to desert, as they had done, and by chance of the massacre of Captain Brown's whole command their crime of desertion was hid unless they themselves exposed it, which they did to Thomas Fruid and I, but I never told anyone about it.

About the 20th of March we concluded to take our horses, mules and wagons, and go to Helena, or on some creek close by, and camp out in our wagons until we could get some freight to haul from Helena to some surrounding mining camp. We would drive slow and be some seven or eight days a-going, and we expected the grass would start early in April or May, and if we had to feed hay we could get it 15 to 20 miles from town, where we expected to camp, and be that much closer to Helena. We bought up some wheat and oats to feed our horses, and I took Jack Farrel, the New Yorker, with me. Thomas Fruid started two days ahead, to get some wheat, by another road, and we were to meet at the Missouri River ferry.

We found the streams awful high yet, and hard to ferry. The main Gallatin was awful swift and high; we had to wait two days to get over. We passed several new gulches where rich discoveries had been made late in the Fall, and I stayed at some of them a few days to look at them, and to see how they prospected, but the water was too much for us, and we went on till we got to Beaver Creek.

[26] Colonel Henry B. Carrington, commanding Fort Phil Kearney, did send out a party to collect the dead. The Indian ambush was too effective and forced them to act otherwise. Carrington was relieved of his command and sent to an insignificant post on the North Platte. The fort was not abandoned until 1868. Details are given in Hebard and Brininstool, *Bozeman Trail*, I, 297–346.

After crossing the Mission Ferry I caught up with Thomas Fruid; he had one of the deserting boys with him, the other two had gone to Virginia City to try and get work. At Beaver Creek, where we stayed over night, we heard that snow still lay on, for over 16 or 18 miles, two or three feet deep. So when we got on a creek called the Spokane, we camped, and found a little old, and some new grass, on the sunny sides of some hills. We camped on the creek, slept in our wagons with double covers on them, and let the horses run on the hills to eat what roughness they could get, and we fed them boiled [wheat] and some hay.

We got at Beaver Creek about the 3d day of April. Thomas Fruid and I went up on the stage sleigh to Helena, the snow from two to three feet deep, and I had to pay $5 to go and return, and $5.75 for meals, lodging and drinks, and I bought some provisions at Wright's Store to take out to our camp as I went back. We left the two soldier boys to take care of our camp and watch the horses and mules. I had but six head left, and Fruid his eight head. The snow went off quite fast, and by April 15th I made arrangements to take a four-horse wagon load of liquors to Virginia City, Montana, for a wholesale firm in Helena named Murphy & Stevenson, on Bridge Street. Jack Farrel had stopped with me, and wanted to go to New York Gulch or Diamond for work. He had no money and I let him have four dollars to pay his way....

When I got all loaded and paid my expense and toll from Helena to Virginia City, I was nearly out of money, having spent during the winter all I had made during the summer, and it looked rather discouraging for Montana. I had so many toll roads and ferries to cross to go to Virginia City that it took off a good deal of the profits of the trip. I had quite bad roads, but got in all right and delivered my freight and got my money.

The merchants there were engaging teams to haul Sutler's supplies and provisions to the Gallatin Valley, for the army of volunteers was located on the West Gallatin near or beyond Bozeman City;[27] two companies from Virginia City; two from Helena. Captain Neil Howie, and Charley Curtis, an auctioneer, was lieutenant,[28] and they were all Vigilantes from Helena. I was offered good pay, or the promise of it when the war was over and the Congress would appropriate money for that purpose, but I thought that would not pay me, so declined.

I could not get any freight to take back with me, so I concluded to go over the divide, go down and cross the Madison River, the middle branch of the three main branches of the Missouri River, then cross over the next divide and come down on the West Gallatin River below Bozeman City,

---

[27] The volunteers were stationed in and around the vicinity of Fort Ellis, just west of Bozeman City and the East Gallatin River, a strategic defense for Bozeman Pass. A sutler is one who is licensed to sell goods, especially foodstuffs, to soldiers.

[28] Niel Howie was appointed 1st major; Charles D. Curtis, lieutenant, in Colonel Thomas Thoroughman's command.

and buy a load of wheat and oats to take back with me to Helena, either to feed myself or sell.

The Governor, Thomas Francis Meagher, was at Virginia City, but he kept full of whiskey while there. In two days he was to take the field and march against the Indians, and he and his staff had a merry time in Virginia City. It was quite a lively town, and had a great name for hanging the murderers and robbers of Henry Plummer and his whole gang, and Alder Gulch had been one of the richest mining gulches and camps in that part of Montana.

The second day after I left Virginia City I lost my five horses one night on a creek of good grass, and lost nearly all day finding them, some five miles off. They had felt good, and had a play and ran away that far, and were quietly feeding when I found them. I had commenced to feel uneasy about them.

That night the Governor and his staff of four or five passed by the place I stopped at, and got something to drink and smoke, and went on. He was a small round-faced light complexioned Irishman. I had already read about him in history, before he came to America, when he was tried by the English for high treason, and I had his engraving framed in my saloon on Althouse Creek and Indian Creek, California, in 1853 and 1855.[29]

I got to the Madison River and ferried over. Next day got to the West Gallatin at Saml. Morse's store, and stopped at Col. Foster's for dinner. I was acquainted with his son John and a young man herding for him when I was wintering on East Gallatin. The young man was named Charley Morrison of Bloomington, Illinois. Next day I bought my wheat within four miles of Bozeman City, and started back for Helena. . . . I got back to Helena on the 4th of May, 1867. . . .

Now here I was about eleven months after getting to Helena, and what I had was not worth as much as I got there with, and nearly a year lost. So I made up my mind that the coming year I would work it different, and not winter the coming winter in the high mountain valleys and spend in the winter for feed all I could make in the summer.

So in a few days I got a load of freight for my four-horse team, partly for a place at White Tail Deer Creek, some $18 worth, and the balance of the load to Argenta, a mining town not far from Bannack City. It was a silver mining town and had a smelting works.[30] I had to go by the Jeffer-

---

[29] Meagher was born in Waterford, Ireland, August 23, 1823. In his early youth he helped found the Irish Confederation. On July 11, 1848, the English authorities arrested him for sedition, brought him to trial in October at Clonmel, and convicted him of high treason. The death sentence was commuted in July, 1849, to banishment to Tasmania. From there in January, 1852, he escaped to the United States. After a lecture career, Meagher joined the Union Army and rose to brigadier general. In 1865, he was appointed territorial secretary for Montana, and at this time was acting governor. [Michael A. Leeson], *History of Montana, 1739–1885*, pp. 248–249.

[30] The 1864 discovery of rich silver deposits in Argenta created the first Montana silver excitement.

son Bridge and toll road, cross the Stinkwater Ferry, and keep up the East side of the Jefferson River to where I crossed back the Jefferson River by a ferry, and delivered my loading at Argenta all right, $82, making, with the $18, about $100 for the load.

I came back by Beaver River toll gate and Big Hole Bridge (toll, too), and kept down the west side of the Blackfoot Range of Mountains, by way of Butte City and Deer Lodge and Blackfoot City, and got to Helena the 27th of May, 1867.

I bought grey horse to mate Pinto and replace the other pinto lost (died), and I paid $65 for him. I bought a fine set of Hill's Concord Stage Harness, of a man named Alma, for $65, for my two bay horses Prince and John, and with my six horses and one wagon I took some passengers to Fort Benton. I had to pay $9 toll for the one wagon and six horses at the King and Gillette toll road, and $7.50 for ferrying at Sun River.

When I got to Fort Benton a steamer had just come in from St. Louis with passengers, and a man and his wife, a German butcher whom I used to know at Walla Walla and Boise; his name was Christopher Walters. I got my large wagon loaded with passengers, and there were still more to go to Helena. I had brought my two extra horses along, for sometimes there were buggies or carriages to haul up that had been shipped to parties in Helena, and I made inquiries and found that the Hazard Powder Company of St. Louis had sent up a new express wagon to advertise their powder. It was a heavy panel box spring wagon, strong enough to carry 1,500 pounds, but the Agent wanted to buy a span of horses to take it up, and he could take five or six passengers and make nearly the price of the horses, besides getting his wagon taken up. There were no horses to buy at Fort Benton [so we arranged that I would haul his wagon].... So I got ready and struck out with the two wagons loaded with passengers at $25 apiece in the spring wagon....

We had some very wet cold rains and some snow going up, and were longer than usual. When I got to Clark's, 25 miles from Helena, my black horse Jack got very weak, and I went in to Clark's to see if I could get some oats or barley to feed him, as we wanted to get through to Helena that day or towards night, but they had no oats or barley, but had some shelled corn to sell at 25 cents a pound, so I concluded to get some of it, so our horses would not give out.... By good care we got into Helena that night, and I got rid of my passengers and the spring wagon. My loads came to $386.82 for the trip, for ten or eleven days....

I bought an iron axle California wagon, $65, and engaged passengers to the amount of $54 for four, and on June 9th took two wagons and got to Fort Benton, and got freight back to Helena for Tutt & Donald's, about $200. I stopped at Helena a few days and old Jerry McKay of Roseburg and Coffee Creek, whom I used to clerk for in the store of Ransom & McKay, brought some horses from Walla Walla to sell. He had them in the

Travis Brothers' stable. He had other company from Walla Walla, Mr. Groville, my neighbor from the Walla Walla River about four miles from our claim and ranch on Dry Creek. He was the Frenchman who had a half-breed Indian wife, and some of his children were nearly white; one girl, a young lady, quite white and goodlooking, and two of her brothers dark as Indians. The girl, or young lady, had married George Harter of the firm of Harter and Keeler, who owned a large pack train of mules, had a ranch near the Hudson Bay Meadows, and lots of cattle. Groville had brought up some cattle with George Harter. . . .

On my . . . trip to Benton June 15th I had an old man, uncle to the Martin Brothers who lived close to us. When I wintered in Gallatin Valley he had bought my bay filly. He and his son were on their way to Kansas; his name was James Martin. . . .

I had two men driving teams for me; one named Edwards, and the other Ben Bowling. On my way up from [this] trip got 150 fire brick for some refinery and freight for Webster & Davis, in all, including one Negro passenger, $230.

On the 20th of July I loaded nine passengers for Fort Benton at $10 apiece in one wagon, and started Ben Bowling with them, and the 24th bought one more horse at Helena, a bay I called Bally (he was white-faced, and two white hind feet), for $63, making me ten head, and I loaded ten passengers to two wagons for Fort Benton at $100, and Edwards drove one team for me, and when we got to Fort Benton there were several steamers ready to go down the river to St. Louis, and Ben Bowl[ing] took a notion to go back home to Covington, Kentucky, where his father run a lot of drays around town and made lots of money, and Edwards got a good chance on a boat to work his passage, so I paid both up and they went down the river together. I got some freight and a few passengers back, and the round trip brought in $293.

I bought the third wagon, a new Schutler, at Helena for $143, and August 3rd I started to Fort Benton for freight with three wagons and ten head of horses. At the Spring, 27 miles this side of Benton, stopped over night. It was a stage station, and they kept one spring boxed up and with a lock and key to it. Water was very scarce, and the other spring was about dry, and I had to pay the stage hostler one dollar to water my ten horses out of their locked-up spring. When I got to Fort Benton one of the teams that was ahead had some passengers to get off, and we lay there some few days for freight.

In the spring, some time in the latter part of May, Governor Thomas Francis Meagher passed through Helena on his way to Fort Benton. It was after the Indian War, and the troops had been discharged, and I do not now remember whether the Governor was going East, or just stopped over at Benton a few days, but he was on a steamer and had been writing some letters. The cabin or his stateroom was too warm or close for him, and he stepped out of the door of his stateroom in the dark, stumbled over a rope cable and pitched headfirst overboard into the Missouri River. The alarm was at once given and some men rushed to the place he fell, and the ship's crew searched, but never saw or heard anything of him, for he never came to the surface of the water. For several days they looked for the body but did not find it; no doubt the undertow or undercurrent kept him under and he was carried far down the river and not found. Some said he had been drinking and staggered and fell over.[31]

The Government was to send out a newly appointed Governor, and he came to Fort Benton on the Steamer *Octavia*, a new boat on her first trip to Benton, and she made the quickest trip that had been made up the river to Fort Benton of any steamer that had come up (to that time).[32] The Governor, Green Clay Smith, was a Kentuckian, and his family and some of his staff were with him. He had a large lot of furniture and household goods. Books and goods were all to be freighted to Virginia City, where the seat of Government of Montana was located.[33]

A merchant named John A. Gaston of Helena, of the firm of Gaston & Simpson, was aboard the steamer with a stock of groceries, just up from St. Louis. While he was gone East, to Leavenworth, Kansas, where his wife and family resided, Simpson had gone into mining speculations, and he had moved a 20-stamp quartz mill he had been operating at New York Gulch, and he had put it up on the Cable Ledge, a new discovery, and somehow his creditors broke him up, and he was on his way down the river, and met John A. Gaston coming up on the *Octavia*. He told John A. Gaston about his failure, and as soon as Gaston got to Fort Benton he sold out the stock of groceries, to keep the old creditors of the firm from attaching them, and went back down the river on the *Octavia* as soon as she unloaded and got ready to go back....[34]

One day a passenger who had come up on the Steamer *Octavia* was on shore in a saloon at Fort Benton, and had been playing cards for the

---

[31] Meagher was drowned July 1, 1867, while on a reconnaissance of the Missouri near Fort Benton. New York *Times*, July 8, 1867.

[32] The *Octavia* was built in 1866. Captain Joseph B. LaBarge brought it up river on its maiden voyage in 1867, arriving at Fort Benton June 20, the fifteenth ship to dock during that season.

[33] Smith, who was serving his second term in the House of Representatives, resigned to accept appointment as governor of the Montana Territory. His commission was dated July 13, 1866. Smith and his family left Richmond, Kentucky, September 1, 1866, and arrived at Virginia City, the territorial capital since February 7, 1865, on the evening of October 3. They traveled on the Overland Stage by way of Salt Lake City, thence by the branch line to Virginia City. He went east to Washington in the early spring and returned on the *Octavia* with his heavy freight. James E. Callaway, "Governor Green Clay Smith, 1866–1868," *Montana Contributions*, V (1904), 108, 113, 120.

[34] Reinhart notes: "I got well acquainted with John A. Gaston afterwards at Sweetwater, South Pass City and Green River, Wyoming Territory, and afterwards again at New Chicago, Kansas, now Chanute, Kansas, 1886."

drinks, and a drunken Irishman who was in the saloon (I do not know if he was in the same game or not) claimed he had been robbed of some $80 or $90, and accused the passenger of the *Octavia* of robbing him, and the man ran to the steamer to keep from being abused by the drunken man and his partner. As he went on board to his family, he passed by the Governor, who was standing on the dock in front of the boat, with some of his staff and friends (I was close by). The drunken Irishman came after the man who had fled onto the steamer, and was cursing him for a thief and a robber, and came up to where the Governor was surrounded by his friends, talking and joking. The Governor asked the Irishman why he was trying to catch up to the passenger I spoke of, and the man told him he had robbed him. The Governor said he did not think the man would do it, and the Irishman commenced on the Governor, and said it was the like of him, and he was a damn thief too, as well as his friend. The Governor just grabbed up a sort of fence stake and struck the drunken Irishman over the head, knocked him down, and would have beat his brains out, but the governor's friends got his stick away from him. I was glad to see the drunken hound get what he did, for he deserved it, and the Governor had not said anything to him to make him curse him and say what he did. Next day it transpired that the drunken hound had just misplaced his money; he found it in some inside pocket where he himself had put it.

After looking for passengers and not finding any to go, for the stage had engaged a large lot, and I could find no freight to Helena, one of the Baker Brothers, commission merchants, asked me if I wanted to take a large lot of furniture and household goods to Virginia City for General Green Clay Smith, the Governor. It consisted of chairs, tables, wash stands (marble tops), bedsteads, bureaus, mostly light bulky loading, and would require about four wagons, not heavy loaded, to carry.

After considering the price, I contracted to take it all to Virginia City. I had only three wagons of my own, and I let a man with a light team take one of the lightest, bulkiest loads, and we took it all. I trailed one of the wagons, and put on six horses and drove the team, saving one driver, and the third wagon I had a man named Roads to drive with four horses. He had accompanied his wife and child as far as Fort Benton on their way to Indiana on a visit, and I got him to drive my teams to Virginia City, where he belonged, and he saved his passage up, and his expense, and I paid him light wages. The other man with the fourth wagon traveled in our company, and I had to help him up some bad pulls, for he had but two horses to his wagon, and in some places four horses had all they could do to make it.

We got to Virginia City all right about August 17, 1867, and unloaded our loading at the Governor's dwelling. His wife was a high-spirited Southern belle, daughter of Basil Duke of Kentucky.[35] The Governor had

[35] Smith's wife was Lena Duke, daughter of James K. Duke of Scott County, Ken-

been a Methodist preacher back in Kentucky, but got into the Union Army and distinguished himself and was a Major General. I believe he had stumped the State of Kentucky the second presidential election of old Abe Lincoln, and I expect was appointed by Andy Johnson after the death of Lincoln, and Thomas Francis Meagher.[36]

I got ready to start right back to Benton; got one family to take at $45, and six other passengers from Virginia City to Fort Benton at $66, and some trunks and freight from the widow of Thomas Francis Meagher, $12.80. She was on her way to New York, and had a light conveyance to carry her. She was the daughter of Dr. Townsend, the sarsaparilla man of New York, a very rich man.[37]

When I got to Helena I got four more passengers and some trunks and potatoes, and loaded up the second wagon, and divided our passengers, and I got to Fort Benton August 31st. Our load brought about $220 dollars down. I got a driver at Virginia City, named Levi Johnson, by the month. September 1st loaded up some freight to the Springs, to Sun River, and some to Clark, all way freight, and 30 stamps and dies for a quartz crushing mill to King & Gillette of Helena, in all $160 up. . . .

October 25th got a load for two wagons to Salmon City, Idaho. Took along two 4-horse teams and one riding horse, and left the two roans, the mule, and one other horse on the ranch. We started October 26, 1867, for Salmon City, a new discovery in the mountains 200 miles above Florence City, discovered in 1863 or 1864.[38] We had to take the same road as far as the bridge on the Big Hole River, cross and go up Red Rock, then go up to Wild Horse Prairie, and over a bad road over a divide, and came down on the Lemhigh Valley, passed the old fort, or adobe ruins of old Fort Lemhigh, a Mormon fort, and in Idaho Territory we kept down the river.[39] Some good farms and stock claims along the valley.

---

tucky. She was a grand niece of John Marshall.

[36] Smith was born in Richmond, Kentucky, July 4, 1826. He saw military service in the Mexican War and rose to the rank of brigadier general in the Union Army during the Civil War. A graduate of Transylvania University, he practiced law and was extremely able in politics. He resigned his governorship April 9, 1869, and moved to Washington, D. C., where he was ordained to the Baptist ministry. He was the National Prohibition Party presidential candidate in 1876. On June 29, 1895, he died in the national capital and was buried in Arlington National Cemetery. *Biographical Dictionary of the American Congress, 1774–1949*, p. 1827.

[37] Elizabeth Townsend, daughter of Peter Townsend, a wealthy Fifth Avenue, New York City merchant, was Meagher's second wife. His first wife, Catherine Bennett, died in Ireland in 1854.

[38] The strike was in the vicinity of the confluence of the Lemhi River and the South Fork of the Salmon River in present-day Lemhi County.

[39] Fort Lemhi was founded as a Mormon Indian Mission June 18, 1855. Originally named Limhi in honor of a *Book of Mormon* figure, the spelling was soon corrupted. Two years later, in early May, Brigham Young visited Idaho's first settlement. Because of an Indian attack and continued hostility, the mission was abandoned in March, 1858. W. W. Henderson (ed.), "The Salmon River Mission: Extracts from the Journal of L[ewis] W. Shurtliff," *Utah Historical Quarterly*, V (1932), 3–28.

We got to Salmon City at the mouth of that stream and our freight had to be packed in to Leesburg, on Salmon River, on a pack train.[40] I got rid of my loading but had to stay all night at the City, and I found Mr. Kroney, whom I knew at Walla Walla, keeping a brewery.... My freight to Salmon City came to about $300. James Buckley and wife were living here in town, and one of the Travis Brothers had a feed stable where I put up my nine head of horses two nights.

[40] On July 16, 1866, the "Discovery Company," composed of Frank B. Sharkey, Elijah Mulkey, Joseph Rapp, William Smith, and Ward Girton, struck gold on Nappias Creek near the mouth of a gulch afterwards named Wards Gulch. Soon afterwards, the town of Leesburg (named for Robert E. Lee) was laid out. Orion E. Kirkpatrick, *History of Leesburg Pioneers*, pp. 23–24.

# 14

## UTAH: MORMONS AND LOST HORSES

(1867–1868)

The day I got [to Salmon City] Ed Hutchinson, our old Indian Creek, California acquaintance, came down from Leesburg with another gambler named Hill, and they wanted to go right down to Salt Lake City, and I found some more passengers who wanted to go to winter there, and in the Spring they wanted to go to the Sweetwater or South Pass City Mines, a new discovery.¹ By looking about I found some freight for Salt Lake City; Col. Noun[na]n of the firm of Noun[na]n & Kiskadin, bankers of Salt Lake City.²

So I concluded to load up the two wagons that I had with me with passengers and freight for Salt Lake City, and let George Woodbury take them down, and then take the eight horses and the two wagons and go to some place and stay and recruit up the horses until I came, for I concluded to winter in Salt Lake Valley that winter....

I kept with them until I got to my turning-off place to go back to Helena. We all stayed at Hall's over night, and old Major Henry Switzer, the passenger who was to drive the teams with George Woodbury, had been stopping here, and we had some way freight for Hall. I bid the boys all goodbye and took my pinto horse with his saddle, and I left George some money for expense, and him to collect freight at Salt Lake City, and I started back towards Helena to get my other horses I had left on the ranch at a little town called Montana....

I got back the 19th of November, and found I could get a lot of passengers to Salt Lake City at a good price if I had a good conveyance, and I

---

¹ For the Sweetwater Mines, see *note* 15, *post*.
² Joseph F. Nounnan, a prominent non-Mormon, was the cashier of the Miner's National Bank, founded in 1866. *Deseret Evening News*, January 8, 1868. He became the first grand secretary of the Masonic Grand Lodge of Utah.

found an 11-passenger heavy thorough braced coach, one of the J. C. Oliver Stage Company's, and bought it for about $300, with some harness, and I bought another set of harness, so as to work six horses on the coach.[3] I got some repairing done to coach and harness and bought a large brown horse of one of the passengers who was to go to Salt Lake City with us, for $75, making me fourteen head. The man was a carpenter, and I took a chest of tools for him. I sold my Schutler wagon for $125, for I had no use for it, and got all ready, with provision for the trip, for we expected to have a chance of being caught in the storms in crossing the Rocky Mountains so late in the season.

My passengers were George Hurt, Alex Williams, Charles Fish, William Richmond, Robert Rippley, Clay Thompson, Willey the carpenter, Frenchy Duvall and Fuller, in all eight passengers, at $246. We got all started and had a good time until we got into the mountains in the snow, and had to do some hard roughing and driving, but got over the mountains to Idaho Springs in some awful snowstorms and snowdrifts, but by all working well we got over the Rocky Mountains all right and got nobody froze or hurt. The boys passed their time in keeping the curtains down and keeping the snow and cold out, and playing cards. I had to sit on the outside and drive, but sometimes one would spell me for a while till I got a little warm inside, but I was dressed warm in a good cavalry overcoat and fur cap, and plenty of buffalo robes to wrap up in. We had so many toll roads, bridges, and ferries that it took quite a little pile of money to pay our way, and when we stopped close to houses in rough weather we would get the privilege of making our beds down on the floor by paying 50 cents per bed near the fire.

December 5th we crossed Snake River at Taylor's Ferry,[4] and the 6th were at Marsden's Hotel, and paid toll through the Portneuf Canyon, where there had been some robberies committed on the Wells and Fargo stages and some of the robbers killed; one a stage driver was in company with them. That night we stayed at a place called the Robbers' Roost, and next day Marsh Valley; December 7th, Peck's at Malad City, a Mormon settlement,[5] stopped at the Dutchman's, a Josephite.[6] Next night Bear

---

[3] Oliver's Stage Company ran a weekly four-horse stage to Salt Lake City. The fare was $50.00 and the trip usually took eight days. J[ohn] L. Campbell, *Idaho and Montana: New Gold Regions. The Emigrant's Guide*, p. 25. Reinhart recruited his passengers for this trip at fares ranging from $25.00 to $40.00 each.

[4] The original ferry across the Snake River was operated by Harry Rickett. In the winter of 1865, James Madison (Matt) Taylor built a toll bridge. This led to the creation of a settlement which in 1872 was renamed Eagle Rock and in 1891, Idaho Falls. M[errill] D. Beal, *A History of Southeastern Idaho*, p. 218.

[5] Henry Peck ran the hotel in Malad City, thirteen miles north of the Utah-Idaho state line. The community was founded in 1855 by a small band of Mormons. In 1866 it became the seat for Oneida County.

[6] Josephites, organized in 1860 at Plano, Illinois, advocated Joseph Smith, Jr., the son

River Toll Ferry (now Cori[n]ne), and next at Oregon Springs; then at Brigham City (Box Elder); next we got to Ogden City, terminus of the Union Pacific and the Central Pacific.[7]

We next got to Case Ward, a small town, next to Sessions, where I found George Woodbury and the eight horses and two wagons.[8] I changed horses and put on my two blacks and two bays, and with the two roans made a spanking six-horse coach team.

We got into Salt Lake City in good style, drove up to the Salt Lake City Hotel, kept by Little, son-in-law of Brigham Young, Sr.,[9] where several of my gambling passengers got off, and Clay Thompson too. The City looked fine to me, and I thought it was the nicest small city I ever saw. The streets were very wide, and the gutters flowing with cool clear water from the mountains above, and the buildings were fine brick, some of them four stories high. The Tabernacle and theatre and Bee Hive Buildings were all very fine, and on the side of the mountains northeast, not over one mile, was Fort Douglas, flying the United States flag, the old Star Spangled Banner—it was a great sight.[10] From the hotel I drove to the Salt Lake Bank of Noun[na]n & Kiskadin, and drew some money and left some packages, and then drove to a livery stable and put up our horses. George Woodbury had hitched up the other horses and came on that same evening, December 17, 1867.

Next day we got all unloaded, bought some provisions, and prepared to hunt up a place for to make it our winter quarters. On December 19th started in the morning with the two four-horse wagons and the six-horse coach to go west some 45 miles to a place called Stockton or Silver Canyon. The first day out we stopped at an English Mormon's. The next day, the 20th, we got to a small town called Tooel[e].[11] It commenced to snow

---

of the martyred Mormon founder, as the head of the church. They also disavowed the practice of polygamy.

[7] The two railroads had not yet been completed. Ogden was made the official terminus by Congress, and became so by the completion of the Union Pacific tracks to there, March 8, 1869, and the subsequent joining of the two railroads at Promontory Point, May 10.

Corinne, Utah was not founded until early February, 1869. The town site was laid out by J. E. House, chief engineer of the Union Pacific. Utah granted a charter to the community on February 18, 1870.

Box Elder was the earlier name for Brigham City.

[8] Case Ward is now Karpville, and Sessions is called Bountiful.

[9] Salt Lake House was operated by Feramorz Little, who was Brigham Young's brother-in-law. It was jointly owned by them. It was Salt Lake City's leading hotel and was purchased in 1865. Andrew Jenson, *Latter-Day Saint Biographical Encyclopedia*, II, 486.

[10] Fort Douglas was erected by Colonel Patrick E. Connor in the fall of 1862 and was officially dedicated on October 26.

[11] Tooele was founded in 1849 and is about eight miles west of Stockton which grew up around ore deposits found by California Volunteers stationed at Fort Douglas during the Civil War.

fearfully that night, but the Mormon we had put up with gave us a good supper. There was a wedding of his daughter, and they danced all night, but we slept well and had a splendid breakfast in the morning. Our landlord was a New Yorker (State), and his wife from Connecticut, and were both regular Yankees and knew how to get up a good meal, and after we went into camp six miles away, sometimes we used to come to get a good meal got up for us. They were fine clever old-fashioned farmers, but Mormons (he had two wives). The last wife a young healthy robust English girl, not over 19 or 20 years old; some of the first wife's children as old if not older, but she seemed to make a good servant for his first wife, and she did not seem to care.

We got a late start on account of the snow still falling, and we got to Stockton, a mining town of 200 or 300 inhabitants, a hotel, two or three stores, two saloons and blacksmiths, butchers, etc. We got some dinner and inquired for a good place to winter our stock, and were told of a place up a canyon, called Silver Canyon, where there was a good spring and plenty of timber for firewood and good shelter for the horses in the hills under the low fir and cedar trees, where the bunch grass grew all winter, and the sun kept the snow melted under the low cedars. So we drove up the Silver Canyon about one and one-half miles and put up our tent on a flat a short distance below the spring, and placed our wagons, one on each side, and the coach in back.

Our big tent had been the hospital tent of the Gallatin Army, when Governor Thomas Francis Meagher went out from Virginia City; I had bought it at auction after the war. It was about 12′ or 14′ by 22′, high wall. The wagons all around made a good shelter, and we built a fireplace at the front, cut cedar boughs and covered the ground and spread blankets and made us bunks, and when we had a good fire we felt quite comfortable. There was an old adobe house a quarter of a mile above us, but it looked too damp and cold to fix up to live in, but two miners came in with a few pack animals, and lived in it, and they were good company for us to pass away the long days and nights. One named Burns, an Irish sailor, was good company, and he and his partner had been in the Blackfoot Mines and at Deer Lodge. We got all straightened up in a few days, and had our Christmas suppers in Stockton at the hotel.

My two drivers were George Woodbury of Worcester, Massachusetts, and Major Henry Switzer of Hartford, Connecticut, who had been a splendid bookkeeper or accountant in New York City before he went to California, then to Idaho City on Moore's Creek, where he still had claims. He stopped all winter with us.

New Year's Day, 1868, George and I went on horseback to our Yankee

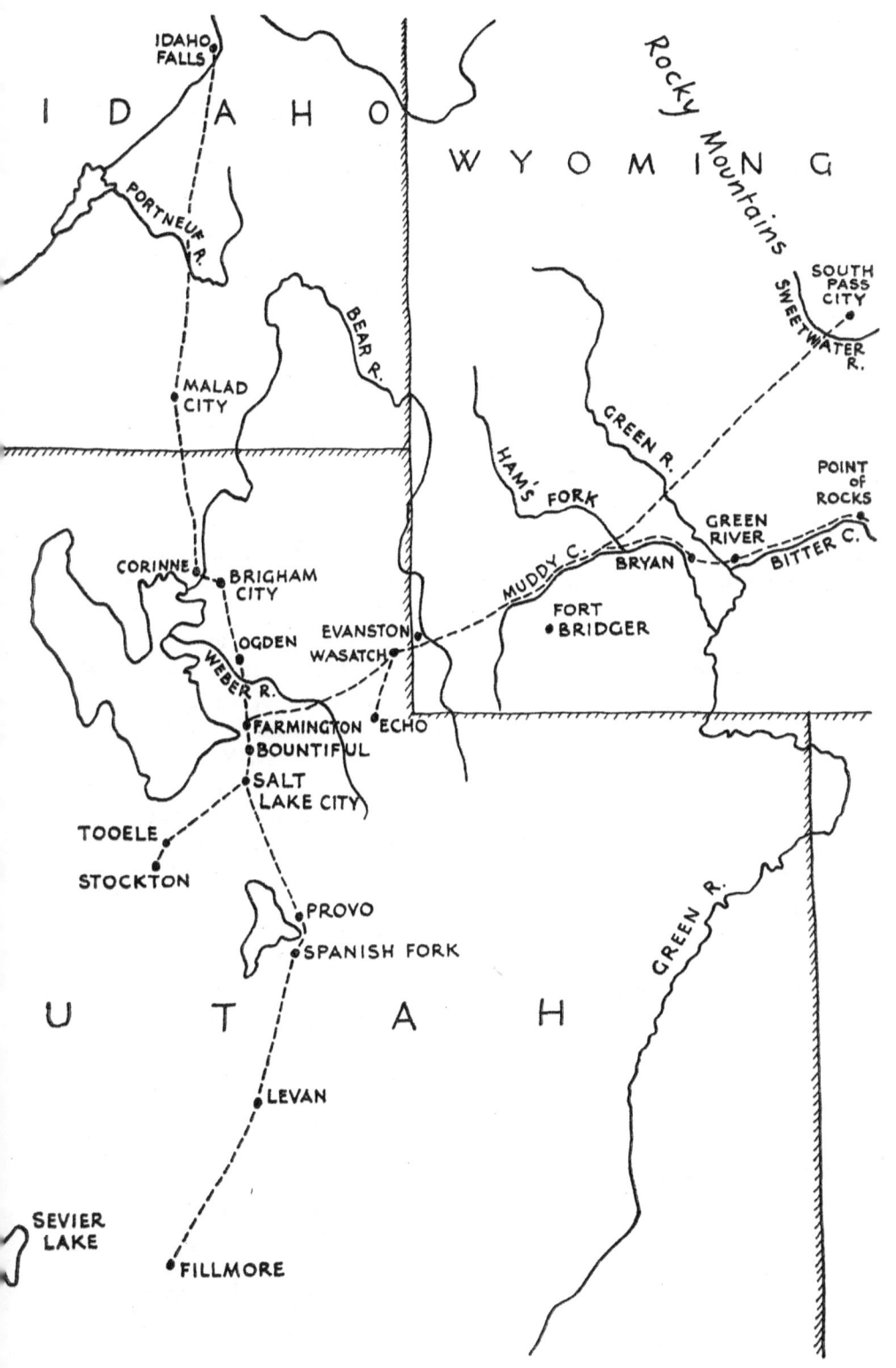

Map 6. Utah and Wyoming

Mormon friends at Tooel[e], and had a good dinner, and I bought a lot of reading matter, papers, magazines and books, for $3 the bunch, and it kept me in good reading for the time. We used to go up on the hills about twice a week, and bring down the horses to make them drink, for they would eat snow and not come down if not drove down. The weather was awful cold, and considerable snow, but somehow the horses kept under the low cedar trees and there was no snow under them, and the hills facing south, it would melt and the grass kept green all winter, and the horses would paw the snow off in some places to get to the green bunch grass, and got fat. When we went in to winter them it was December 25th and by March 10th they were in splendid working order, and I had not fed them over a bushel of oats (and that to give them some worm medicine in—rosin and copperas).

Some four or five years before, some soldiers or California Volunteers (miners) with General Connor,[12] had found some silver ore, prospecting in this canyon, and there was great excitement about it, and some New York company, and some German company and several different ones had built up smelting works close to Stockton (the town being named after the California town of Stockton, General Connor's home), and the whole country was staked off and hundreds of shafts sunk, and tunnels run into the hills, and found very rich.[13] But the cost of transportation to the East was too high, for the ore would have to be hauled in wagons to the Missouri River, and then on trains by rail to New York, and on to Wales, for there was no works to separate the silver from the lead, and transportation from Salt Lake City to Omaha was worth 25 cts per pound. It was found they would have to wait until the railroads would be built before it would pay. I heard old General Connor had about broke himself up prospecting and then building the smelting works. . . .[14]

The weather got fine and I got some neatsfoot oil and Ivory Black, took all my harness apart and repaired, washed, oiled and spunched them, all ready to go on the road; put all my halters, feed sacks, wagon breaks, whips, and all ready. Major Switzer took a trip to Salt Lake City to see

---

[12] Patrick E. Connor was born in Ireland, March 17, 1820. After service in the Mexican War, he went to California, eventually settling at Stockton. When the Civil War broke out, he was made Colonel of the 3rd California Volunteer Infantry and ordered to Utah to guard the overland mail route. In 1865, he was given command of the vast Plains District, and was mustered out the following year. He died at Salt Lake City, December 17, 1891.

[13] Connor organized the first Utah mining district at Bingham Canyon. He discovered ore in Little Cotton Canyon and developed mining and smelting operations at Stockton. The Knickerbocker and Argenta Mining and Smelting Company of New York City exploited Rush Valley.

[14] Andrew L. Neff, *History of Utah, 1847 to 1869*, p. 643, notes that mining venture profits for 1865 to 1870 totaled no more than $597,129, or $589,933 for gold and $7,196 for silver. Mining became profitable only with the coming of the railroad to Utah.

if he could get some store to clerk until May, and another man named Sol Norton (From McKokety[?], Iowa) came and staid with me. He wanted to go to South Pass City, to the Sweetwater Mountains in the spring, and I thought I would take up passengers, and if the mines did not suit me, to try and get loading for some other point, Idaho or Montana.

I had partly made up my mind to run in eggs, butter, and peaches as soon as they got ripe, by fast stages, and stationing my teams so as to carry passengers back and keep supplying Helena and Virginia City and all other mining camps. I had made a calculation that by investing my profits into more teams and wagons I could clear about 8000 dollars the coming year of 1868. My expense for provisions and repairing was fearful high, and I had to take some medicines myself. I felt nervous and asthmatic and I had to take Hostetter's Stomach Bitters at $3 per bottle, and all we bought was two or three prices, but I could not help it.

March 24th I got some five of my horses newly shod. I heard that there would be a rush to the Sweetwater Mines as soon as the Spring opened up, and I had just the outfit to carry passengers and fast freight at good prices....[15]

April 1st, 1868, got some more of my horses shod, and we broke up camp and started again for Salt Lake City. Our horses were in fine fix, as if they had been kept on plenty of grain, all nicely shed off, everything in tiptop order. Drove to within four miles of the City and camped at a place called Brighton, put up our tent, and wagons all around, as we had in Silver Canyon....

I went into the city and found a good many passengers to go to South Pass City as soon as the waters would permit, and Noun[na]n the banker said he would engage passage on my coach at $40 or $45 apiece and sixty pounds of baggage to each passenger, and I could take a lot of freight and passengers on the two four-horse wagons, so as to make it up to $900 to

[15] The *Sweetwater Mines*, March 24, 1869, reported:

Gold in the Sweetwater district was first discovered in 1842 by a Georgian who came here with the American Fur Company for the recovery of his health. After remaining a year he started for home, intending to organize a company and bring them here to work the mines. He never reached his home, however, and was supposed to have been killed by the Indians.

The real boom began in the fall of 1867 when, on June 8th, H. S. Reedall, J. W. Lawrence, and Harry Hubbel, among others, hit the Carissa lode (originally discovered in 1865 by Tom Ryan who later uncovered the Carter lode). South Pass City was laid out that fall. Thus did a short-lived fame come to a historic Oregon Trail landmark. Rossiter W. Raymond, *Mines and Mining of the Rocky Mountains* . . ., pp. 237–238. Newspaper extract quoted in C[harles] G. Coutant, *The History of Wyoming*, p. 637. (The *Sweetwater Mines* newspaper is extremely scarce. The Bancroft Library, University of California, Berkeley, holds all known copies. The above cited issue is not in their holdings.)

$1,000 for a trip. I was to start in about ten days. When I got home I told the boys and they were glad to get off so soon.

One day a man with two ponies rode up to the tent, and I knew him. I had known him at Coffee Creek, Oregon, Roseburg, Jacksonville, and at Walla Walla. His name was James Wilson. He was the man who had taken the Thomas Kav[a]naugh horse, and missed the discovery he had made of Gold Hill. I was surprised to see him. He told me he had wintered four or five miles west of Stockton at a sawmill. He had heard I was at Stockton but had not seen me. He had been at Tooel[e] some, too. I asked him where he was going; he said to South Pass City, Sweetwater, that he had been there last Fall and had taken up some claims and he wanted to get in as soon as possible. I asked him to stay over night, which he did.

Next day I drove in my six horses on my coach, and he rode in with me. He left his ponies at my camp; one he had packed, and the other he rode. I looked about town, and saw the passengers, to see how soon to start, and engage more, and let them see my team and coach. It was on a Monday, and I made my arrangements to start on the following Monday (a week from then), and I went back to camp. Wilson left next day with his two ponies; he carried a rifle and wore a coonskin cap.

Wednesday evening we had up our horses and salted them all, and then hobbled four of the largest leaders and kept Whitey or Pinto staked close to camp. I took my night glass, or field glass, and just dusk could see the thirteen head grazing off west towards the lake. We had been cautioned not to let our horses go on to some wheatfields north and east of us, where a German Mormon named H[i]skey lived.[16] He was a fine old Pennsylvania farmer, and we got our milk and butter from him, and we were on good terms with him and all the Mormons close by, so we took pains to keep our horses from molesting them.

I considered they were all right and away from the wheatfield direction, so we went to bed, all quiet. Next morning early George or Sol went out for the horses while I was getting our breakfast. I got all ready, when they came in; they should have taken the saddle horse staked close by, but they found he had pulled up his picket pin and rope and was gone, and they thought he had gone on to where the others would be, but found no trace or saw no horses.

So we all got breakfast and two of us struck out on foot. I found the tracks which led off South, and after going some four or five miles lost track of them, there were so many other tracks. Our horses were all fresh shod all around. I kept on and could see horses away off, but when I got there they were not mine, and I saw cattle that I would take to be horses

---

[16] This should be Benjamin Hiskey.

until I would get nearly to them before I would see that they were cattle, and I got off ten or twelve miles and it commenced to get dusk.

I found a cabin but no one in it, nor could I open it, and I was so tired that I could go no further so I laid down in a shed. I had no overcoat, and the night still very cold, and I had to pass a cold night, and I had a scant early breakfast. I was much troubled with my ankles; I would wrench them every little while, they seemed weak, and I suffered from them. When it came daylight I took a look as far as I could see, and looked at the tracks, but there had been cattle and horses all over the country, and my eyes got weak and watery in looking long distances on the plains or prairies, and I struck for our camp, about 12 miles off. I was in hopes the other boys would find them while I was gone, but when I got to camp there was no sign or news of them.

Some of us thought that some of the Mormons had drove them into the hills so that they would not get on their wheat fields. The hills were 7 or 8 miles west and we thought they might have struck back to Silver Canyon, where we had wintered, and Sol Norton had gone back a ways. Some thought they had been stolen.

I got my breakfast, and I was nearly used up. I told George to go to where I had left off, and get out of the valley where all the cattle and stock tracks were, and strike a ridge beyond the valley and he could examine the tracks, for our horses were all sharp new shod, and fourteen head would surely make a plain trail. I went to the city to make inquiries, but could hear of none. I got back to camp at night and George had returned but no news of the horses; he had gone over the same ground I had gone over already, instead of going beyond the regular stock grazing part of the valley. I began to get uneasy and thought they might have been stolen, and I wrote a letter to South Pass City to James Wilson to look out for my horses up there, as my whole fourteen head were gone, and he would know my horses, as he had stopped in my tent at Brighton several days and saw all the horses. I told him to look for them and let me know at Salt Lake City. I addressed the letter to South Pass City, Wyoming Territory.

We were all on foot and could not get horses to ride, for the farmers were all busy with their horses plowing, so we had hunted our stock afoot. It was tiresome, and we could not get far away to look, so I went to Salt Lake City to the livery stables to get horses to ride and hunt mine on, but they made all kinds of excuses, that they had no saddle horses, and did not like to let their driving horses go to hunt horses on, as they would be rode hard and fast; so I could get none. Everybody had his opinion, and advice to offer, that the Mormons had stolen them, and the Mormons said it might be that the notorious Bill Hickman and his brother George had stolen them and drove them into the mountains southwest of

Salt Lake City, some 20 or 30 miles, and kept them there hid, as there were plenty of gulches to hide them in, and a stranger would not find them in a year.[17]

I went to the United States Judge in Salt Lake City,[18] and he said I had better get a good hunter or mountaineer, who understood the country, and get him to hunt for me. I telegraphed west and north, and I found two young men, Mormons, who kept a general store in the City, named the Shaw Brothers, and I got two ponies to hire at $5 per day, and I had to promise to keep it secret, as the livery stable keepers would have prosecuted them for doing livery business without a license.

So I got George to take one horse, and Norton the other, and they took along blankets, and some provisions to last them two days and nights. I kept making inquiries in town of all those passing through the city. I went to see Brigham Young, and he could only refer me to a stockman who had a big stock ranch about 180 miles west, at Point of Rock, close to Nevada. On that road he had herders herding his stock, and if the horses were stolen and drove west they might see or hear of them, for seven or eight were large American work horses about 16 hands high. I went to see the man.

He was the famous, or infamous, Porter C. Rockwell of the Destroying Angels; he was their leader when they massacred the immigrants in the Mountain Meadows in 1857.[19] He made me a fair offer, better than any

---

[17] The Hickman brothers were former Mormons who had turned to a number of extra-legal activities. [William A.] Hickman, *Brigham's Destroying Angels: Being the Life, Confession, and Startling Disclosures of the Notorious Bill Hickman, The Danite Chief of Utah*, passim.

[18] Utah Territory was in a judicial turnover at this time. Associate Justice Thomas J. Drake was in the East, as was his colleague, Solomon P. McCurdy. The appointment of Chief Justice John Titus expired in January, 1868. McCurdy was appointed as his successor, while Enos D. Hoge, Salt Lake City, was appointed associate justice. Perhaps Reinhart approached the latter since McCurdy was still in the East. Earl S. Pomeroy, *The Territories and the United States, 1861–1890*, p. 117; *Deseret News*, January 29, 1868.

[19] On September 8, 1857, about 140 California-bound emigrants from bitterly anti-Mormon districts of Arkansas and Missouri, were attacked by Mormons and Indians at Mountain Meadows in southern Utah. On the morning of September 11, the party surrendered under a flag of truce only to be deliberately slaughtered—only seventeen very young children were spared. The participating Mormons were led by John D. Lee, who twenty years later was executed for his part in this butchery. Juanita Brooks, in *John Doyle Lee*, and in *Mountain Meadows Massacre*, treats this incident in detail.

Orrin Porter Rockwell, to whom Reinhart should refer, served first as a major, then as a colonel, in the Nauvoo Legion (Utah militia) in the "Utah War," 1857–1858. Part of the tactics of the Utahans against what they felt was the U. S. Government's aggressive designs was to destroy U. S. Army supplies and outposts in a "scorched-earth" policy. A balanced treatment of the 1857–1858 events is given in Neff, *History of Utah*, pp. 410–484, and the important "Utah War" documents have been published by LeRoy R. Hafen and Ann W. Hafen (eds.), *The Utah Expedition, 1857–1858*.

Rockwell was chief of the Danites, the Mormon church's "Destroying Angels," a secret police force established in 1838. He also served as deputy marshall of Salt Lake City for a number of years. His biographers label him "the Terror of the Plains." He

Gentile or white man in Salt Lake City. He said he would take one of his horses, let me have one, and we could put a pack of blankets and provisions on the third, and he would go with me to his place at the Point of Rock, 180 miles, and hunt all the way there, and then get his herders to start from there to hunt my horses. He would be to all expense and if we got the horses I was to let him have a choice of two horses for his trouble. He seemed to act perfectly square and like a liberal gentleman, (if he was a Mormon). I was to let him know in a few days if I should conclude to take his offer, so as to prepare for our trip.

I kept inquiring of all stockmen in the valley but could not hear anything, and I had about run out of money in fitting out my teams and my expense had been high, and the passengers that I had engaged began to feel that I would not be able to go, for I had not heard of the horses. After being out two days George came in. He had heard of some horses being seen at Bingham Canyon, a new discovery about 27 or 30 miles in the west mountains, and he had gone over there and found some horses, but not mine. Sol Norton came in, and he had heard of nothing.

While we were getting supper a man with a load of hay, on his wagon, came up to camp, and George Woodbury said he was Brigham H. Young, the nephew of old Brigham Young, and George knew him.[20] George had been in Salt Lake City a year before, and had got acquainted with him (he, George, was a distant relation of Brigham H. Young on his mother's side). George spoke to Mr. Young, and asked him to stay all night with us, as it was still four miles to the city and getting dusk. So he stayed with us, and as we had supper about ready, eat with us.

After supper George told me to tell Brigham H. Young how I was situated—my horses all stolen and I was about out of money to hunt them with (I had only some seven or eight dollars of gold specimens left). We told him about the livery men not letting us have any horses to hunt ours with, and one of us had to stay and watch camp while the others hunted the horses. He told me I could have my coach and two wagons hauled to his barnyard in town, and we could put our seven sets of harness in his granary, and he had a small house close by the barn with a stove in it where we could cook and stay, and he would loan me $50 on one of my wagons, and if I got back my horses I could pay him back; if not, he would pay me the balance of the wagon, $100 more, and keep it, and I might then dispose of my harness, coach, and other wagon, and any provisions I needed I could get from him, such as flour, meat, vegetables, potatoes, butter, milk, eggs, and I could settle when I got ready to leave.

---

is described by them as a "murderer, [who] robbed, and terrorized; he burned and scourged; he perjured and lied." He did not, however, take part in the Mountain Meadow Massacre. Charles Kelly and Hoffman Birney, *Holy Murder: The Story of Porter Rockwell*, pp. 3-4, 21-24, *et seq.*

[20] Brigham Hamilton Young was Brigham Young's nephew.

Now I was surprised at his splendid offer right in time of need, and I was awful thankful, and next day moved all my things into the city to his place, and found everything just to suit me, and we got our meals in the little house. I sent George Woodbury out again and Norton in another direction. I got the $50 from Mr. Brigham H. Young, and let George and Norton have some money along to pay their expenses while out.

Mr. Young had two wives and families in separate houses, all well fixed. His first wife was a fine lady. They were all from Massachusetts. He had a daughter, a fine player on the piano, and their house was furnished all first class, and his children well raised. They lived in houses one on each side of our little house; each one had a fine house, and a quarter block of a lot to their houses, all in good style.

Now I felt quite bad to be here at Salt Lake City with but seven sets of harness and my coach and wagons, and no money to buy horses with, and if I should have to sell all I had left would not have brought over seven hundred dollars.[21] I was owing George about $80 or $90 and I was under obligation to take George and Norton to South Pass City anyhow, but here I was, after being away from home eighteen years, not heard of any of my relatives for fourteen years, had spent the best part of my life roughing it, and taking all kinds of risks of my life for the eighteen years, and after having got property to the amount of about $2,500, to lose it all at once, when I had the prospect of making from $5,000 to $8,000 the coming year. It was discouraging, and I felt very revengeful toward who must have stolen my horses, and it seemed as if I could kill them on sight. I had come to the conclusion to sell the harnesses, coach, and other wagon, pay George off, and get both a passage to South Pass City Mines, take what money I would have left, buy me a horse to ride, and make it the one thing to do to find the man or men that stole my horses and no matter how long, or how far I would have to go, to get them and kill them, if I had to hang for it. The next minute I felt desperate.

I found that there had been more stock lost or stolen; mules belonging to a Mormon bishop, and he came to see me and got a description of my horses, and he gave me a description of his mules, and while we were all hunting we might come acrost one another's stock, and we would make a more thorough hunt for it; he had telegraphed all around, too. . . .

One evening George Woodbury got back from the hunt and said he came acrost a man and his son who had moved some sheep and some young cattle from the valley west some 25 miles and he went along a track with the sheep and cattle in the evening, and next morning came back a horseback over the same road, and saw that during the night a lot of fresh shod horses had been drove over the trail after his sheep had

---

[21] While at Silver Creek Canyon, Reinhart inventoried the property he held and valued at $2,724.00 in gold coin. *Reinhart Manuscript.*

passed along and said it was some twelve miles from our camp, and it was on Wednesday night, he thought two weeks ago, and George went to where the man and boy said, and saw the horse tracks quite plain yet, and he was positive they were my horses' tracks, but being as it was nearly two weeks ago, thought no use to go after them until he let me know about it, for by that time they would no doubt be 200 or 300 miles away.

Now where George saw the fresh shod tracks was on a straight line to the hills or ridge from our camp, and the first night I was out I had been within three or four miles of the place, and when I came in that morning, tired and hungry, I told George to go acrosst the place I had been and strike to the high land beyond, where the stock had their regular grazing grounds, and to go acrost that to the rise of the ridge, and had he done so, he would have got on the track that very day, not over four miles from where I slept under the shed. After that we could have kept on and got to the horses in two or three days. So by a fatal neglect to take my advice we had lost the horses.

Sol Norton came in. He had been back to Silver Canyon and had not heard or saw anything of the horses. So we let the ponies rest and it being Saturday, we went to the theatre to see Stark, the American tragedian, play "King Lear," and one of Brigham Young's daughters played Cordelia in the same piece, and a man named Kimball made a "Leap for Life."[22] The theatre looked like a New York theatre, only smaller, and the pit, or parquette, was full of Mormons, the women knitting and sewing and embroidering between acts. Us Gentiles had the boxes at 50 cents apiece, or seat, and we were well entertained. They had good stock actors, all Mormons, and kept good order. The Gentiles were very liberal and patronized the theatre well.[23]

Sunday morning we had laid off our route for Monday, and I was going to telegraph south towards New Mexico, or Parranegut, a mining camp near a Mormon settlement at or beyond St. George.[24] After breakfast I went around to a livery stable and as I walked along a man said the Chief of Police of Salt Lake City was looking for me, and I walked up

---

[22] James Stark, a noted and successful actor in San Francisco and much of the West, played the theater, January 7-21, 1868. An examination of the *Deseret Evening News*, January 1-21, 1868, discloses that Stark performed Shakespeare's, "The Merry Wives of Windsor," taking the role of Falstaff, on the 14th. His other presentations were either Shakespearean adaptations or popular fare: "Brutus," "Richelieu," "Money," "Money and Misery," "Katherine and Petruchio," and "Jack Cade." "King Lear" was presented on May 7, with George Waldron and his wife. *Ibid.*, May 7, 1868. In 1864, Stark served as a member of Nevada's state constitutional convention. George D. Pyper, *The Romance of an Old Playhouse*, p. 181; San Francisco *Golden Era*, August 23, 1863; January 31, 1864.

[23] Brigham Young sanctioned the building of the Salt Lake Theater in 1861. Construction began on July 1 and was completed on March 5, 1862. It was a very active thespian center. Pyper, *Old Playhouse*, p. 85, *et seq*.

[24] Reinhart probably refers to Pahranagut, a mining town in Nevada.

towards the police office and met the Mormon Bishop who had lost the mules and had a description of my horses, coming towards me with another gentleman. He introduced me to him, and asked how many horses I had lost; I told him fourteen head, and the Chief of Police told me he had got a dispatch that morning from the Sheriff of Fillmore City, 160 miles south, that they had arrested two men on suspicion, with a lot of stolen horses, fourteen head in all.[25] I asked him if they had the names of the men, and he said one's name was Jackson and one Anderson, but they looked like father and son. I asked if we could send a message right down to the sheriff, but Burd said not till evening when they could call the roll, as all the employes would be off, as it was Sunday, and for me to come around in the evening and we could send some messages.

After a little the Sheriff of Salt Lake City, named Burton, came to me to see about getting the horses, and he saw Burd, the Chief of Police, but he did not give him any satisfaction. The Sheriff did not like it because the Sheriff at Fillmore did not telegraph to him (so that he could get his fingers in the pie on the horses). Burton was a bad man, and all the Gentiles hated him. He was head of the Destroying Angels after Porter C. Rockwell was out of that office.[26] I went around to the Police office and we asked the names of the men held in custody at Fillmore, and they said Jackson and Anderson, and one claimed to be my partner, and that he had a wife and family at Tooel[e], that he was a Mormon. I told them it could not be, for I knew no one at Tooel[e], a Mormon, but we telegraphed back to hold the two prisoners until we would come, and to take good care of the stock, and the Chief of Police said he would send his deputy chief, named Calder, with me, and go Monday morning in the 9 o'clock stage, and we would get to Fillmore on Tuesday or Wednesday.[27] I got two saddles to take along for us to ride. George Woodbury and Sol Norton and Brigham H. Young were glad to hear the good news.

Now I will go back a few days to show how near a person can come to

---

[25] I have been unable to identify with certainty the sheriff of Fillmore. Guy L. Robins, Millard County clerk, in reply to my query, wrote: ". . . the oldest citizen here [in Fillmore] says Peter Huntsman was sheriff sometime near 1868 but he isn't certain just when."

On April 4, 1868, the *Deseret Evening News* reported: ". . . a couple of horse thieves were caught in Millard county and taken into Fillmore, with fourteen head of horses." The *Deseret News* reported the same news item, May 5, 1868, giving warning to local citizens of such occurences.

[26] Jenson, *Biographical Encyclopedia*, I, 238–241, gives a detailed sketch of Robert T. Burton, who was sheriff. He held many civic and elective positions, including President Lincoln's appointment as collector of internal revenue. He discharged his peace officer duties with integrity and zeal. Burton served in the "Utah War" as a colonel of the Mormon militia. Andrew Burt was the chief of police at this time.

Porter apparently retained his leadership over the Danites, or "Destroying Angels," having only been challenged by William A. Hickman. Burton reportedly had a hand in the murder of Dr. J. King Robinson. Kelly and Birney, *Holy Murder*, pp. 228–230.

[27] I have been unable to identify the Deputy Calder mentioned by Reinhart.

finding things out and not do so. One evening while I was still camped at Brighton, I had been in town and stopped at a house on a creek not far from town, between the River Jordan; there was a man and his wife, and I asked if they had seen or heard of any stray or lost horses, and they said not. They asked on what night I had lost them; I told them Wednesday, and the old woman said someone had stolen a long stake rope off of one of their horses and they had tracked the person with two ponies off toward our camp at Brighton, and they thought they knew who got their rope, that a man with two ponies, riding one and leading the other, had camped on Tuesday close by their house, and he had gone to town and came back where he had left one of his horses, and another young man with him, and camped. The oldest man who had come first had a Mississippi Rifle (or Yorger by some called),[28] and he had on a fur cap, and his front teeth were out, and one called himself Jackson; the other, a young man of about 19 or 20, said his name was Anderson, but the old folks thought he was the son of the old one. They said they were going to South Pass City, to the Sweetwater Mines, and started next morning towards Salt Lake City together, and that night the rope was stolen off of their horse, and the man had tracked the man who took the rope towards Brighton, and saw two pony tracks going that way.

Now when I got the dispatch from Fillmore, and the names, I began to see that they were the parties who had stolen the stake rope, and then rode over to where our tent was and pulled up the picket pin and led off Pinto, and then took the hobbles off of the four American horses and all started South with them, and so much stock running at large had tramped out their horseshoe signs, and I lost them and came back. If I had been on horseback no doubt I would have gone on, clear of the stock range, and would have struck their track on the ridges beyond.

So Monday, April 28, the Deputy Chief of Police Calder and myself took the stage. There were several officers and lawyers on their way to Provo, where the District Court was being held, and we had good company, and got a good dinner at a hotel kept by a Mormon widow. That night had supper at Provo, quite a nice looking town. The peach and apple trees were in blossom, and I never saw finer views of villages; they looked from a distance like large lilac patches. Some fine farms in good cultivation. All along the route we passed many villages.

---

[28] "In 1841, [Army] Ordnance adopted a brand-new rifle with a percussion lock, the last to fire the traditional round ball. . . . It was generally known as the 'Yager rifle,' a misspelling of the German *Jaeger,* or as the 'Mississippi Rifle,' as a result of its performance in the hands of Jefferson Davis' Mississippi Regiment during the heroic stand at the Battle of Buena Vista."

It appears that "Yorger" is a further misspelling of *"Jaeger,"* or that it was pronounced this way by this time. Larry Koller, *The Fireside Book of Guns,* p. 65. (My young friend, J. Thomas Dixon, provided me this reference.)

On the 29th had breakfast at Chicken Creek and dinner at Round Valley. While the men were changing horses I asked the new driver if he had seen the men who they had in custody at Fillmore, and he said yes, that the old one was a hard case, and claimed to be a Mormon of family from Tooel[e], named Anderson. I asked him to give me a description of him, and he said he was tall, slim, and had lost his front teeth, had on old Yorger, and wore a fur cap. As soon as he said "fur cap," I all at once knew it must be Wilson, who had left us at Brighton on Tuesday morning for South Pass City.

Now I could account for the two men at the old man's house where they had taken his picket rope. I was dumbfounded to think that after knowing Wilson some eight or nine years, and letting him stay in our tent and eat with us he would do so mean a trick on me, and taking every horse, and leaving me worse than afoot, when he had two ponies and provisions to take him to the mines, where he said he had good claims. It was inexcusable.

We got to Fillmore about 11 o'clock A.M. and drove up to the hotel kept by McBride, where both prisoners were kept. The horses were in a lot back of the hotel. There was quite a crowd at the hotel and the Sheriff had the two men in charge upstairs.

As soon as I got the dispatch at Salt Lake City on Sunday morning, and heard the names, I knew they were the two men with the two ponies that had stolen the picket rope, and I knew that they should have sixteen head, with my fourteen horses, and they must have sold two or divided with some other horse thieves which I thought there must be, and had the Bishop's mules and others, and I was afraid that some of mine were not with that lot. But I had no idea that it was my old acquaintance Wilson until the driver said "fur cap." The hotelkeeper McBride showed me the horses, but the two Chapman roan horses were not among them; twelve of mine, and the two ponies, were all there was.

I went upstairs with the Police Chief and saw James Wilson with handcuffs on his wrists. He was very much ashamed of himself, or pretended that he was when he saw me. I asked him how I had ever misused him that he should serve me such a mean trick. He said he was awful sorry that he had done so, but he did not know what came over him; at first he was jealous because I had so many nice big horses and such a fine outfit and he had but the two ponies, and he thought if he could run off seven or eight of the best and take them down to New Mexico he would make quite a raise and I could soon make up my loss again. The other young fellow (he said) had persuaded him to take them all and they would be more likely to get aw[a]y with them, for if they left me any I would be apt to follow on and I would be more likely to overhaul them, but by taking all the horses we would be left afoot and not have the means at hand to track and follow them. He laid the blame mostly to Anderson, and he begged me to give him a chance to repay me my loss, and I might take the two ponies to pay part of my expenses so far.

At first I felt like taking my pistol and blowing his (Wilson's) brains out on the spot, but after considering that I had got nearly all my horses back again, by the merest chance in the world, I then got a more forgiving mood on me, and he was in the hands of the Sheriff, and I would not be justified in killing him, by law. So I got calm and cool and asked him what he had done to the two bobtail roan horses, the Chapman roans; he said they had drove on Thursday night after they started, all night, and in the dark had lost them two and did not discover their loss till daylight, and then they were afraid to go back to where they must have lost them, for fear we might be close behind them or we would catch up that much while they were hunting the two lost, so went on without them. I did not believe them, and still thought they were in with a gang of horse thieves who had the two roans, and the mules lost, and other stock missing in the valley, but they assured me it was the truth.

Mr. Calder had an order from Chief Burd of Salt Lake City for the prisoners, and from Judge Smith, one of the three head judges of Utah Territory,[29] and we were to leave in a short time to take our prisoners and horses back to Salt Lake City. Mr. McBride and the herder then told us how they came to arrest the two prisoners on suspicion of having stolen the horses, and Wilson helped explain how he got where he did. To commence with Wilson, they used to drive at night and lay by in some gulch in the daytime, and kept away round in the edge of the mountains, outside of all travel until they got to S[e]v[i]er Lake, and they calculated to keep on the west side of the lake and not strike the road to Parranegut or St. George, 300 to 400 miles ahead, but in the dark Wilson rode Pinto ahead and Anderson drove the horses after him. By chance Wilson in the dark got on the east side of S[e]v[i]er Lake and went on and had to cross the S[e]v[i]er River, running into the lake on the east side. The river was quite high, and the horse Wilson was riding had to swim. Wilson had his gun tied to the horn of the saddle and a sack containing some provisions to last them to the settlements in New Mexico when the horse began to swim. He strangled, and in the struggle Wilson had to jump off, and in doing so broke loose the gun and sack of provisions, and the horse came out all right, so did Wilson, but the provision was gone, and nothing to eat, and no gun to shoot anything with to live on.

So they were compelled to strike in for the settlements to get some provisions and when they got close to Fillmore they had the horses in a canyon out of sight that day, intending to go into Fillmore that night

---

[29] Elias Smith was probate judge of Salt Lake County, an office he held until 1882. In this period the probate court was practically the peoples' court of justice. Jenson, *Biographical Encyclopedia*, I, 722.

after the provision, but a herder that was out looking for some stray stock saw the fresh horse tracks, and saw they were going South, contrary to all the travel at this time, which was all going North toward Salt Lake City. He rather thought it strange, and saw the horses hid in a gulch, and the men, one riding a pack saddle and the other a riding saddle, keeping the stock out of sight. He came into Fillmore and told McBride and the Sheriff and a lot more men around town, and they concluded to go out and look at the men and the horses. So they got the herder to get up their own horses close to town, but the herder could only find two or three of them, and the sheriff's horse was not among them. So he and most all the others got out of the notion to go, but McBride and the herder and a cowboy had their horses and said they would go, just for the fun of a little ride. The others tried to persuade them not to go, that they might be some cattlemen from New Mexico on their way home from Montana or Idaho, but the landlord of the hotel, McBride, told the two others to come, and away they went.

When they got to the place where the herder had seen them last, they were not there. They had moved on, and the three men followed the tracks a short distance and lost it, and all rode ahead a while, but no signs, and they were just about to turn back when one of them saw a fresh cud of tobacco on the rocky road or gravel, and saw some tobacco spit along the ground, and they just kept on a short distance and struck the fresh track, and directly saw the horses in a gulch and the men had camped. McBride rode up and spoke to Wilson and asked him where he was going. He said to New Mexico, but he was waiting for his partner whom he had left at Salt Lake City with more stock. The other young man was up the gulch with the horses, so McBride told the two men to stay with Wilson and he rode up to Anderson and asked him the questions he had just asked of Wilson, and he told a different story about it.

So McBride asked the young man to come down to where Wilson and the others were, and when he got there he sent the herder to drive down all the horses, and they packed up and brought them all down to Fillmore City. The crowd got around to asking them questions, and Wilson got to telling too many stories, and the men saw that he was the head, or leader, and put a rope around his neck and pulled him up to a tree, and he just confessed all about it, and told them the horses belonged to Herman F. Reinhart, and how they had stolen them.

This was on a Saturday evening, and the Sheriff just telegraphed to the Chief of Police in Salt Lake City on Sunday morning. That night they both tried to bribe the guard to let them go, offering them four of the best horses, and they had made offers to McBride when he first brought them in, but McBride was an old thief-catcher and Fillmore was a regular thief-catching place, nearly out of the Salt Lake settlement and most of the men belonged to the Vigilantes, and at Salt Lake City they told me if I had just said it they would have hung them for me in a short time. But Wilson let on and claimed to be a Mormon and that he had a wife and four children at Tooel[e], and the Mormon women at Fillmore City sympathized with him in his troubles. Calder told me if they had known he was a Gentile they would liked to have helped hang him.

We got out our saddles and saddled up two of the best for Calder and me to ride, and put the other two saddles, one a packsaddle, on two not so fast for the two prisoners. We took off their handcuffs and made one go ahead and the other to drive the horses. We bid the folks at Fillmore goodby and started about two o'clock back for Salt Lake City. When we got to Round Prairie, or Valley, where we had changed horses before, we stopped and got them to get us supper. We wanted the horses to graze a little.... [We] saddled up and rode until after dark to a little town. We put up at a Mormon Bishop's and were treated like brothers, good beds and good meals. We made up a shakedown near our bed for the two prisoners, and when they got into bed Calder put handcuffs, one on each of their wrists, and we had a good night's sleep. We rode and drove about 50 miles the first day out from Fillmore,[30] and my black horse Jacks... always took the lead, and he could trot as fast as the others could lope. The prisoners complained of getting sore (one was riding a pack saddle), and we made them take turns riding it. We made them do the driving of the horses, and we followed on behind; sometimes we would be back over a half a mile, but they were afraid, for we always kept the fastest horses, and if they had made a break to escape we could have soon overhauled them, and we could have called on all the Mormon help we wanted, and Wilson knew it. Calder told me if I really felt like killing Wilson I could easy do so, and I would not be molested by the authorities; all I would have to do would be to let Wilson get a start, and say he was trying to escape, and I could fill his body full of holes, for I had two large revolvers with me, and so did Calder. But I felt I could nearly forgive Wilson for his thieving, for I was so thankful to get back the most of [the] horses, which I had all but given up as lost forever, and I knew Wilson would be convicted and go to the penitentiary for ten or fifteen years, and that would be punishment for him anyway. He kept asking and begging us to have it made as easy for him as possible, and he would some time replace all my loss when he would get free again, and he acted very sorry all the time on the road.

Next day we got an early start, and stopped at noon at a place called Spanish Fork with some Mormon Bishop,[31] and we had good dinners and our horses fed, and at night we again stopped at a Mormon bishop's, and we were treated as if we were Mormon brothers, and got the best they had to eat, and good beds. I told the Bishop about the losing of our two

---

[30] The town may have been Levan which is some fifty miles north of Fillmore.
[31] Bishop John K. Thurber lived at Spanish Fork.

best horses, and Wilson described the place to him, and the Bishop said he had heard of cattle and young stock running not far from the place Wilson had lost the two horses and would have his herder to hunt them up. I told him if he found them and brought them in all right to Salt Lake City I would pay him $25.

Next day we stopped at Provo, had dinner at the place we had stopped on our way down. There was to be a big Mormon conference held at Provo, and all the big high church officials and dignitaries would be at Provo, and in the afternoon we went on, and within twenty miles of Provo we met a fine carriage and two or three persons inside. When they got up near us Calder saluted the gentleman in the carriage, and he told me it was President Brigham Young.[32] He motioned for Calder to come to the carriage, and they spoke a while, and Calder told him about the finding of my horses, and there were the thieves. Brigham Young said they deserved hanging, and they bid us good evening and drove on.

We got within twelve or fifteen miles of Salt Lake City the third day out from Fillmore. We found stabling that night; it rained. We had good suppers, breakfasts and beds, and before noon we got into Salt Lake City. We took the two prisoners to the police station and they took them to a jail for safekeeping. I took the horses to Brigham H. Young's place, where our boys were, and they were awful glad to see the horses again.

Burton the Sheriff came to see me, and talked like I ought to pay all expenses and trouble, he thought about $300, but the Chief of Police Burd, and Deputy Calder said to not mind what Burton the Sheriff said, that Judge Smith had told him to tell me to come to his residence to dinner next day. I done so, and he said he was glad I had got the horses back, as I must be prejudiced against living with the Mormons, but I must overlook it, they could not help it, and if the thieves had been Mormons they should have punished them severely right off, in fact, he thought they would have been hung. But as it was, they were both Gentiles they should have a fair trial, as there had been ill feeling between the Mormons and the Government in regards to the assassination of Dr. Robinson. So the Mormons did not wish to displease our Government officers, who had said General Connor would hold President Brigham Young responsible in the future. He invited me to dinner and he told me he was sorry that I had lost so much time and money in the stealing of my horses, but that I should not pay one cent of the expense the authorities had been to, that it was their place to get back my property after being taken in hand at Fillmore by the Sheriff there, and he hoped I would not have any hard unpleasant feeling against the welfare of Salt Lake City or Utah, and hoped I would return again and winter my stock, and not have them stolen again. I told him about Burton, the Sheriff, and he told me that

---

[32] President Young left Salt Lake City May 1, for Provo, probably for a Stake Conference. *Deseret Evening News*, May 1, 1868.

was all a get-up of Burton's, and for me to pay him nothing. I told him that Wilson the thief had offered me the two ponies to pay part of my expenses, but I would turn them over to the Chief of Police and the city authorities could do with them as they pleased.

So I left Judge Smith, one of the Mighty Three, with a favorable impression, and of the Mormons in general.[33]

Now all my passengers that I had engaged for my coach and wagons for South Pass City, Sweetwater Mines, had all gone some two weeks. I had a big expense and borrowed money to raise to pay Brigham H. Young his money, and for provisions, and for the two ponies I hired from the Shaw Brothers. I could not get many passengers or freight for South Pass, so concluded to sell one of the wagons and some harness and the big tent, so as to pay up my debts and get money to fit up for a trip as soon as possible. I sold the wagon to Shaw Brothers for $110, the tent and harness for $55, to some party in the city. . . .

I got what horses need shoeing shod up again, and on the 11th day of May, 1868, a man came to me from the Bishop, where we had stayed all night on our way up from Fillmore. He had got his herder to look for my two roan horses that Wilson had lost, and he found them not far from where they had been lost, and they were fat and in splendid order. He brought them to me and I paid him the $25 reward I had offered the Bishop to find them. I could have sold them right off for $350, but I did not want to part with them yet. They had been undisturbed for over three weeks and on good grass, and they felt and looked well.

So I got all ready to go to South Pass City, paid up all my bills, and gave Brigham H. Young ten dollars more than he asked me; he had made his bill so very moderate. I had my stage fare to Fillmore to pay, and all my expenses, but I found five passengers for the coach, at $118, and some baggage at some $10, and 2,000 pounds of flour in the wagon as freight to South Pass City, at $120, so in all it came to $248. One of the passengers was Kimball, the performer at the theatre, who had made the Leap for Life.

So I bid all my Salt Lake City friends goodbye, and left on the 15th of May, 1868. I saw Wilson a few days before, in Jail. His trial was to be on the first Monday in June, and I got excused from having to wait that long to appear against him as the prosecuting witness, for it would have detained me two or three weeks longer at a big expense to me for my men and horses. Judge Smith (one of the vice presidents of the State for Desserette for Utah)[34] said that Mr. H[i]sy[e]y, the German who owned the

[33] Elias Smith was president of all the high priests in the church from 1870 to 1877, and later president of the high priests' quorum in the Salt Lake Stake of Zion, 1877–1888. Reinhart's reference, "one of the Mighty Three," probably refers to the fact that Judge Smith was one of Joseph Smith's close relations living in Utah and was a power in the church. Jenson, *Biographical Encyclopedia*, I, 719–721.

[34] Smith was elected chief justice of the State of Deseret, April 16, 1862. The state-

farm where I was camped when my horses were stolen and saw Wilson in my tent at the time, could be the main witness against Wilson and Jackson....

---

hood effort failed. Utah later replaced the name of Deseret.

# 15

## HAULING FOR THE UNION PACIFIC

(1868–1869)

The first night out we stopped between the Sessions[1] and Farmington; the second night in Weber Canyon, and had to pay $6 toll, and we had toll roads, and toll bridges, and toll ferries, until I was sick of them. At Echo Canyon the water was very high, and all the way, clear to Green River, it was the same; Yellow Creek, Bear River, Sulphur Creek, Muddy Bridge, Smith's Fork, and Ham's Fork of Green River—about $40 in tolls to South Pass City, Sweetwater Mines. We got in May 30th, and found a lively mining town and many old acquaintances.[2] John A. Gaston of Helena was running a grocery store. . . .

I put up my horses at a ranch and had them herded while I was in the mines and unloaded my freight and passengers. I looked around the mines, but the water was still too high to mine or prospect, and George

---

[1] Peregrine Sessions founded the community in 1847. First called Session's Settlement, today it bears the name Bountiful.

[2] John W. Clampitt, special agent, Post Office Department, Salt Lake City, writing to the postmaster general, observed:

> . . . South Pass City is situated within eighteen miles of the South Pass of the Rocky Mountains, and distant from Cheyenne, D[akota] T[erritory], two hundred and fifty miles, and from Fort Bridger one hundred and forty miles. . . . It has a population of one thousand *bona fide* residents, who have laid off the town in streets and squares and erected a large number of houses. . . . The population of South Pass City in one or two months from the present date will be at least three thousand, and judging from other reports presumed to be reliable by the 4th of July, there will be a population of ten thousand persons to celebrate, at that point, the nation's anniversary.

*Sweetwater Mines*, March 28, 1868. (It should be noted that the Territory of Wyoming was not created until July 25, 1868 when President Andrew Johnson signed the Congressional act. At this time, Wyoming was divided between the territories of Utah and Dakota.)

and Norton did not like it. I looked around to see if I could find any passengers for Idaho or Montana, but could find but four or five Chinamen who wanted to go to Virginia City, at $80 apiece, but that would not justify me to go back 800 miles.

The big excitement and rush to South Pass City was over, and all had returned to Utah, Idaho, and Montana, and the different mining places they had come from, but if I had not lost my horses I could have got into the Sweetwater Mines with my first load of passengers, baggage and light freight, and cleared over $1,000; unloaded, and could have got a load for Salmon River, Virginia City, or Helena, all passengers for my three-horse teams and maybe bought a fourth team if I could have made it pay.[3] My passengers and loading for the three or four teams for Montana would have cleared me about $2,000, and I should have been in Montana when the great White Pines Mines were discovered in Nevada.[4] The big rush would have given me a trip for all the teams I should have bought with the money I had cleared going up, and what I should have had left from my South Pass City trip, and I would not be exaggerating the trip to White Pines from Montana, Idaho, and way passengers to say four or five thousand dollars, and it would have brought me to White Pines in July, and if the mines had not suited me I could have put my teams all on the railroad (the Central Pacific) work at from seven to eight dollars per day per span. But now that was past and lost.

So I concluded to take the present and the future. I had got sort of discouraged by the horses being stolen and the direct consequence of it to me of the loss of about eight thousand dollars, and I considered it one chance out of a thousand that I ever got them back. So I concluded to take what passengers or loading I could find for Green River, where the Union Pacific Railroad was surveying and working and contracting for on the grades.

When I unloaded my flour at South Pass City, the consignee was deputy United States Collector of Internal Revenue for Utah, and I paid him $10 for my license for the year of 1868.[5] So then I found three of my

---

[3] The Sweetwater Mines quickly turned out to be poor. For a first-hand and splendid view of this short-lived mining boom, see Lola M. Homsher (ed.), *South Pass, 1868: James Chisholm's Journal of the Wyoming Gold Rush*.

Reinhart would have made more profit in transportation than in mining since Wells, Fargo and Company had not yet established a stage line from Cheyenne, the end of the Union Pacific Railroad tracks at this time, to South Pass City. The *Sweetwater Mines*, March 28, 1868, gave voice to the hope that: "As soon as the spring dawns, Wells Fargo & Co. will place a line of stages from the end of the Railroad to South Pass City and from thence to connect at Hams Fork with the great overland route, over which the United States Mails are now being transported...."

[4] The White Pine Mining District around Hamilton, later created as White Pine County, April 1, 1869, was organized on October 10, 1865. In the fall of 1867 a discovery was made at Treasure Hill which created a short-lived rush in mid-1868. Rossiter W. Raymond, *Mines and Mining of the Rocky Mountains*, pp. 145–185.

[5] Robert T. Burton was appointed deputy collector of internal revenue by President

other passengers from Salt Lake would go with me to Green River, and George Woodbury and Sol Norton went along. They did not care about mining. I thought at Green River I could either carry passengers or work teams until Fall, sell out, and take a trip back to Illinois, Wisconsin and New York, and see if I could see or hear from my sisters from whom I had not heard for fourteen years. Woodbury and Norton wanted to go, too, after making a little money, and go back East to their folks. My passengers paid me $15 (just for company).

About June 3d, 1868, I got to a place newly laid off as Green River City, at the mouth of Bitter Creek, down which creek the railroad came; it was being built, grass sod and some frame and log houses, mostly saloons and a few provision stores. Lots were held quite high. The town was just above the ferry, but it was uncertain where the future Green River City was to be, above or not. After we stopped a few days we located our camp and took up some lots about one mile above. The surveyor had run off the grades and located bridges at different points. I put my horses on a ranch where they were herded down the river bottoms. The town was all excitement, taking up lots, and building at different parts. We stayed around town and looked to get into work for my teams.[6]

A contractor named Jocelyn who was station agent above the ferry one mile kept the railroad supplies and had a large storehouse on the west side of Green River, took a contract from the river west, and promised me work for my teams as soon as they had commenced operation on the bridge. Mr. Jocelyn was a fine little gentleman and we thought we would wait on him to go to work, but I got tired of the expense while laying by. The town was full of gamblers and drinking saloons, and gambling and fighting and shooting was a common thing.

The roughs got up a prize fight on an island in the river. Kimball put up his ropes and made a Leap for Life, and performed on tight and slack ropes, and many got drunk, fought, and were beat out of their money.

I ran quite low on funds to pay our expense. I had to sell what few gold specimens I had been carrying for a long time. About the 20th of July I got my teams to work on the Jocelyn contract, at $5 per day per team. I had seven teams, but some were too free and high-spirited to work on scrapers on a 16-foot fill to commence on.[7] They had been used to staging

---

Lincoln, serving from 1862 to 1869. *Deseret Evening News*, April 25, 1868. For a brief sketch, see Chapter XIV, *note* 26, *ante*.

[6] In early June, 1868, H. M. Hook, the first mayor of Cheyenne, and Frank B. Gilbert, formerly of Salt Lake City, staked out a city site on the eastern bank of the Green River "about a mile above the stage crossing. . . ." When the railroad came through, a property-right row resulted, and the railroad laid out its own town on the western bank of the Green. Since the townsite company acted without railroad authorization, its investment was impaired by the creation of Green River City. *Sweetwater Mines*, June 3, 1868; *Frontier Index*, July 3, 1868; Mary L. Pence and Lola M. Homsher, *The Ghost Towns of Wyoming*, pp. 197–198.

[7] Scrapers prepared the way for the building of the rail track bed.

and wagons, and the scraper work was too slow and hard to pull in the sage brush and willow roots, but I got in from four to six teams most of the time. Caldwell the foreman bought a half interest in one of my teams (the 6-horse leaders, Frank and Captain). I could not make them work coolly, so we got a man to drive them on a scraper by themselves, but after the second day Frank was so hot and fiery that the man took them to the river in the evening to cool them off, and he poured cold water on his back while he was all of a sweat, and next morning Caldwell and I had a dead horse, between us. I bought back Captain and worked him with one of the others who was cool and slow. I had sold the two for $225, and bought Captain back at $100. (It was very careless in the man to throw water on him in that condition.) I had some six or seven drivers at $2 per day, and I cleared $3 on each team.

Sol Norton took sick with the mountain fever, and had to lay off work, but after a short time Jocelyn gave him a job herding the horses and mules nights. The balance of our crowd of teamsters were nearly all Mormons, and the railroad camp above us were all Irish (Cheseborough & McKees),[8] and the Irish graders used to steal our scraper doubletrees, and I lost several of mine, but when they found out that I was not a Mormon, left them alone and took the Mormon boys' only. I had for drivers George Woodbury, Sol Norton, Tim O'Neil, James O'Neil, E. Davis, Dinsmore, John Rushman, John Williams, and sometimes three peons or Mexicans. . . .

After we had got nearly through with that contract, George and Norton and Ward, another driver, told me that the work was too hard on the horses and I had better carry passengers with my coach and wagons from Green River to the terminus of the Union Pacific Railroad on Bitter Creek, at the Point of Rock, 40 miles [east], and passengers paid five dollars apiece, coming or going.[9] So I concluded to draw off my teams when payday came, and I had worked up to September 10th, forty-eight days, which came to nearly eleven hundred dollars, including expense (from Jocelyn).

I fixed up my teams and wagon to haul, and on the 14th of September took four passengers to the end of track, about 45 miles, at $20. When we got down to the railroad it made me feel homesick, and George Woodbury and Sol Norton concluded to go East. Norton was not quite strong of the mountain fever and the night air was bad for him, and Jocelyn got him a ticket to Omaha free, and I gave him $20, and George was to see Norton home, in Iowa, near Clinton. I paid off George, over $200, and

---

[8] Cheesebrough and MaGee were contractors who supplied railroad construction laborers.
[9] Point of Rocks is seventy miles directly south from South Pass City. The Union Pacific did not reach Green River City until October 2. Chicago *Tribune*, October 5, 1868.

saw them on the cars. Oh, how I longed to go, too! I had not seen a railroad or cars since I left Chicago, in August, 1850, eighteen years—and they looked so comfortable to ride on. But I could not go just then, so I bid the boys goodbye. George, after seeing Norton home, was going to Worcester, Massachusetts.

I loaded up twelve passengers and some freight, in all, up and back, about $120, and I went the second trip, and found I could not buck against Wells, Fargo & Company Stage Company carrying passengers, for those from the East were already engaged, or had their passage paid to Green River; those from the West I had my share. But the stages carried the United States Mail, and could carry their passengers free and still make big pay for the company out of the Government. So I took what passengers I could, and then loaded up the wagon with fast freight....

My third trip I loaded with furniture and household goods for a fast woman named Mollie Shepherd, from Chicago to Cheyenne or Laramie; I got $125. When we got back to Green River we got a load of passengers to Bryan, to a horse race; it came to $40. The horses run were Jack Gillmore's and the Missouri roan of Rob Wade.[10] James Ward drove for me (his family lived at Green River City).

September 29th I got loading for Bryan, $45. October 3rd, six passengers to end of track, $15. October 9th loaded up with 6,000 pounds of shelled corn for Warner & Whitman, contractors; got $185. In going through a steep banked creek, my coach had on 6,000 pounds and I could not pull it out of the banks of the creek. I put on twelve horses but could not get a fair pull. There was a man hauling hay with six yoke of cattle, and I asked him to pull up the bank. He thought he could pull it out easy with but four yoke, but when he tried he could not, so he put on the six yoke of cattle, and I had to put on four horses, the best, to pull it out. He charged me $5 for his pulling.

I hired a driver named Herman Dustin, and I got a lot of groceries and goods to take to Bear River City, a new laid off town on Sulphur Creek two miles from the Bear River crossing, for Durrand & Company.[11] It came to $285. When we got down there, there was quite a little town doing big business, and several large railroad camps close by—Carmichael's, Cheseborough & McKee, and O'Neil.[12] There was a tie site close by, where the

---

[10] Bryan was about ten miles west of Green River City on the Union Pacific roadway. The *Frontier Index*, September 22, 1868, reported that a horse race would be held in Bryan, September 30, between Jack Gilmore's "Missouri Roan" and John Bulwer's "Soda Springs horse." The outcome was related in the October 6th issue.

[11] Bear River City, more frequently called Bear Town or Beartown, was about ten miles from present Evanston, Wyoming. It was a typical "quickie" boom-to-bust railroad construction town. End of track reached the settlement November 29, and plans reputedly called for the city to be winter headquarters. Since the railroad had not laid it out, the settlement had a brief existence. *Sweetwater Mines*, December 2, 5, 1868.

[12] The construction camps were named for the contractors who provided and supervised the labor force.

ties were brought and unloaded, from the hills and mountains from 15 to 25 miles off. We had just left our lots at Green River, or abandoned them, and Mr. Durrand, secretary of the Bear River Town Company, was very anxious to sell me some lots, but I put him off a while.[13] Mr. Ward bought him one, and he put up a tent for his family, who moved down from Green River. Their lot was right back of the Southern Restaurant. We made our camp close by Ward's camp.

We got into Bear River City the 19th of October. I found some old friends, and amongst them was Jeff Standifer, who was keeping a saloon and doing good business selling whiskey and gambling.[14] He had been nearly shot and cut to death at Green River, just below where we worked on the grade. He had a falling-out with a barkeeper in a log building near our camp, and the barkeeper got away with Jeff.[15] He was drunk and quarrelsome, just like his friend Ferd Patterson used to be.

The most of the buildings put up at Bear River City were of logs got in the mountains 12 to 15 miles off, and there was a demand for them. So I concluded to cut and haul logs and poles or lumber from the mountains. I bought a third wagon of a man named John Race, an old Indian Creek, California, acquaintance, and I bought some new chopping axes and log chains, and got provisions to take with the chopper, and the teams were to make a round trip a day, and the chopper camped in the timber, but the first time they did not get back till next day, and at that rate it would not pay me for the four men's wages and horse feed. I stayed at camp and sold the logs. So they tried the second trip, and lost the horses, so they did not get back until the third day, having lost one day hunting the horses. So I concluded to haul lumber from a mill, but it would pay but little over expenses.

I had bought a lot to build on, of Durrand, for $125, and took some of our logs and laid a foundation (my wagon and coach boxes formed our camp). I laid up some five or six logs on each side for the house, which made it seven or eight feet high, and had put up a lot of large poles to make ridge poles for it. I expected to finish the building after I had hauled a lot of logs and lumber and make a bakery or restaurant out of it. I sold logs from one to two dollars apiece, and a wagon load of roofing for $20.

A railroad man named Harper came to me and asked me to put my teams into hauling and distributing ties from the tie site above on the line west, and he offered me $20 per day for four-horse teams. I had three four-horse teams. I made some inquiries and was told that Harper

[13] The *Sweetwater Mines*, December 30, 1868, reported that in Bear River City, "Business is dead. A number of citizens have gone to Evanston and Echo City."

[14] Jeff Standifer opened a saloon, "The Mountaineer," in partnership with Billy Huston. *Frontier Index*, November 13, 17, 1868.

[15] Jimmy McGuire shot Standifer in the right hand and left arm in a fracas on August 21, at a trading post a mile from Green River City. His wounds were not serious, however. *Frontier Index*, August 25; September 11, 1868.

had at one time belonged to an association called Davis, Sprague and Harper, or Sprague, Davis and Associates at North Platte River or City, and they had defrauded many men and teams out of their pay, so I was not very anxious to work my teams and not get my pay. So [I] kept on with my logs and lumber hauling.

A week or two after, Harper asked me again, and raised the offer to $22 per day per four-horse team, and told me that he was engaging me for S. B. Reed, the superintendent of construction,[16] and I could draw my pay monthly of Reed's paymaster, Mr. Frost and Winters, at their store on Bear River at the crossing, and I could get supplies at the same place on account, such as flour, bacon, sugar, corn, syrup, coffee, and tobacco. I saw two Mormon brothers named Farnsworth, and they had two teams hauling, and they told me the pay would be all right, so I took my three teams and commenced to distribute ties along the grade.

I got ready the morning of the 5th of November, 1868. I sold the wagon I had bought of John Race, it was too light; I got $65 for it, and I bought a large new Schutler wagon for $155. After settling up my log hauling and the drivers had considerable blacksmithing to do on my coach and wagon and shoeing, and axle setting on coach. John Race wanted to go down to Echo City[17] to get on the works down there, asked me for an order for $40 worth of goods provision of Harper's on Bear River, and as I owed him the amount on the wagon I gave him the order. I was up at the tie siding at the time.

On the 25th of November I drew $200 from Winters on my contract, but he was not the regular paymaster of the construction company, and I could not draw the balance until Smith, the paymaster, paid off for the company. I still had over $900 due me.

Mr. Ward, one of my drivers, went to Salt Lake City to winter and I got George Dustin to drive in his place. I was not very well myself, but kept camp, and drove some when some of my drivers laid off. I made my time count for $22 per day per team, or $66 for my three teams, was the highest I had ever made, regular pay, and very easy on the horses.

We took along our blankets, bedding, and cook box, and we camped at one or the other end every night; at the tie siding we stopped in a cabin; at or near the other end we had left our wagon boxes with bows and wagon covers on them, to sleep in. It was getting quite chilly nights, and it was uncertain how soon the winter would set in, for we were right in the Wasatchor part of the Rocky Mountains,[18] and the season was so far uncommon open. My drivers asked higher wages to stay, as they took

---

[16] Samuel B. Reed started out as a division chief, but soon was taken from the field and appointed superintendent of construction for the Union Pacific. Later he became general superintendent for the railroad.

[17] Echo, Utah, is located approximately forty miles southwest of Evanston, Wyoming.

[18] Reinhart refers to the Wasatch Mountains in northeastern Utah.

risks in staying, for if the winter was to set in rough, I should have to pull up my camp and take my teams to or beyond Salt Lake City to winter until Spring again, when I could get work again. I had to pay a big price for corn to feed my horses. I had one extra to spell some out of the three 4-horse teams. A sack of corn shelled, a large gunny sack, weighed about 160 to 170 pounds, at 13 cents per pound came to from $20 to $22 per sack, and I need to take from three to five sacks of shelled corn at a time, and the provision was quite high, so that after paying my drivers' wages, and feed and provisions, and blacksmithing, I still cleared about $1,400 or $1,500 per month.

I have distributed ties along that Union Pacific Railroad track between Bear River and Evanston and Wasatch City that cost the company $5 apiece, from the mountains until they were put on the grade at or near Evanston.[19] First they had to pay to have them cut out in the timber, in the mountains; then Christmans Brothers' six-mule teams hauled them from the mountains to the tie siding. They got $36 per day for each 6-mule team. The roads were bad and they only took from 20 to 28 to the 6-mule team, and only made four or five loads a week. I then took them from the tie siding, and I put on from 28 to 30 ties to a 4-horse team, and sometimes it took me two days to make one load to each team, at $22 per day per team. For about two weeks some of Mr. Carmichael's teams were sent up from the grading camp; they got $12 per day per span to work on the scrapers, and were sent up to the mountains to help haul out ties to the siding or to the foot of the mountains eight miles from the tie siding, and when the ties were loaded on cars and sent on to Wasatch I was sent to put my three teams in the Carmichael outfit to help them. A Mr. Reynolds was their timekeeper, and he kept my time with his the two weeks, and I was allowed $24 per day per two 4-horse teams, same as the others, making me $72 per day for the three teams.

The first Sunday Reynolds sent his teams to the mountains for ties. Whenever his teams had worked on Sunday on the grade they had been allowed double pay, or $24 per day per span, and I sent mine along, too. But on Monday we were told that they did not wish us to haul on Sundays, for they were not so much pressed for ties, and had only paid double pay on the grade on the scrapers, as they had run onto some contractors that had not got their grade finished up, and were in a hurry to have it done quick, so as not to delay the construction train laying tracks.

For every day General Jack Casement's train lay idle, or could not lay track, the contractor who had the grade unfinished had to pay the expense of the construction train, and it came to over $1,000 per day.[20]

---

[19] Wasatch City, Utah, lies between Evanston, Wyoming and Echo, Utah.
[20] General Jack Casement and his brother, Dan, had the supervising contract for laying 1,000 miles of track from central Nebraska west. He earned his miliatry rank after conspicuous service with Ohio troops during the Civil War. His specific job was

Whenever Casement ran onto a grader, the company put on all the force that could be put on, and paid the biggest prices for teams, and the contractor had to foot the bill for all the extra teams. It had been the means of breaking up many a contractor between North Platte and Evanston.

Hall and Noun[na]n & Kiskaden of Salt Lake City had a contract of over 100 miles from Bear River west, and they failed.[21] Col. Noun[na]n had wanted me to haul off from Harper & Reed and go on their works, but I was afraid they would not be able to pay up their obligations and I would have to lose my pay, so I kept refusing him and kept my present place. The Noun[na]n Brothers had commenced suit against the Union Pacific Railroad Company for heavy damages in not delivering their tools and railroad supplies as they agreed to when they took the big contracting for grading, for it caused them to fail in having their contract finished as per contract and Gen. Jack Casement put on the big force at high pay, and the laying by of the construction train in waiting on the Noun[na]n & Kiskaden, and damaged Noun[na]n & Company $200,000 or $300,000. (I never knew how they settled the thing in court, for I left shortly after.)[22]

Bear River City had got to be a hard place. The cities and towns farther east had been overrun with hard cases; murderers, robbers and cutthroats had been run off and some hung at Cheyenne and Laramie, and they had come west. Our paper at Bear City had notified all the roughs to leave, or they would be served as some had been at Laramie or Cheyenne, and the paper talked and approved of the actions of the Vigilantes down there. Some of the roughs left, but some stayed for spite, and braved it out.[23]

---

on the construction end of the line; his brother, Dan, at the rear forwarded necessary supplies.

[21] Reinhart's memory may be at fault here. Joseph F. Nounnan, a non-Mormon, was let one of three major subcontracts for grading. Brigham Young negotiated the U. P. contract for much of the Utah construction. *Deseret Evening News*, May 27, 1868. Other major subcontractors, in addition to Nounnan, were Joseph A. Young, a son of the church president, and John Sharp, assistant to the superintendent of public works. Leonard J. Arrington, *Great Basin Kingdom*, pp. 262–263 (although he spells Nounnan, Nounan).

An advertisement for 1,000 laborers "to grade the railroad from Quaking Asp mountain to head of Echo Cañon," ran in the *Sweetwater Mines*, May 27, 1868, and was signed, "J. F. Nounnan & Co." Apparently this was a separate Nounnan venture and is probably relevant to Reinhart's recollection, for the subcontracts for the work west of the head of Echo Cañon were not let by the Mormons until June 11. *Frontier Index*, June 26, 1868.

[22] I have been unable to verify this.

[23] The paper Reinhart refers to was the *Frontier Index*, a traveling end of line newspaper edited by the two brothers, Legh R. and Fred K. Freeman. Legh R. Freeman became editor of the so-called "vanguard edition" of the paper on August 11, 1868, when the rail construction reached Green River City. The paper reached Bear River by October 13. The editor wrote in the November 13 issue:

We have never been connected with the vigilantes at any time though we do heartily

There had been some shooting at the Southern Restaurant, where Jimmy Ryan shot and got shot a few days [later]. We had our coach box on my lot and the log foundation was right back of the Southern Restaurant on the next street, and the jail was right acrosst the street from my building. One night an old German came to Bear River City and was looking for a boarding and lodging house to stop over night, and he asked a young man, or boy of 18 or 19, named O'Neil, to show him one. O'Neil took him on a back street and put a pistol to his head and asked the old man to shell out, or he would blow his brains out. The old German was so badly frightened that he just grabbed the boy's arms and called "Police!" and "Help!" The police came up and took O'Neil right in the act of highway robbery, and took him to jail, where two other young men were in for shooting and highway robbing. Some of them were the ones that had been drove from Cheyenne or Laramie by the Vigilantes.[24]

One night while I and my teams were down near Evanston with ties, a lot of Vigilantes came up from Laramie on a train, got off at some railroad outfit camp at Quaking Asp, and there they took horses and came to Bear River City, broke open the jail and secured the jailer, took out the three prisoners, all young fellows not over 22, and swung around one of my building logs, and put a fork of two logs tied together under one end, and hung the whole three of them.[25] Herman Dustin, one of my drivers, had been sort of sick for a few days and not at work, and he was sleeping in a bed that he had in the coach box, not twenty feet from where the men hung. When he awoke the next morning he looked out and saw the three dead men hanging there; he said he heard nothing of the taking of

---

endorse their actions in ridding the community of a set of creatures who are not worthy of the name of men, and who have caused our town to be shunned by thousands of honest laborers in the timber and on the railroad grade who would otherwise come here to spend their money and enrich our tradesmen.

[24] Reinhart has confused two crimes. Jimmy Powers was caught in the act of robbing John Anderson by the city marshal who heard the cries for help. Jack O'Neil and C. C. Jones were arrested for beating and robbing Tom Moylen (a German) in Weaver & Bailey's Saloon. *Frontier Index,* November 3, 10, 1868.

[25] Reinhart's account is partly garbled. A vigilance committee was formed in Bear River after continuous lawlessness. Charles Deloney and Russell Thorp, Sr., headed the group. On the night of November 11, 1868, the committee executed three jailed culprits. The *Frontier Index,* November 13, reported the provocation and set its seal of approval on the act in these words:

Little Jack O'Neil, one of the trio hung Tuesday night did, together with a confederate—Jones—knock down horribly mutilate and rob a man in broad daylight in the saloon of Weaver and Bailey on Uintah street and Jimmy Powers (not the clog dancer) was caught in the act of demanding the "money or life" of a man a few nights prior, and Jimmy Reed (not the prize fighter) garroted a party, and had to be badly beaten before he would surrender himself to the officers. When such open and high handed acts as these are committed every hour of the night or day, by men who follow murder and robbery for a livelihood, we not only justify the people in administering a sure and speedy retribution but we say that we are in favor of hanging several more who are now in our midst.

the jail or of the hanging. He must have slept very sound or the crowd must have been very orderly to hang them and he not know anything about it, so nearby. There was a heavy frost on the ground and one of the boys hung was barefooted and had a gunny sack laying under him, where he had stood when they pulled him up. An empty glass stood on the sack. It had had brandy or whiskey in it, and they first gave them a drink before hanging them. I saw them in the morning. They had been cut down and lay in jail on the floor, and during the morning were buried by some of their railroader friends.

Young O'Neil had an uncle in Cheseborough & McKee's outfit.[26] They were all three Irish boys (I have forgotten the names of the other two with O'Neil). Next day or the day after, when I got back to the supply store at the crossing of Bear River, about one or one and one-half miles from Bear River City, with my teams, they told me at the store that there was a red hot time up at Bear River City—that Cheseborough & McKee's outfit had come down and burned up the house and tent where the paper was published that had upheld and took part with the Vigilantes of Laramie and Cheyenne, destroyed the press, and come over to the jail and took out a railroader of the Carmichael outfit in for getting drunk and being disorderly, and then got on the jail, tore the roof off, and set fire to it, and one of the ringleaders got on top of the jail and called on the crowd or mob to burn up the "damn town."

But he fell off, shot dead by some men or citizens in a log store a few doors below the Southern Restaurant with a Winchester rifle. The crowd from the jail then went up to town and commenced firing on the citizens, who got into the log houses or stores and fired into the mob with their rifles and shotguns. The store where the first shot came from and killed the leader on the jail roof was kept by some men who used to belong to the Vigilantes at Cheyenne or Laramie; one of them had been a sheriff there. The party in that store had 15 or 20 Winchester rifles and they done the most execution.

Thomas Smith, a gambler who lived in Bear River City, was an Irishman, and he took up with the railroaders and seemed to lead them. He stood in open view in front of the log store windows and door with his revolver and fired into the store ten or twelve shots, and the men from inside cut his hat and coat and shirts with their Winchester rifles and only wounded his arm until his pistols were empty and his gambler friends got him away from the front of the building. Thomas Smith was smuggled onto the stage and run in towards Salt Lake City by some of his city friends. The railroaders were beat off after leaving ten dead on the ground, and they started back to camp to get reinforcements to take

---

[26] One account states that Jimmy Reed's brother was a member of McGee's camp. *Sweetwater Mines*, November 25, 1868.

and burn up "the whole damn town," so they said.[27]

By good luck for the city, Carmichael's and O'Neils two big camps had been sent on 15 or 20 miles west the day before, or they might have got the whole of them, 800 or 900 men, to help the mob, but some of the contractors would not let their men come to town to fight us, but the citizens felt uneasy for fear of another attack with great numbers, and a dispatch was sent to Fort Bridger for a company of dragoons and one company of artillery, and the dragoons came that night, about 60 miles off. The citizens lost one man wounded, and he died next day. The contractor McKee sent down some men for the bodies of the rioters, and ten were buried just above the tie siding where I hauled my ties from. The citizens stood guard day and night for several days, until relieved by the soldiers from Fort Bridger.[28]

There were hot times all along the road, killing and shooting every day and night, and most of the towns were nothing but tents to sleep in, and stoves were very unsafe when tents were made of ducking. . . .

It got to be after December 20th, and I expected every day to have to draw off and go into the valley and go into winter quarters but the winter was extraordinarily open, and no snow right in a part of the Rocky Mountains where there had been 15 to 20 feet of snow fell and laid on generally until April. Our work would not last but a few weeks longer, but Col. Noun[na]n, the banker of Salt Lake City, wanted me to go to Bear Lake Valley and put my teams into getting out railroad ties, and I was to have the same price per day I got then, for all winter and spring, and the Farnsworth boys were agoing too.

About this time I felt as if I was going to have the mountain fever, something like typhoid fever, and I was not in good plight for work, for the storm might set in any minute, and I might be sick among strangers, and no telling, I might again have my horses stolen or get accidentally shot in some broil or in my tent or wagons, and I had not heard for over fourteen years from any of my sisters. The Farnsworth boys asked me if I would sell out my teams, or I would leave them in charge of someone

---

[27] Thirty-nine casualties and one death resulted from the riot. Douglas C. McMurtrie (ed.), *The History of the* FRONTIER-INDEX . . . *Reproduced from the Butte City* UNION FREEMAN *of June 24, 1883*, p. 9.

[28] Further wholesale arrests followed the vigilance action of November 11. This resulted in growing confusion.

> On [the morning of] November 20, armed men from the railroad camp came into the town, released the prisoners confined in the jail and then applied the torch to the building. They next visited the *Frontier Index* office . . . applied the match and the building and its contents were consumed. The town was abundantly supplied with a police force, but these were helpless in this emergency.

A call was sent to Fort Bridger, and on the following morning, November 21, one hundred troops arrived at 8 o'clock in the morning and order was quickly restored. "In this riot no one was killed but several were badly injured." Coutant, *Wyoming*, p. 683. *Sweetwater Mines*, November 25, 1868, gives a full account.

and I could go back East for a few months. I thought that would be too risky, for someone might abscond with my whole outfit.

So after studying a little, I offered to sell out the whole outfit for $2,500, and I could go back to Chicago, or Fond du Lac, where my two sisters and their husbands lived, when I last heard of them in 1854. The boys offered me $2,200. I studied over that. If I should sell now and go East; if I was not suited I could, if I wished to, go to teaming again. I could go to Canada and buy some Canadian horses good and cheap (and some of mine were getting old) and I could ship them from Canada to Evanston, and be ready in the Spring to go to work again. If I did like to stay East, I had nothing to take me back west, where for over eighteen years I had spent an uncertain, dangerous life.

So I concluded to take the $2,200 for my thirteen horses, three wagons, one coach, seven sets of harness, and all my camping outfit, bedding, buffalo robes, and all, and I was to get my pay at Echo City when the paymaster came around.

So on Christmas Day, December 25, 1868, I sold out to the Farnsworth Brothers, and they kept on the work we then had. I got to Echo City and found I would have to wait for the December pay until January 15, 1869. So I went back. I had drawn the November pay for myself and some for Farnsworth Brothers, and I settled up all my drivers' wages, and they kept on for Farnsworth's. . . .

I got awful uneasy staying at Echo City at Jinks Hotel, for fear I would be delayed by the snow blockade east on my way to Chicago. Paymaster Smith came on the 15th of January and paid me only November time, and said he would have to go to Ogden and pay off there, that he had not enough currency with him to pay all up to the 15th of January. I asked him if he could not settle my balance, of $2,600, as I had been waiting twenty days to go East. He said he could not, but I might make some arrangements with his assistant, that Morton & Co., a banking house in New York, gave checks on New York for a certain discount. I asked him what the discount on $2,600 would be. He said 4%; it came to $104.

Now I was uneasy after getting all ready to start. I would have to wait until the Paymaster got back, or take the discount, and I was afraid that the Paymaster might be delayed and not come back to Echo to pay until next month. The company at one time was three months' pay behind when I was at Bear River City. Some men had time to sell and got uneasy and some bankers took advantage of their fear, and it was nearly a panic. At one time the President Duran[t] was stopped at Piedmont and they would not let him go until they had paid off a large lot of hands on some contract. The Government was owing the Union Pacific R. R. some six million dollars.[29]

[29] Thomas C. Durant, who died in 1885, was the builder of the Union Pacific. In early July, 1868, he was practically made dictator over the road by the company's exec-

So I concluded to take the discount and pay $104 rather than risk the chances of waiting there a month. So I took the check on New York for the 2600 dollars. My time hauling for the Union Pacific Railroad Construction Company came to 50 days, including expenses, $3,288, or in all, with Jocelyn's 48 days including expenses $1,070.16, from July 20, 1868, to December 25, 1868 (five to seven span of horses), a total of $4,358.16, including the price of teams sold to Farnsworth Brothers, $2,200.00 equaled $6,558.16.

In February, 1868, at Silver Canyon, Stockton, Utah, I had made a cash estimate of the teams and wagons and coach, with one more horse, and the tent, at $2,724. But my expenses were so high that a person not posted on the prices of feed and provisions, wages and blacksmithing would hardly believe it. So I will give my bills nearly in full so that anyone can judge for themselves, and I will put down the days I hauled with what time I put in, and what it came to each day, and you can see I kept them at it day in and day out, from when I first went to hauling ties for the construction company until I sold out and turned over my teams to Farnsworth Brothers, near Evanston. They had two or three 4-horse teams of their own or working them for some parties at Utah Valley, and they took all the teams with some of my drivers on to Bear River Lake on January 10, 1869.

I paid out and bought, from April 10th, 1866, the time I dissolved partnership with my brother, Charles T. Reinhart, at Dry Creek, Umatilla County, Oregon, up to March 31, 180: Received, $16,308.83; paid out expense, $10,780.00, leaving clear about $5,528.83, in the three years or less. The four horses and wagon I had for my share were worth about $700. But I had left my brother four other mares or horses and two wagons, and our farm of 320 acres of land, about 100 acres improved, and good houses, barns, some cows, colts, hogs, chickens, plows and tools, everything to do with as he wished, and we had divided up all the debts each was to pay, which I did, and some of *his* over my amount . . . and I left him there at his farm and never saw him after April, 1866, when I started on my way to Montana loaded with the freight for Virginia City or Helena, Montana, for the Woolf Brothers. . . .

So after settling up all my affairs at Echo City and buying me a suit of clothes, a fur cap of beaverskin, and a pair of nice embroidered buckskin otter gauntlet Salt Lake City gloves, and after dinner on the 15th of January, 1869, I bid all friends goodby at Echo City and got into a stage sleigh and had a sleigh ride to Wasatch, where the cars on regular trains started for the East. . . .

---

utive committee. At the time, financially, the U. P. was in solvent condition, although the government was in arrears of its support to the road.

# EPILOGUE

On returning to the East, Reinhart spent the better part of two years in travel. It was in his blood: it was a habit. He even visited Niagara Falls —alone. But the East was too tame for a man who had spent twenty migrant years on the frontier. Rootless and tired of wandering though he was, the frontier still beckoned him.

In 1871, he settled at New Chicago, later renamed Chanute, in Kansas. On those broad plains he was to live out his remaining years. He had found his permanent and final home: Kansas soil would claim his body. Until the day of his death, January 14, 1889, he wandered no more. The reason may well lie in the fact that on an 1871 visit to his married sister, Emma Musgat, in Fond du Lac, Wisconsin, he met Margaret Brown. On June 16th they were married by a justice of the peace and he brought his young bride to Kansas where their children were born. This was his root and he accepted his responsibility.

Reinhart's last eighteen years were spent in keeping the "Pioneer Livery Feed & Sale Stable" and in operating a restaurant in Chanute: his trade skills still served him as his livelihood. As an early settler of his community, he found himself called to public service: councilman, city marshal, street inspector, fire warden. He proved to be a solid citizen and a good provider for his family.

His epitaph could well read:

"Here lies Herman Francis Reinhart, the American mining frontier's *common man*, indeed *everyman*."

# APPENDIX

[The following document was apparently prepared by Reinhart on March 20, 1888, for that is the last entry noted. The manuscript gives in brief a resumé of his yearly movements from his German birthplace, Jena, to his deathplace, Chanute, Kansas. In summary, it is a history of his travels and ventures.]

The whereabouts [of] Herman Francis Reinhart at the different years: a record of every year from 1833 to 1888.

1833   at Sena or Jena, Saxony, now Prussia, Germany. Father kept Bakery. I was born here.

1834   at Gabestadt, Saxony, Now Prussia, Germany. Father [kept] a large Hotel and Pleasure Garden.

1835   at same place—it was on the Main King's Highway at the edge of the village [so] named.

1836   At Sena or Jena again. Same place and occupation [of his father].

1837   at Echarts Burge or Bergs—at large Castle on the hill. I went to School.

1838   at same place. Father off to the Great Fair at Leipzig with horses.

1839   went with my Mother to Halle, Noumburg, Mershelburg & Saxony.

1840   with my Mother & 2 Sisters and Brother to Wienhanover, Breman on our way to New York, U. S. Am.

1841   at New York City. Went to School on Bleeker Street. My Father at Kingland Bakery.

1842   253 Centre Street. My Father bought out Bakery & Confectionary of Johnson, the Temperance Preacher.

1843   231 Centre St., corner Grand. Father built Bakery & Sold out 253 to H. Rose.

1844   Same place.

1845   at Mrs. Bennets on Grand Street be[tween] Mott & Mulbery learning my trade [as a] Baker. D[id] work for E. L. Carmen, No. 19 & 20 Butcher Stalls Centre Market. Residence of Mr. Carmen, 19th Street near 9th avenue. Slaughter House 18th Street between 8th & 9th avenue, New York City.

1846   at Brooklyn on Atlantic Street at Baking.

1847   went to Newark, New Jersey until July, then with my Father and Sisters & Brother went to Buffalo & Chicago, Illinois to stay.

1848   was at our home in Fremont Center, Lake County, Illinois at farming.

1849   went to Clerk at Wind Mill Flour Store, R. Morley, South Water Co.

*Appendix* 309

1850 at Andrews Baker, South Clark Street, Chicago & in fall through Illinois, Iowa, and Ran Threshing Machine near Dubuque, Iowa. Bur[n]t out same Fall.

1851 Started acrost the Plains through Nebraska, Utah, Wyoming, Nevada, California, & Oregon; Yreka and Humbug, California.

1852 Klamath River, Yreka & Jacksonville, Oregon, Canyonville, South Umpqua, Cow Valley, Oregon.

1853 Althouse Creek, Josephine County, Oregon—Mining & Bakery & Ball Alley at Sucker Creek.

1854 To Crescent City, Cal., to Gold Beach, Prattsville, Port Orford & Pacific Ranch. Build house & locate 320 acres Land.

1855 at my Pacific Ranch. Start to Happy Camp, Klamath River, Del Norte County, California.

1856 Klamath River Mines, and Back to Indian Creek—mine there. Build the Eldorado Saloon & Bakery—go off.

1857 to Sucker Creek & Bowling Creek Mine and build Ball Alley at mouth of Bowling Creek and sell out in part.

1858 Through Oregon, British Columbia and Shuswap Lake, down Fraser River to Victoria, Vancouver Island, and back on the Pacific Ocean to Crescent City, California.

1859 to Roseburg, South Umpqua River and keep Burnett House and am Postmaster. Go to Coffeeville, clerk in store.

1860 Wolf Creek & Kioty [Coyote] Creek, Ore. Whipsaw and go all through Willamette Valley, Ore.: Winchester, Oakland, Eugene, Albany, Corvallis.

1861 Through Oregon to The Dalles, Portland, Walla Walla, W[ashington] T[erritory]. To Idaho Mines or Nez Perce and Clearwater, Lewiston. Rhodes Creek and mine some.

1862 Come back to Walla Walla, W.T., and on my Ranch, Dry Creek, Umatilla Co., Ore. [now Washington], and fr[oze] out—my horses die, & I go to Walla Walla.

1863 at Walla Walla, W.T. Keep Saloon and Bakery until Fall and go back [to] my Ranch on Dry Creek and improve it.

1864 Teaming at Placerville, Centerville, Idaho City, Boise City and Hogem, or Pioneer City.

1865 Team to Boise City, Virginia City & Helena, Montana and to Fort Benton, Montana.

1866 Fort Benton, Helena, Gallatin Valley, East & West; the Jefferson & Madison Rivers, Stink Water, Big Hole, Deer Lodge, etc.

1867 Helena, Blackfoot, Mont., Salmon City, Idaho, Lemhi Valley and Salt Lake City, Utah.

1868 Stockton, Utah, Salt Lake City, Provo, Brigham City, Ogden, Fillmore, Box Elder, Utah. South Pass City, Fort Bridger, South Pass City, Sweetwater, Wyoming.

1869  Echo City, Green River, Bryan, Cheyenne, Omaha, Chicago, Canada Niagara Falls, Newark, Philadelphia, Pittsburgh, Baltimore, Washington, Cincinnati, Buffalo, Milwaukee, New York, Brooklyn, and Fond du Lac, Wis.
1870  Wisconsin, Illinois, Iowa, Missouri, Kansas. Kansas City—Leavenworth, all through. Bought and keep Pioneer Stable park of Kansas and settle at New Chicago, now Chanute.
1871  New Chicago, Erie, Osage Mission, Fort Scott, Thayer, Humboldt, Kansas. Keep Pioneer Livery Feed & Sale Stable yet.
1872  Still keep Pioneer Stable and Bon To[n] Restaurant same town. Elected to the City Council of New Chicago.
1873  Chanute, Pioneer Stable & Bon To[n] Restaurant, 4th Street near Malsome.
1874  Lyons Building.
1875  Baileys Blue Front.
1876  City Marshal and City Hotel. Street Inspector & Fire Warden.
1877  Winfield or Baileys old corner.
1878  Jackson, opposite Baileys Bank.
1879  Corner Centre & 5th & Pioneer Stable.
1880  Eigansatz [House].
1881  lot 6, Block 33, 1st house west of my Pioneer Stable.
to
1888
1888  March 20th. [Reinhart dies, January 14, 1889.]

# BIBLIOGRAPHY

## Primary Sources

Beall, Thomas J. "Recollections of W[illia]m Craig," Lewiston *Morning Tribune*, March 3, 1918.
Burnett, Peter H. *Recollections and Opinions of an Old Pioneer.* New York: D. Appleton & Co., 1880.
Campbell, J[ohn] L. *Idaho and Montana: New Gold Regions. The Emigrant's Guide.* Chicago: J. L. Campbell, 1865.
Clark, Dan E. (ed.) "Some Episodes in the Early History of Des Moines [Selections from the Autobiography of John A. Nash]," *Iowa Journal of History and Politics*, XIII (1915), 175–237.
Crawford, Medorem. "Journal of the expedition organized for the protection of emigrants to Oregon . . . ," *U. S. Senate, Executive Documents*, 37 Congress, 3 Session, Document 17 [Washington, D. C.: 1863 (?)].
Delano, Alonzo. *Across the Plains and Among the Diggings.* New York: Wilson-Erickson, Inc., 1936.
Dunn, Mary M. *Undaunted Pioneers.* Eugene, Oregon: Valley Printing Co., 1929.
Epner, Gustavus. *Map of the Gold Regions in British Columbia.* Victoria, British Columbia: Gust[avus] Epner, 1862.
Frost, Robert. "Fraser River Gold Rush Adventures," *Washington Historical Quarterly*, XXII (1931), 203–209.
Giffen, Guy J. *California Expedition.* Oakland, California: Biobooks, 1951.
Goulder, W[illiam] A. *Reminiscences: Incidents in the Life of a Pioneer in Oregon and Idaho.* Boise, Idaho: T. Regan, 1909.
Hafen, LeRoy R. and Ann W. (eds.). *The Utah Expedition, 1857–1858.* Glendale, California: Arthur H. Clark Co., 1958.
Hammond, Otis G. (ed.). *The Utah Expedition, 1857–1858: Letters of Captain Jesse A. Gove.* Concord: New Hampshire Historical Society, 1928.
Henderson, W. W. (ed.). "The Salmon River Mission: Extracts from the Journal of L[ewis] W. Shurtliff," *Utah Historical Quarterly*, V (1932), 3–28.
Hickman, [William A.] *Brigham's Destroying Angels: Being the Life, Confession, and Startling Disclosures of the Notorious Bill Hickman, The Danite Chief of Utah.* New York: Geo[rge] A. Crofutt, 1892.
Homsher, Lola M. (ed.). *South Pass, 1868: James Chisholm's Journal of the Wyoming Gold Rush.* Lincoln: University of Nebraska Press, 1960.
Howay, Frederic W. (ed.). *The Early History of the Fraser River Mines.* Victoria, British Columbia: C. F. Banfield, 1926.
Lockley, Fred. "Impressions and Observations of the Journal Man [The Recollections of Elijah C. Hills]," *The Oregon Journal*, December 30, 31, 1932.
Lyman, H[orace] S. (ed.). "Reminiscences of Daniel Knight Warren," *Oregon Historical Quarterly*, III (1902), 296–309.
Maloney, Alice B. (ed.). "Some Oatman Documents," *California Historical Society Quarterly*, XXI (1941), 107–111.
McCormick, S. J. *The Oregon and Washington Almanac, 1862.* Portland, Oregon: S. J. McCormick, 1862.
McMurtrie, Douglas C. (ed.). *History of the* FRONTIER-INDEX *('the Press on Wheels')* . . . *Reproduced from the Butte City* UNION FREEMAN *of June 24, 1883.* Evanston, Illinois: n. p., 1943.
[Metlar, George W.] *Northern California, Scott & Klamath Rivers . . . by a Practical Miner.* Yreka, California: Yreka Union Office, J. Tyson Printer, 1856.
Moore, James. "The Discovery of Hill's Bar in 1858," *British Columbia Historical Quarterly*, III (1939), 215–220.
Morgan, Dale L. (ed.). *The Overland Diary of James A. Pritchard.* Denver, Colorado: The Old West Publishing Co., 1959.

Partoll, Albert J. (ed.). "Anderson's Narrative of a Ride to the Rocky Mountains in 1834," Historical Reprint No. 27 of *Sources of Northwest History*. Missoula, Montana: n.d. [Reprint from *Frontier and Midland*, XIX (1938).]
Raymond, Rossiter W. *Mines and Mining of the Rocky Mountains, The Inland Basin, and the Pacific Slope*. New York: J. B. Ford & Co., 1871.
Reid, Robie L. (ed.). "To the Fraser River Mines in 1858," *British Columbia Historical Quarterly*, I (1937), 243–253.
Riddle, George W., *History of Early Days in Oregon*. Riddle, Oregon, 1920. [Reprint from the *Riddle Enterprise*.]
———. "Recollections," *The Oregonian*, March 29, 1925.
Rockwood, E. Ruth (ed.). "The Letters of Charles Stevens," *Oregon Historical Quarterly*, XXXVII (1936), 137–159; 241–261; 334–353.
Skavlan, Margaret (ed.). "Over the Westward Trail [The 1847 Journal of Lester Hulin]," *The Oregonian*, May 3, 1931.
Stratton, Royal B. *Captivity of the Oatman Girls*. San Francisco: 1857; reprinted by the Grabhorn Press, 1935.
Tichenor, Captain W[illiam]. "Among the Oregon Indians." Manuscript. Bancroft Library, University of California, Berkeley.
*Transactions of the Oregon Pioneer Association, 1917; 1919*. Portland, Oregon: Chausse-Prudhomme, Co., Printers, 1917; 1919.
U. S. Senate, *Executive Documents*, 32 Congress, 1 Session, Document 1. Washington, D. C.: A. Boyd Hamilton, 1851. 3 vols.
Victor, Frances F. *All Over Oregon and Washington*. San Francisco: J. H. Carmany & Co., 1872.
Waddington, Alfred. *The Fraser Mines Vindicated: or, The History of Four Months*. Victoria, British Columbia: P. De Gano, 1858; reprint; Vancouver, British Columbia: Robert R. Reid, 1949.
Winther, Oscar O., and Rose D. Galey (eds.). "Mrs. Butler's 1853 Diary of Rogue River Valley," *Oregon Historical Quarterly*, XLI (1940), 337–366.
Woody, Frank H. "From Missoula to Walla Walla in 1857 on Horseback," *Washington Historical Quarterly*, III (1912), 277–286.
WPA. *Told by the Pioneers*. [Olympia, Washington]: State Printing Office, 1936–1938. 3 vols.

## Secondary Sources

*Books*
Arrington, Leonard J. *Great Basin Kingdom*. Cambridge, Massachusetts: Harvard University Press, 1958.
Babbitt, Charles H. *Early Days at Council Bluffs*. Washington, D. C.: Press of Bryon S. Adams, 1916.
Bancroft, Hubert H. *History of British Columbia*. San Francisco: History Publishing Co., 1887.
———. *History of California*. San Francisco: A. L. Bancroft & Co., Publishers, 1886–1890. 7 vols.
———. *History of Oregon*. San Francisco: History Publishing Co., 1886–1888. 2 vols.
———. *History of Utah*. San Francisco: History Publishing Co., 1890.
———. *History of Washington, Idaho, and Montana*. San Francisco: History Publishing Co., 1890.
———. *History of Nevada, Colorado, and Wyoming*. San Francisco: History Publishing Co., 1890.
———. *Popular Tribunals*. San Francisco: History Publishing Co., 1890. 2 vols.
Bagley, Clarence B. *History of Seattle*. Chicago: S. J. Clarke Publishing Co., 1916. 3 vols.
Baxter, James P. *The Introduction of the Ironclad Warship*. Cambridge, Massachusetts: Harvard University Press, 1933.
Beal, M[errill] D. *A History of Southeastern Idaho*. Caldwell, Idaho: Caxton Printers, Ltd., 1942.

# Bibliography 313

———, and Merle W. Wells. *History of Idaho.* New York: Lewis Historical Publishing Co., 1959. 3 vols.
Begg, Alexander. *History of British Columbia.* Toronto: W. Briggs, 1894.
*Biographical Dictionary of the American Congress, 1774–1949.* Washington, D. C.: Government Printing Office, 1950.
Brooks, Juanita. *Mountain Meadows Massacre.* Stanford, California: Stanford University Press, 1950.
———, *John Doyle Lee.* Glendale, California: Arthur H. Clark Co., 1961.
Burlingame, Merrill G. *The Montana Frontier.* Helena, Montana: State Publishing Co., 1942.
———, and K. Ross Toole. *A History of Montana.* New York: Lewis Historical Publishing Co., 1957. 3 vols.
*California Blue Book, or State Roster, 1909.* Sacramento, California: State Printing Office, 1909.
Carey, Charles H. *A General History of Oregon Prior to 1861.* Portland, Oregon: Metropolitan Press, 1936. 2 vols.
———. *History of Oregon.* Chicago: Pioneer Publishing Co., 1922. 3 vols.
Caughey, John W. *Gold Is the Cornerstone.* Berkeley and Los Angeles: University of California Press, 1948.
Chittenden, Hiram M. *History of Early Steamboat Navigation on the Missouri River.* New York: F. P. Harper, 1903. 2 vols.
Corning, Howard M. *Dictionary of Oregon History.* Portland, Oregon: Binfords & Mort, 1956.
Coutant, C[harles] G. *The History of Wyoming.* Laramie, Wyoming: Chaplin, Spafford & Mathison, Printers, 1899.
[Dick, William B., pseud., "Trumps"]. *The American Hoyle.* 18th edition. New York: Dick & Fitzgerald, 1907.
Dimsdale, Thomas J. *The Vigilantes of Montana.* 3rd printing. Butte, Montana: W. F. Bartlett, 1915.
Drury, Clifford M. *Marcus Whitman, M.D.: Pioneer and Martyr.* Caldwell, Idaho: Caxton Printers, Ltd., 1937.
Dunn, J[acob] P. *Massacres of the Mountains: A History of the Indian Wars of the Far West.* New York: Harper & Bros., 1886.
Eells, Myron. *Father Eells.* Boston and Chicago: Congregational Sunday-School and Publishing Society, 1894.
Ernst, Alice H. *Trouping in The Oregon Country.* Portland, Oregon: Oregon Historical Society, 1961.
Evans, Elwood. *History of the Pcaific Northwest: Oregon and Washington.* Portland, Oregon: North Pacific History Publishing Co., 1889. 2 vols.
Fisher, Vardis. *Idaho: A Guide in Word and Picture.* Caldwell, Idaho: Caxton Printers, Ltd., 1937.
———. *The Idaho Encyclopedia.* Caldwell, Idaho: Caxton Printers, Ltd., 1938.
Fuller, George W. *A History of the Pacific Northwest.* New York: A. A. Knopf, 1931.
Furniss, Norman F. *The Mormon Conflict, 1850–1859.* New Haven, Connecticut: Yale University Press, 1960.
Gard, Wayne. *Frontier Justice.* Norman: University of Oklahoma Press, 1949.
Gaston, Joseph. *The Centennial History of Oregon, 1811–1912.* Chicago: S. J. Clarke Publishing Co., 1912. 4 vols.
Gilbert, Frank T. *Historic Sketches of Walla Walla, Whitman, Columbia and Garfield Counties, Washington Territory.* Portland, Oregon: A. G. Walling, 1882.
Gudde, Erwin G. *California Place Names.* Berkeley and Los Angeles: University of California Press, 1960.
Hafen, LeRoy R., and W[illiam] J. Ghent. *Broken Hand: The Life Story of Thomas Fitzpatrick, Chief of the Mountain Men.* Denver, Colorado: The Old West Publishing Co., 1931.
———, and Francis M. Young. *Fort Laramie and the Pageant of the West, 1834–1890.* Glendale, California: Arthur H. Clark Co., 1938.

Hebard, Grace R., and E. A. Brininstool. *The Bozeman Trail.* Cleveland, Ohio: Arthur H. Clark Co., 1922. 2 vols.
Heitman, Francis B. *Historical Register and Dictionary of the United States Army.* Washington, D. C.: Government Printing Office, 1903. 2 vols.
Hodge, Frederick F. *Handbook of American Indians.* Washington, D. C.: Government Printing Office, 1910.
Howay, F[rederic] H., W. N. Sage, and H. F. Angus. *British Columbia and the United States.* Toronto: The Ryerson Press, 1942.
Hunt, Aurora. *The Army of the Pacific.* Glendale, California: Arthur H. Clark Co., 1950.
Hutchison, Bruce. *The Fraser.* New York and Toronto: Rinehart and Co., Inc., 1950.
Jackson, W. Turrentine. *Wagon Roads West.* Berkeley and Los Angeles: University of California Press, 1952.
Jenson, Andrew. *Latter-Day Saint Biographical Encyclopedia.* Salt Lake City, Utah: Andrew Jenson History Co., 1901–1936. 4 vols.
Jones, J. Roy. *Saddle Bags in Siskiyou.* Yreka, California: News-Journal Print Shop, 1953.
Johnson, Allen, and Dumas Malone (eds.). *Dictionary of American Biography.* New York: Charles Scribner's Sons, 1928–1936. 20 vols.
Kelly, Charles, and Hoffman Birney. *Holy Murder: The Story of Porter Rockwell.* New York: Minton, Balch & Co., 1934.
Kendall, John S. *The Golden Age of the New Orleans Theater.* Baton Rouge: Louisiana State University Press, 1952.
Kirkpatrick, Orion E. *History of Leesburg Pioneers.* Salt Lake City, Utah: Pyramid Press, 1934.
Koller, Larry. *The Fireside Book of Guns.* New York: Simon and Schuster, 1959.
Langford, Nathaniel P. *Vigilante Days and Ways.* Boston: J. G. Cupples Co., 1890. 2 vols.
[Leeson, Michael A.] *History of Montana, 1739–1885.* Chicago: Warner, Beers & Co., 1885.
MacMinn, George R. *The Theater of the Golden Era in California.* Caldwell, Idaho: Caxton Printers, Ltd., 1941.
McArthur, Lewis A. *Oregon Geographic Names.* 3rd edition. Portland, Oregon: Binfords & Mort, 1952.
McNeal, William H. *History of Wasco County, Oregon.* [n.p., n.d. (mimeograph).]
Meany, Edmond S. *Origin of Washington Geographic Names.* Seattle: University of Washington Press, 1923.
Montgomery, Richard G. *The White-Headed Eagle: John McLoughlin, Builder of an Empire.* New York: Macmillan Co., 1935.
Morice, A[drien] G. *The History of the Northern Interior of British Columbia.* 3rd edition. Toronto: W. Briggs, 1905.
Murray, Keith A. *The Modocs and Their War.* Norman: University of Oklahoma Press, 1959.
Neff, Andrew L. *History of Utah, 1847 to 1869,* edited and annotated by Leland H. Creer. Salt Lake City, Utah: Deseret News Press, 1940.
Ormsby, Margaret A. *British Columbia: A History.* Vancouver, British Columbia: Macmillan Co. of Canada, Ltd., 1958.
Paden, Irene. *Prairie Schooner Detours.* New York: Macmillan Company, 1949.
———, *The Wake of the Prairie Schooner.* New York: Macmillan Company, 1943.
Pence, Mary L., and Lola M. Homsher. *The Ghost Towns of Wyoming.* New York: Hastings House, 1956.
Peterson, Emil R., and Alfred Powers. *A Century of Coos and Curry: History of Southwest Oregon.* Portland, Oregon: Binfords & Mort, 1952.
*Portrait and Biographical Record of Western Oregon.* Chicago: Chapman Publishing Co., 1904.
Pomeroy, Earl S. *The Territories and the United States, 1861–1890.* Philadelphia: University of Pennsylvania Press, 1947.

Pyper, George D. *The Romance of an Old Playhouse.* Salt Lake City, Utah: Deseret News Press, 1937.
Rich, E[dwin] E. *The History of the Hudson's Bay Company, 1670–1870.* London: Hudson's Bay Record Society, 1958–1959. 2 vols.
Sabin, Edwin L. *Building the Pacific Railway.* Philadelphia, Pennsylvania: J. B. Lippincott Co., 1919.
Scott, Harvey W. *History of the Oregon Country.* Cambridge, Massachusetts: Riverside Press, 1924. 6 vols.
Siskiyou County Historical Society. *Yearbooks, 1946–1960.* Yreka, California. 3 vols.
Swanton, John R. *The Indian Tribes of North America.* Washington, D. C.: Government Printing Office, 1952.
Trimble, William J. *The Mining Advance into the Inland Empire.* Madison: University of Wisconsin Press, 1914.
Victor, Frances F. *The Early Indian Wars of Oregon.* Salem, Oregon: F. C. Baker, 1894.
Wagner, Henry R. *The Plains and the Rockies.* Revised by Charles L. Camp. 3rd edition. Columbus, Ohio: Long's College Book Co., 1953.
Walbran, John T. *British Columbia Coast Names.* Ottawa: Government Printing Bureau, 1909.
Wallace, W[illiam] S. *The Dictionary of Canadian Biography.* Toronto: Macmillan Co. of Canada, Ltd., 1926.
Walling, A[lbert] G. *Illustrated History of Lane County, Oregon.* Portland, Oregon: A. G. Walling, 1884.
———. *History of Southern Oregon, Comprising Jackson, Josephine, Douglas, Curry and Coos Counties.* Portland, Oregon: A. G. Walling, 1884.
Warren, Sidney. *Farthest Frontier: The Pacific Northwest.* New York: Macmillan Co., 1949.
Wells, [Harry L.]. *History of Siskiyou County, California . . .* Oakland, California: D. J. Stewart & Co., 1881.
Winther, Oscar O. *The Great Northwest: A History.* New York: A. A. Knopf, 1958.
———. *The Old Oregon Country.* Bloomington: Indiana University Publications, [n.d.]
WPA. *Oregon: End of the Trail.* Portland, Oregon: Binfords & Mort, 1940.
———. *Montana: A State Guide Book.* New York: Viking Press, 1939.
———. *The Oregon Trail.* New York: Hastings House, 1939.
———. *Utah: A Guide to the State.* New York: Hastings House, 1941.
———. *Washington: A Guide to the Evergreen State.* Portland, Oregon: Binfords & Mort, 1941.
———. *Wyoming: A Guide to Its History, Highways, and People.* New York: Oxford University Press, 1941.
Wright, E[dgar] W. (ed.). *Lewis & Dryden's Marine History of the Pacific Northwest.* Portland, Oregon: Lewis & Dryden Printing Co., 1895.

*Articles*
Anderson, Bernice G. "The Gentile City of Corinne," *Utah Historical Quarterly*, IX (1941), 141–154.
Anonymous. [Obituary for General Persifer F. Smith], *Harper's New Monthly Magazine*, XVII (1858), 258.
———. "Steamboat Arrivals at Fort Benton, Montana, and Vicinity," *Contributions to the Historical Society of Montana*, I (1876), 317–325.
Bagley, Clarence B. " 'The Mercer Immigration': Two Cargoes of Maidens for the Sound Country," *Oregon Historical Quarterly*, V (1904), 1–24.
Baker, Hugh S. C. "The Book Trade in California, 1849–1859," *California Historical Society Quarterly*, XXX (1951), 97–116; 353–367.
Berelson, Bernard, and Howard F. Grant. "The Pioneer Theater in Washington," *Pacific Northwest Quarterly*, XXVIII (1937), 115–136.
Briggs, John E. "The Removal of the Capital from Iowa City to Des Moines," *Iowa Journal of History and Politics*, XIV (1916), 56–95.
Brown, William C. "Old Fort Okanogan and the Okanogan Trail," *Oregon Historical Quarterly*, XV (1914), 1–38.

Burlingame, Merrill G. "John M. Bozeman, Montana Trailmaker," *Mississippi Valley Historical Review*, XXVII (1941), 541–568.
Callaway, James E. "Governor Green Clay Smith, 1866–1868," *Contributions to the Historical Society of Montana*, V (1904), 108–127.
Chadwick, S. J. "Colonel Steptoe's Battle," *Washington Historical Quarterly*, II (1908), 333–343.
Colvig, William H. "Indian Wars of Southern Oregon," *Oregon Historical Quarterly*, IV (1903), 227–240.
De Loney, Burton. "Press on Wheels," *Annals of Wyoming*, XIV (1942), 299–314.
Dillard, W. B. "The Beginnings of Lane County," *Oregon Historical Quarterly*, V (1904), 133–138.
Elliott, Russell R. "The Early History of White County, Nevada, 1865–1887," *Pacific Northwest Quarterly*, XXX (1939), 145–168.
Gray, Mary A. "Settlement of the Claims in Washington of the Hudson's Bay Company and the Puget's Sound Agricultural Company," *Washington Historical Quarterly*, XXI (1930), 95–102.
Hacking, Norman R., "'Steamboat 'Round the Bend': American Steamers on the Fraser River in 1858," *British Columbia Historical Quarterly*, VIII (1944), 255–280.
Hafen, LeRoy R. "Mountain Men—William Craig," *Colorado Magazine*, XI (1934), 171–176.
Hermann, Binger. "Early History of Southern Oregon," *Oregon Historical Quarterly*, XIX (1918), 53–68.
———. "Port Orford Homecoming," *Oregon Historical Quarterly*, XXV (1924), 311–328.
Johnson, Walter W., "List of Officers of the Territory of Montana to 1876," *Contributions to the Montana Historical Society*, I (1876), 326–333.
Jones, Idwal. "The Man From Scott's Bar [Anton Roman]," *Westways*, 40 (1948), 8–9.
Martig, Ralph R. "Hudson's Bay Company Claims, 1846–69," *Oregon Historical Quarterly*, XXXVI (1935), 60–70.
Nixon, Robert J. "Sheriffs of Siskiyou County," *The Sisikiyou Pioneer and Yearbook*, III (1959), 66–71.
Oviatt, Alton B. "Steamboating Traffic on the Upper Missouri River, 1859–1869," *Pacific Northwest Quarterly*, XL (1949), 93–105.
Platt, Robert T. "Oregon and Its Share in the Civil War," *Oregon Historical Quarterly*, IV (1903), 89–109.
Prosch, Thomas W. "The Indian War in Washington Territory," *Oregon Historical Quarterly*, XVI (1915), 2–23.
———. "The Indian War of 1858," *Washington Historical Quarterly*, II (1908), 237–240.
Rice, William B. "The Captivity of Olive Oatman," *California Historical Society Quarterly*, XXI (1941), 97–106.
Sage, Donald. "Gold Rush Days on the Fraser River," *Pacific Northwest Quarterly*, XLIV (1953), 161–165.
Scammell, Col. J. Marius. "The Siskiyou Cavalry (1861–66)," *The Siskiyou Pioneer and Yearbook*, III (1959), 72–74.
Scott, Leslie M. "The Pioneer Stimulus of Gold," *Oregon Historical Quarterly*, XVIII (1917), 148–166.
Stern, Madeleine B. "Anton Roman," *California Historical Society Quarterly*, XXVIII (1949), 1–18.
Talkington, Henry L. "Story of The River: Its Place in Northwest History," *Oregon Historical Quarterly*, XVI (1916), 181–195.
Van der Zee, Jacob. "Forts in the Iowa Country," *Iowa Journal of History and Politics*, XII (1914), 163–204.
Victor, Frances F. "A Knight of the Frontier [Benjamin Wright]," *The Californian*, IV (1881), 152–162.
Wheat, Carl I. "Ned, The Ubiquitous," *California Historical Society Quarterly*, VI (1927), 3–36.

Williams, George H. "Political History of Oregon From 1852 to 1865," *Oregon Historical Quarterly*, II (1901), 1–35.
Woodward, Walter C. "The Rise and Early History of Political Parties in Oregon," *Oregon Historical Quarterly*, XII (1911), 245–263.
Wyman, Walker D. "Omaha: Frontier Depot and Prodigy of Council Bluffs," *Nebraska History Magazine*, XVII (1936), 143–144.

*Newspapers*

British Columbia
    Victoria *Gazette*
California
    Los Angeles *Star*
    Red Bluff *Beacon*
    San Francisco *Chronicle*
    San Francisco *Daily Alta Californian*
    San Francisco *Examiner*
    San Francisco *Golden Era*
    Shasta *Courier*
    Yreka *Journal*
Idaho
    Lewiston *Morning Tribune*
Illinois
    Chicago *Tribune*
New York
    New York *Times*
Oregon
    Coos Bay *News*
    Oregon City *Oregon Spectator*
    Portland *The Oregonian*
    Portland *The Oregon Journal*
    Port Orford *News*
    Salem *Oregon Statesman*
    Salem *Statesman*
Utah
    Salt Lake City *Deseret Evening News*
    Salt Lake City *Deseret News*
Washington
    Walla Walla *Statesman*
Wyoming
    *Frontier Index*—published at the "end-of-track" on the Union Pacfic.
    *Sweetwater Mines*—published at several different Wyoming towns.

# INDEX

*Active*: and boundary survey, 139 n
Ada County, Idaho: Dave Updike hanged in, 250 n
Adams, Miss: and Warty, 47
Adams, Dr. [boardinghouse-keeper]: 27
Adams, Charley: 194
Adams, Henry: 47 n
Adams, John: and killing of Warty, 47, 56
Adams, John Jr.: 47 n
Adams, Mary: 194
Addams, Augustus A.: 204 n
adultery: among Indians, 45
Agnew, James: and Samuel Driggs, 227
Albany, Oregon: mention of, 10, 176; Reinhart passes through, 111, 177, 182, 309
Alder Gulch, Montana: gold discoveries at, 211, 211 n, 238; hanging of Plummer gang at, 240, 258; mention of, 264
Alexander, Dr.: Reinhart consults, 177
Alexander, Idaho: 9 n
Allbecker, [hotelkeeper]: 197
Allen, Beverly S.: as Indian commissioner, 34 n
Allen, Charles: murder of, 239 n
Allen, Captain James: and Des Moines, Iowa, 2 n
Allis, Samuel: 3 n
Alma, [unidentified]: Reinhart buys horse from, 265
Alpawa Creek, Washington: 188 n
Althouse, Phillip: discovery of gold by, 57 n
Althouse Creek, Oregon: mining on, xxi, 58, 60–61, 67, 68, 70; mention of, 57, 59, 75, 77, 79, 81, 82, 86, 90, 91, 93, 94, 97, 99, 100, 105, 118, 134–135, 144, 149, 150, 164, 165, 168, 172, 177, 184, 192, 199, 220; Reinhart at, 58, 59–60, 60–63, 69, 78, 148, 309; Democrat Creek near, 60; politics at, 69; law enforcement at, 70, 72–75
Althouse Creek saloon: *See* Browntown saloon
Amberson, Jim: and Reinhart, 215
American Baptist Home Mission Society: missionaries of, 2 n
American Bar, British Columbia: 134, 134 n
American Fur Company: and Hudson Bay Company, 57; and Fort Benton, 252 n; mention of, 277 n
American Hotel: in Victoria, 141
Anderson, Alexander C.: 133 n
Anderson, George: 185, 194

Anderson, John: robbery of, 302 n
Anderson River, British Columbia: 135 n
Andrews Baker, Chicago: Reinhart works for, 309
Applegate, Charles: 163 n
Applegate, Jesse: 163
Applegate, Lindsey: 163 n
Applegate Creek, Oregon: gold discoveries on, 40; and Rogue River War, 97
Applegate party: 23 n
Applegate Road: Cornelius J. Hills on, 10, 10 n; Oregon wagon train passes, 22 n
Argenta, Montana: freight hauled to, 264; silver deposits at, 264 n
Arkansas: Judge Young from, 69
Armer, Henry: 208
army: deserters from, 99, 155, 170, 261–262
Ashcraft, [blacksmith]: 45, 49
Aspinwalls, the: 252
Astoria, Oregon: 213
Auburn, Oregon: gold discoveries at, 211, 211 n
Aulfrey, [ferryman]: 199–200
Avery, Joseph C.: as founder of Marysville, 54 n

Baker, Dr. Dorsey S.: and Oakland, 53 n
Baker, Edward D.: and Joseph Hooker, 159; background of, 159 n; in Mexican War, 160
Baker, Oregon: *See* Baker City, Oregon
Baker Brothers: 268
Baker City, Oregon: Reinhart stops at, 236
Baker County, Oregon: Powder River in, 211 n; Auburn in, 211 n
bakeries: mention of, xx; at Minersville, 28; at Browntown, 63–64, 65, 66, 67–68, 75, 77–78; at Sucker Creek, 70, 71, 75, 77–78, 97; at Prattsville, 81, 86, 95; at Indian Creek, 93, 97, 99, 144, 309; at Sucker Creek second, 99, 103, 104–105, 147
ball alleys: *See* bowling alleys
Ball's Bluff, Virginia: Edward D. Baker killed at, 159 n, 160
Baltimore, Maryland: C. Miller from, 261; Reinhart at, 310
Bannack, Montana: gold discoveries at, 211, 238; hanging of Plummer gang at, 240, 240 n; prices in, 248 n; mention of, 252; freight hauled to, 264
Baptist Home Mission Society: missionaries of, 2 n
Barnes, [sawyer]: 60, 61

Barnett's [in Idaho]: Reinhart sells pigs at, 224
Barrow, Major John E.: and Council Bluffs Indian Agency, 3 n
Bartholomew, Parodi: and Reinhart, 103, 108; on Fraser River trip, 108; injury of, 116–117
Bartholomew, Sidney: See Bartholomew, Sidolia
Bartholomew, Sidolia [Sidney]: on Fraser River trip, 108, 112, 128; languages spoken by, 108; horses of, 117
Bates, [unidentified]: 65 n
Bean, Jim: and McCoy and McCloud, 73
Bear Creek, Oregon: Indians near, 23 n
Bear River, Wyoming: soldiers stationed at, 9; ferry on, 273; Reinhart crosses, 293; Bear River City near, 297
Bear River City, Wyoming: Reinhart hauls freight to, 297; location of, 297 n; mention of, 298 n, 300; description of, 301; vigilantes at, 301, 302, 302 n; riot in, 303
Bear River Town Company: 298
Bear Town, Wyoming: See Bear River City, Wyoming
Beaver Creek: 262, 263, 265
Beckney, John C.: winters with Reinhart, 215; in Pioneer City, 228–229; in Idaho City, 235
Beckworth, Charles W.: in Oregon wagon train, 3, 4 n; and cattle stampede, 12–13; and Indians, 15; marriage of daughters of, 147, 149
bedbugs: at Round Prairie, 159–160
Bee Hive Buildings: in Salt Lake City, 273
Beerup, Charley: character of, 61; on theft of gold, 66
Beerup, Daniel: mine of, 61
Begbie, Matthew B.: law enforcement by, 143 n–144 n
Beidler, John X.: 257 n
Belknap, [juggler]: 66
Bell, [miner]: partnership of, 59 n
Bell, Dr. [swindler]: 81
Bellevue, Nebraska: Indian school at, 3 n
Benecia, California: 28 n
Bennet, Mrs.: Reinhart works for, 308
Bennett, Catherine: 269 n
Benton, Fort: See Fort Benton, Montana
Benton, Fort: See Fort Benton [no. 2]
Benton County, Oregon: Reinhart in, 54, 177; Corvallis in, 54, 111, 134, 175
Berry, Jerry: robbery of, 240 n
Berry, Joseph: robbery of, 240 n
Biernbaum, Joseph: on trip to Montana, 250
Big Bar, Oregon: 41

Big Hole River, Montana: Reinhart on, 250, 269, 309; bridge across, 265
Big Horn Mountains, Wyoming: 260 n, 261 n
Big Horn River: Fort Benton [no. 2] on, 254 n
Biglow, [farmer]: 216
Big Luckiamute: See Luckiamute, Big
Bill, Boston: See Latham, William
Bill, Dancing: See Clark, Bill
Bill, Dancing: See Fountain, William
Bill, Poker: See Willowby, William
Bill, Red Face: See Willowby, William
Birch Creek, Oregon: 185 n
Bitter Creek, Wyoming: Green River, Wyoming on, 295; Union Pacific Railroad on, 296
Bitterroot Mountains, Idaho: 191 n
Black, Dr.: 71–72, 77, 91
Black Canyon, British Columbia: 137 n
Blackfoot City, Montana: freight hauled to, 258; Reinhart passes through, 265, 309
Blackfoot Indians: on Bozeman trail, 260
Blackfoot Mines, Montana: 248, 274
Blackfoot Mountains, Montana: 265
Black Rock Canyon, Nevada: Oregon wagon train passes, 18 n, 20 n; finding of liquor near, 21 n
Black Rock Desert, Nevada: Oregon wagon train passes, 20 n
Black Rock Springs, Nevada: Oregon wagon train passes, 21, 22 n
Black Tail Deer Creek, Montana: Reinhart passes, 250
Blanco, Point: See Point Blanco, Oregon
Bland, John: marriage of, 149; Reinhart stops with, 151, 152
Bland, Mrs. John: 149, 152
Bleeker Street, New York: Reinhart goes to school at, 308
Blood Indians: at Fort Benton, 255
Bloody Point, Oregon: Oregon wagon train passes, 20 n; Indian massacre at, 22–23, 28
Bloomington, Illinois: Charley Morrison from, 264
Blue, Amos: mine of, 40
Blue Mountains, Oregon: emigrant trails across, 184; and Umatilla Indian Reservation, 185 n; streams from, 186; gold discoveries in, 211 n; mention of, 214, 233 n, 236; freight hauling through, 218, 219, 220
Blue Mountain Wagon Road Consolidated: 215, 215 n, 218
Blunt County, Tennessee: James D. Burnett from, 56 n
Bob, Cherokee: See Talbot, Robert

Bodega, California: 28 n
Boicy Brothers: 219
Boise, Idaho: mining at, xxi, 229; mention of, 211 n, 229, 248, 252, 265; Reinhart passes through, 226–227, 228, 250, 309; vigilantes in, 250
Boise Basin, Idaho: gold discoveries in, 211 n, 214; mention of, 215, 243; hauling trips to, 218–221, 223–225
Boise Mountains, Idaho: Reinhart crosses, 220
Boise River, Idaho: Hay Press Ranch on, 227; mention of, 236, 250
Bolton, Dick [Cribbage Dick]: 199–200
Bonaparte River, British Columbia: 129, 129 n
Bon Ton Restaurant: kept by Reinhart, 310
books: read by Reinhart, 33, 83, 158, 259, 276
Boonesville, Idaho: 211 n
Boonville, Missouri: people from, 29
Boston, [gambler]: 144
Boston, Massachusetts: people from, 69, 111, 195 n, 213, 225
Boston Bar, British Columbia: 135, 135 n
Boston Bill: See Latham, William
Bountiful, Utah: See Sessions, Utah
Bourbon County, Kentucky: 4 n
Bowen, Louisa: marriage of, 155
Bowling, Ben: works for Reinhart, 266
bowling alley, Sucker Creek: See Sucker Creek saloon and bowling alley; Mayflower Saloon & Bakery & Bowling Alley
bowling alleys: in Browntown, 66, 69
Bowling Creek, Oregon: gold discoveries on, 98, 101; Reinhart at, 148, 309; mention of, 211, 222
Box Elder, Utah: See Brigham City, Utah
boxing: exhibition of, 90, 257–258
Boyles, [cattleman]: gold claims of, 180
Boyles, Ned: 211
Bozeman, John M.: xxii, 259–260, 259 n–260 n
Bozeman City, Montana: mention of, 260; soldiers near, 263, 263 n
Bozeman Pass: defense for, 263 n
Bozeman Trail: travellers on, 259–260; Indians on, 260; Fort Phil Kearney on, 261
Brady, James: Reinhart meets with, 188–189; mention of, 197
Brady, Washington: 189, 197
Bramson, George: in Oregon wagon train, 4 n, 40 n
Bramson, Newtown: in Oregon wagon train, 4 n; mine of, 40, 40 n
Braziel, [miner]: and Emmanuel, 61

Breckenridge, John C.: and General Joseph Lane, 111
Breed, Bertha, as Reinhart's sister, 81, 111
Breed, Orson: as Reinhart's brother-in-law, 111, 148
Breman, Germany: Reinhart in, 308
Brickley, James: 171
Bridger, Fort: See Fort Bridger, Wyoming
Bridge River, British Columbia: 130 n
Bridge Street, Helena: 263
Brigade Trail, British Columbia: Reinhart travels on, 131, 133, 136
Briggs, Almira: in Oregon wagon train, 4 n
Briggs, David: 4 n
Briggs, George E.: 64
Briggs, Samuel B.: in Oregon wagon train, 3, 4; finding of liquor by, 20; toll bridge run by, 63, 147–148; mention of, 63 n
Briggs, Susanna Phillips: in Oregon wagon train, 4 n
Briggs, William: in Oregon wagon train, 4 n
Briggs and Watkins: 78
Briggs Creek, Oregon: prospecting on, 67
Brigham City, Utah: Reinhart passes through, 273, 309
Brighton, Utah: Reinhart stops at, 277, 286
British Columbia: Reinhart's trip to, 108–145, 309; Hudson Bay Company in, 56–57; ships from, 87; gold discoveries in, 108 n; law enforcement in, 143–144
Brock, Catherine C.: marriage of, 207
Brock, Vineyard C.: 207
Brooklyn, New York: people from, 89–100, 225; Reinhart in, 308, 310
Broughton, John: on Walla Walla River, 185, 194
Brown, [storekeeper]: 33
Brown, [merchant]: death of, 199–200
Brown, [miner]: mine of, 101
Brown, [teamster]: on trip to Montana, 250
Brown, Captain Frederick H.: killing of, 261–262
Brown, Henry H. [Webfoot]: and founding of Browntown, 60 n; Reinhart's conflict with, 63
Brown, James: 134
Brown, John: hanging of, 36–39
Brown, Margaret: marriage of Reinhart to, 307
Brown, Peter: 90, 91
Brown, William D.: Missouri River ferry of, 3 n
Brownlee, [Jed Powers' partner]: ferry

operated by, 178, 183; and Bill Bunton, 179, 185; at The Dalles, 183
Brown's butcher shop: in Browntown, 74
Brownsville, Oregon: 60 n
Browntown, Oregon: Democrat Creek near, 60; the Reinhart brothers in, 63–78; description of, 70; law enforcement at, 70, 72–75
Browntown saloon: building of, 63–64; prices at, 65, 66; operation of, 65, 67; and George Fetterman, 65, 67–68; costs of, 75; abandoning of, 77–78
Brush Creek, Oregon: crossing of, 86
Bryan, Wyoming, 297, 310
Buchanan, Colonel Robert C.: 95 n
Buck, Bill: as partner of Reinhart, 97
Buck, John: as partner of Reinhart, 97
Buckley, James: as sheriff of Walla Walla, 188, 188 n; mention of, 197, 270; as southern sympathizer, 205; in Boise, Idaho: 227; and Dave English, 239
Buckley's Ferry, Idaho: 188 n
buckskin coat, Reinhart's: 9, 17, 21–23
Buena Vista, Mexico: Joseph Hooker at, 160
buffalo: killing of, 11
Buffalo, New York: mention of, 99; Reinhart in, 308, 310
buffalo chips: cooking with, 9
Bulwer, John: horse of, 297 n
Buncombe County, North Carolina: General Joseph Lane from, 52 n
Bunton, William: hanging of, 52; and Brownlee, 179; and Plummer gang, 240 n
Bunton, Eleza: 52–53
Bunton, Mrs. Eleza: on Walla Walla River, 185
Bunton, Elijah: and Brownlee, 179
Bunton, Minerva: mention of, 52; marriage of, 185
Buntons, the: on Whiskey Creek, 194, 194 n
Burd, [chief of police]: See Burt, Andrew
Burgess, [unidentified]: marriage of, 41
Burgess, Tony: marriage of, 41
Burk, Bill: gambling of, 100
Burn, Jim: 189
Burnett, Colonel James D.: in Round Prairie, 56, 56 n; hotel of, 158, 158 n; and the Reinharts, 161
Burnett, Josiah: 161
Burnett, Molly: 56, 158, 159, 160, 166
Burnett House: Reinhart at, 158–161, 309; James Wilson at, 166
Burns, [miner]: winters near Reinhart, 274
Burnstein, [merchant]: 199
Burnt River Valley, Oregon: mention of, 209, 226, 227; stage robbery in, 229
Burpee, James S.: in furniture business, 171 n
Burpee and Linn: furniture shop of, 171
Burt, Andrew: and recovery of Reinhart's horses, 284, 287, 290
Burton, Robert T.: and recovery of Reinhart's horses, 284, 290; as collector of internal revenue, 294 n–295 n
Butte, Montana: Reinhart passes through, 265

Cain, A. J.: 186 n
Calapooya Mountains, Oregon: Reinhart crosses, 53, 55, 111, 175
Calaveras County, California: gold from, 109
Calder, [deputy]: and recovery of Reinhart's horses, 284, 285, 287
California: mention of, 1, 2, 3, 10, 53, 56, 70 n; gold in, 18; Indian department of, 24; boundary of, 25; Thomas Smith in, 43 n; farming in, 54; timber in, 54; fish in, 54–55; Chinook jargon in, 56–57; vigilantes in, 105; Mexican War in, 189 n; Reinhart in, 26–36, 78–79, 90–98, 145, 146, 309
Calvin, [storekeeper]: 210
Cambria County, Pennsylvania: Captain McDermit from, 28 n
Campbell, Archibald: 139 n
Campbell County, Kentucky: Thomas Smith from, 43 n
Canada: James W. Nesmith from, 54 n. See also British Columbia
Canadian-United States boundary survey: 138–139, 138 n–139 n
Canby, General E. R. S.: killing of, 23, 23 n
Canemah Falls: See Falls Canemah, Oregon
Canoe Country, British Columbia: gold discoveries in, 128, 131 n; Reinhart's trip to, 131–133
Canoe Creek, British Columbia: 131 n
Cantlen, Oliver: mining activities of, 28
Cantonment Loring: soldiers stationed at, 10 n
Canyon City, Oregon: gold discoveries at, 211, 211 n
Canyon Creek Canyon, Oregon: Reinhart in, 41 n–42 n
Canyon John Kitchen and Co.: mine of, 44
Canyonville, Oregon: Charles Beckworth at, 4 n; Samuel Briggs at, 4 n; Joseph Knott at, 42, 43, 46; Ashcraft from, 45; mention of, 82, 147, 148, 149, 158, 208;

Reinhart at, 111, 175, 309; supplies brought from, 151
Cap-a-lil Creek, Washington: 188 n
Cape Alava, Washington: 82 n
Cape Blanco: *See* Point Blanco, Oregon
Captain Jack [Indian]: *See* Captain John [Indian]
Captain Jack's Company: Reinhart works for, 28
Captain Jane: 41
Captain John [Indian]: and Rogue River War, 94, 95
card games: of miners, 33, 67
cardplaying: by Reinhart, 60, 77, 83, 99–100, 104–105, 170–171, 173–174
Caribou Mines, British Columbia: 137
Caribou River, British Columbia: 131
Carissa Lode, Wyoming: 277 n
Carmen, E. L.: Reinhart works for, 308
Carmichael, [railroad contractor]: 297, 300, 304
Carrington, Colonel Henry B.: and Fetterman disaster, 262 n
Carson Desert, Nevada: *See* Carson Sink, Nevada
Carson Sink, Nevada: Oregon wagon train passes, 20 n
Carter, Aleck: and Plummer gang, 240, 240 n–241 n
Carter Lode: 277 n
Cartwright, B. D.: 53 n
Cartwright's Crossing: of Calapooya Mountains, 53
Cascade Falls: on Columbia River, 182
Cascade Mountains, Oregon: 25 n, 112–113, 116, 135, 182
Casement, Dan: and Union Pacific Railroad, 300 n–301 n
Casement, General Jack: and Union Pacific Railroad, 300–301; account of, 300 n–301 n
Case Ward, Utah: Reinhart passes through, 273
cattle: in Oregon wagon train, 7–8, 10–11, 12–13, 19; of Thomas Richmond, 183, 187; of Caleb Grover, 187, 200–201, 210; prices of, 210; of Reinhart, 9, 84, 214, 216
Cayuse Indians: and Fraser River miners, 114; and Umatilla Indian Reservation, 185 n; killing of Whitman by, 213 n
Cedar Rapids, Iowa: Reinhart in, 1
Celilo, Oregon: 200
Centerville, Idaho: Reinhart in, 221, 309
Central, Montana: founding of, 211 n
Central City, Nebraska: 6
Central Pacific Railroad: at Ogden, 273

Chadwick, S. F.: and trial of Joseph Knott, 47 n
Chalmer, Horace: murder of, 239 n
Chalmer, Robert: murder of, 239 n
Chanute, Kansas: prices at, 136; mention of, 196, 267 n; T. C. Jones in, 255 n; Reinhart in, 307, 310
Chapman, [restaurant owner]: 64
Chapman, Artinecia: in Oregon wagon train, 4, 4 n; mention of, 59; marriage of, 63
Chapman, Dr. J. A.: 196 n
Chapman, John: in Oregon wagon train, 4 n
Charles, S. M.: as Charley Reinhart's partner, 86; revolver of, 109, 192; treachery of, 113
Charley, Dutch: *See* Dutch Charley
Chase, Albert: and joke on Coom, 154
Chavner, Thomas: and Gold Hill Mining Company, 175 n
Cheesebrough and MaGee: 296, 297, 303
Cherokee Bob: *See* Talbot, Robert
Cheseborough & McKee: *See* Cheesebrough and MaGee
Cheyenne, Wyoming: Reinhart at, 310
Cheyenne Indians: on Bozeman trail, 260
Chicago, Illinois: James Clugage from, 33–34; the Orson Breeds in, 81, 111; Reinhart in, 88, 308, 310
Chicken Creek, Utah: Reinhart in, 286
chickens: raised by Charley Reinhart, 104; hauled by Reinhart, 228
Childs, Nat: mine of, 61
Chile: Emmanuel from, 61
Chimney Rock, Nebraska: Oregon wagon train passes, 18
Chinese: in Yreka, 163
Chinese miners: mining by, 103, 147; characteristics of, 103–104; and other miners, 104; employed by Reinhart, 108, 147
Chinook jargon: 56–57, 80, 138, 143
Chinook wind: effects of, 201
cholera: on Oregon trail, 12
Chrismas, Campbell: marriage of, 4 n
*Christian Science Monitor:* accounts of Reinhart in, xxiii
Christmans Brothers: mule teams of, 300
Cincinnati, Ohio: Reinhart at, 310
Civil War: Reinhart on, xxii: participants in, 4 n, 28 n, 159 n, 269 n, 276 n, 300 n; effects of, 195, 204–207, 242, 243; enlistment practices during, 262
Clackamas County, Oregon: Oregon City in, 182
Clark, [unidentified]: Reinhart stops with, 151
Clark, Bill [Dancing Bill]: prospecting

by, 45, 46; and Sucker Creek bowling alley, 75
Clark, Lester: in Oregon wagon train, 3
Clarke, General Newton S.: and Indians, 114 n
Clark's [near Helena]: 265, 269
Clarkson, Washington: 190 n
Clearwater Mines, Idaho: 181 n, 186, 211
Clearwater River, Idaho: gold discoveries on, xxi, 181, 184, 184 n; mention of, 117, 202; Reinhart at, 309
Clevenger, John: murder of, 95
Clinton, Iowa: Sol Norton from, 296
clothing: prices of, 35, 62
Clubfoot George: and Plummer gang, 240
Clugage, James: pack trains of, 33–34; mine of, 40; and Thomas Fruid, 252
Cochran, William: on Fraser River trip, 117–118, 119, 120–121, 133–144; in Yreka, 163, 166; mention of, 238
Cocoran's Ranch: McCloud at, 74
Coeur d'Alene, Idaho: 186
Coeur d'Alene Indians: and Colonel Steptoe, 114 n
Coffee Creek, Oregon: gold discoveries on, 148–149; naming of, 152; mention of, 154, 166, 179, 180, 207, 265, 278; Reinhart at, 149–154, 156, 309
Coffee Creek, Washington: Reinhart crosses, 188
Coffeeville, Oregon: *See* Coffee Creek, Oregon
Coffin, Fred: and Charley Johnson, 28
Cole, Rennick: prospecting by, 46
Coleman, William: rides with Reinhart, 256–257
Cole's Valley, Oregon: H. B. Flournoy in, 57 n
Collins, [storekeeper]: mine of, 79
Collins, [mulatto]: and Joseph Hooker, 160
*Colonel Wright:* on Columbia River, 191 n
Columbia River: mention of, 99, 120, 125, 299 n; cities on, 112, 116; Reinhart crosses, 117; Mount Hood on, 163; driving of horses up, 182–183; description of, 183, 186
Columbia River Indians: and Fraser River miners, 114
Columbus, Nebraska: 5 n
Colusa, California: 118, 205
Colville, Washington: 114 n, 127, 195, 210
Comstock Lode, Nevada: discovery of, 180
Connecticut: Henry Switzer from, 274
Conners, [miner]: in Captain Jack's Company, 28
Connor, Jim: rides with Reinhart, 258
Connor, General Patrick E.: mention of,

273 n; discovery of silver by, 276, 290; account of, 276 n
Constant, Isaac: and Cornelius J. Hills, 10 n
Cook, John D.: 164 n
Coom, [butcher]: joke played on, 153–154
Cooper, Cal: *See* Reinhart, Wilson and Company (first)
Cooper, John: and Plummer gang, 240
Coos County, Oregon: 77 n, 83 n
Coover, Thomas W.: and John M. Bozeman, 260 n
Coquihalla River, British Columbia: Fort Hope on, 106 n, 138 n
Coquille River, Oregon: gold discoveries on, 46, 77, 81, 82; mention of, 76, 83 n, 145
Corinne, Utah: 273, 273 n
Cornelius, Colonel Thomas R.: 195 n
Corvallis [Marysville], Oregon: Reinhart passes through, 54, 111, 175, 309; mention of, 134, 179, 207, 222, 227; vigilantes near, 177–178
Cosbey, John D.: *See* Cosby, John D.
Cosby [Cosbey], John D.: Reinhart hires, 163–164, 166, 168; account of, 164 n; and trial of Abel George, 168
Cottonwood Creek, California: 26, 36, 163
Council Bluffs, Iowa: Mormons at, 2
Council Bluffs Indian Agency: 3 n
Cove Valley, Oregon: 219
Covington, Jim: 100
Covington, Kentucky: Ben Bowling from, 266
Cow Creek, Oregon: Charley Reinhart's land claim on, 35, 39; Reinhart's land claim on, 40, 309; mention of, 41 n, 43 n, 51, 58, 62, 75, 79, 111, 147, 149, 172; description of, 43; ferry on, 43, 51
Cowen, Samuel: and McCoy and McCloud, 73
Cow River, Montana: Indians on, 260
Coyote Creek, Oregon: 173, 173 n, 309
Crabb, Jack: 116
Craig, William: 181 n, 190, 190 n, 194
Craig Mountain, Idaho: Florence near, 202 n
Cranmer, Major: 150
Cranston, James: at Browntown saloon, 65, 69; at Sucker Creek saloon, 70, 75; death of, 164
Crawford, Captain Medorem: emigrant train escorted by, 195–196, 195 n–196 n, 261
Creamy, Steven: in Captain Jack's Company, 28
Creighton, Captain John: mention of,

## Index

236; killing of Freelon Schnebley by, 246
Crescent City, California: mention of, 53, 96, 103, 144, 174, 241, 243; Reinhart at, 78, 90, 145, 309; pack trains from, 79 n; soldiers from, 94–95
Cribbage Dick: *See* Bolton, Dick
Cripple Creek, Colorado: John M. Bozeman in, 259 n–260 n
Crook and Cosbey: 164
Crossroad Jack: *See* Elliot, Jack
Crow Indians: and Bozeman trail, 260 n
Crutchfield, Joe: at Coffee Creek, 152–153
Culbertson, Major Alexander: and Fort Benton by, 252 n
Cummings, Dr. J. Lyle: 28 n
Curry County, Oregon: 77 n, 83 n
Curtis, Charley: 263, 263 n
Cushing, Milo M.: hotel of, 116, 116 n
Cushman, Walter: ferry operated by, 178, 183

Dallas, Oregon: Reinhart goes to, 177
Dalles, The: *See* The Dalles, Oregon
Dancing Bill: *See* Clark, Bill
Dancing Bill: *See* Fountain, William
Danforth, John: gold claims of, 150; at Coffee Creek, 152–153
Daniels, James: and Reinhart, 105, 138, 231; on Fraser River, 105–106, 137, 138; on Umatilla River, 184, 231
Danites: 280 n
Danzig, Germany: Joe Nitsell from, 29
Davenport, Joe: and Reinhart, 177
Davidson, Erk: 192, 193
Davis, Ann: and Indians, 19 n
Davis, E.: works for Reinhart, 296
Davis, Jefferson: 242
Davis, John: works for Reinhart, 217
Davis, Vincent: at Roseburg, 151–152; and McPherson, 155–156; and minstrel party, 158
Davis, Walter: mention of, 185, 216; as southern sympathizer, 195; account of, 195 n; and government cattle, 196
Davis, Sprague and Harper: 299
Day, Dora: 162
Day, George: 162
Day, Mrs. Pat: marriage of, 162
Deady, Judge Matthew P.: mention of, 47 n, 70 n; and McPherson, 157; background of, 157 n
Dearborn River, Montana: Reinhart crosses, 254
Deer Creek, Oregon: 169
Deer Lodge, Montana: hanging of Plummer gang at, 240; gold discoveries near, 248; Reinhart passes through, 265, 309; mention of, 274
Deer Lodge County, Montana: New York Gulch in, 258
Deer Lodge River, Montana: 250, 259
Dejarlais, A.: at Minersville, 27, 27 n; Reinhart's meeting with, 36. *See also* Dejarlais Brothers
Dejarlais, N.: at Yreka, 27 n. *See also* Dejarlais Brothers
Dejarlais, O.: at Scott Bar, 27 n; death of, 36. *See also* Dejarlais Brothers
Dejarlais Brothers: stores of, 27; pack train of, 27; fate of, 36, 36 n. *See also* Dejarlais, A.; Dejarlais, N.; Dejarlais, O.
Delano, Alonzo: 21 n
delirium tremens: of George Williams, 41
Del Norte County, California: mention of, 53, 78; Crescent City in, 53; Reinhart in, 309
Deloney, Charles: in vigilantes, 302 n
Democrat Gulch, Oregon: Northcut Brothers on, 59, 60; McCloud at, 74; George Fetterman in, 76
Denny, French Joe: and Reinhart, 105; death of, 135–136, 137
Denny, John: on prospecting trip, 68
Deschutes House: 200
Deschutes River, Oregon: 183, 199
Deseret: as name for Utah, 292
Des Moines, Iowa: 2, 2 n
Des Moines River, Iowa: Reinhart passes, 2; Des Moines, on, 2 n
Desmond, Jack: mine of, 101; killing of, 211–212
Destroying Angels: members of, 280, 284
devil-fish: killed by Reinhart, 141
Devil's Back, Oregon: 88–89
Devil's Gate: Oregon wagon train passes, 18
Diamond, Montana: 263
dianna: playing of, 194
Ding, John: Reinhart sells oxen to, 216; Reinhart's lawsuit with, 216–217
Dinsmore, [teamster]: works for Reinhart, 296
Discovery Company: 270 n
District Court: in Yreka, 32
Dixon, Jacob: hanging of, 250 n
Dog River, [Oregon?]: 183
Dolly [horse]: 247
Donahue, Thomas: *See* Donnahough, Thomas
donation claims: *See* land claims
Donehough, [miner]: 171
Donehough and Ferguson: 171

## 326 Index

Donnahough, Thomas: and Ferd Patterson, 243, 244–245; fate of, 245
Donovan, Thomas: *See* Donnahough, Thomas
Dooling, James: mine of, 101; killing of, 211
Doolittle, Henry: hotel of, 93–94
Douglas, James: in Oregon wagon train, 4, 24
Douglas, James: as governor of British Columbia, 130, 142, 142 n
Douglas, Fort: *See* Fort Douglas, Utah
Douglas, Port: *See* Port Douglas, British Columbia
Douglas County, Oregon: settlers in, 4 n, 42 n, 158 n, 161; Charley Reinhart's land in, 35 n, 39; commissioners of, 41 n, 43 n; county seat of, 51; mention of, 79; creeks in, 154, 228
Douglas Portage: travel via, 133 n
Down, John: as clerk for Brown, 33
Downer, Douglas: in Oregon wagon train, 3
Downer, James: in Oregon wagon train, 3, 24; in Humbug City, 29; bowling alley of, 66, 69, 72, 73; mention, 173
Doty, George: marriage of, 149
Drake, Thomas J.: 280 n
Driggs, Samuel: and James Agnew, 227
drinks: *See* liquor
Driscoll, Jack: and McCoy and McCloud, 73, 74, 172; and volunteers, 172; and Bob Williams, 172
Drum, John: and Thomas Fruid, 252
Drummond, Montana: mining at, 256
Dry Creek, Oregon: Reinhart's farm on, 53, 187, 193–194, 207, 208, 209–210, 212, 214, 215–216, 217, 225, 237, 247, 306, 309; description of, 185; Reinhart crosses, 185, 188; Caleb Grover on, 187; meadows on, 212
Dryden's Livery: in Idaho City, 234, 235
Dubuque, Iowa: 83
Dudrick, Bob: 60
Duke, Basil: *See* Duke, James K.
Duke, James K. [Basil]: daughter of, 268
Duke, Lena: marriage of, 268
Dunlap & Co.: and Browntown saloon, 67
Dunn, Bill: 220
Dunsmith, [unidentified]: shot by McPherson, 156
Dupuy, [Charley's partner]: and Sucker Creek saloon, 70–71, 75, 77; and McCoy and McCloud, 73; as vigilante, 75
Durant, Thomas C.: and Union Pacific Railroad, 305
Durrand, [secretary]: 298
Durrand & Company: Reinhart hauls freight for, 297

Dusenberg, Moses: mine of, 61
Dusenberg Bar, Oregon: Reinhart's mine claim on, 61, 63
Dusenberg farm: Schertz boys on, 99
Dustin, George: works for Reinhart, 299
Dustin, Herman: works for Reinhart, 297; and vigilante hangings, 302–303
Dutch Boy's Claim: 92
Dutch Charley: murder of, 207–208
Dutch John: *See* Schertz, John
Dutch Louis: movements of, 41
Dutcher, [constable]: and Dr. Black, 71, 72; and McCoy and McCloud, 73, 74
Duvall, [miner]: gambling by, 77
Duvall, Frenchy: rides with Reinhart, 272

Eads, George: and Reinhart, 177
Eagle Creek, Oregon: Philip Foster on, 182 n
Eagle Hotel: Reinhart passes, 220
Eagle Rock, Idaho: *See* Idaho Falls, Idaho
Eals, Dr.: *See* Eells, Reverend Cushing
Eastman, Chester: and Reinhart's saloons, 69, 70, 75, 97
Ebert, Charles: and Walla Walla Hotel, 197
Echarts Burge, Germany: Reinhart in, 308
Echo, Utah: 299, 300 n, 305, 306, 310
Echo Canyon, Utah: Reinhart passes through, 293
Eddy, [butcher]: 83
Eddy & Pratt: 83, 84
Edson, [hotelkeeper]: 83
Edwards, [teamster]: works for Reinhart, 266
Eells, Reverend Cushing: 213, 213 n
Egerton, Sidney: as governor of Montana, 260 n
Eightmile Creek, Oregon: 185
Eldorado Saloon [in Pioneer City]: run by John C. Beckney, 228
Eldorado Saloon and Bakery [Indian Creek]: 93, 97, 99, 144, 309
elections: Reinhart on, xxii, 69–70, 226
Elizabethtown, Oregon: 80, 90, 94
Elk City, Idaho: 239 n
Elk Horn, Idaho: gold discoveries at, 211 n
Elkhorn River, Nebraska: Oregon wagon train crosses, 5
Elk Mines, Idaho: gold discoveries at, 211
Elko, Nevada: 10 n
Ellensburg, Oregon: 79 n, 86, 145
Elliff, Hardy: Reinhart visits, 41, 148; mention of, 208
Elliot, Jack [Crossroad Jack]: hanging of, 222
Ellis, Fort: *See* Fort Ellis, Montana
Ellrod, John: death of, 134

Elpowa River, Washington: 189, 190
Emigrant Road: 215, 215 n, 218
Emmanuel, [Chilean miner]: 61, 62
Emmanuel, [merchant]: Reinhart hauls for, 231, 234
Emory Bar, British Columbia: 136, 137 n
Empire City, Oregon: mention of, 76, 145; gold discoveries at, 77, 81; Spanish Mary's saloon at, 83
Eneas [Indian]: See Enos [Indian]
English: in Chinook jargon, 56; spoken by Reinhart's neighbors, 80
English, Dave: murder of Scott by, 239, 240 n
English Navy: at Victoria, 142
Enos [Indian]: murder of Benjamin Wright by, 22 n; Reinhart's meeting with, 85, 95; background of, 85 n; in Rogue River War, 94
*Enterprise:* on Fraser River route, 137 n
Erie, Pennsylvania: Reinhart at, 310
Esquimault, British Columbia: English Navy ships at, 142, 142 n
Estes, Oregon: Reinhart in, 111
euchre: played by miners, 33, 67; Reinhart's skill at, 60
Eugene, Oregon: mention of, 54; Reinhart passes through, 111, 175, 309
Evans, Davis: ferry operated by, 41, 41 n
Evans Bridge, Oregon: Oatmans at, 166
Evanston, Wyoming: 297, 300, 300 n
Ewing, Johnny: and Mike Mitchell, 189
Express Ranch: 220, 226

Fairchild, John B.: marriage of, 166 n
Fairchilds, [unidentified]: 210
Fairchilds, Lucius: as governor of Wisconsin, 210
Fairfield, Oregon: Reinhart visits, 112
Falls Canemah, Oregon: Reinhart passes, 112
Fandango Creek: Oregon wagon train passes, 18 n, 20 n
farming: by the Reinhart Brothers, 158–161, 162, 165, 169, 175, 193–194, 207, 208, 209–210, 212, 214, 215–216, 217, 225, 237, 247
Farmington, Utah: Reinhart at, 293
Farnsworth Brothers: and railroad, 299; mention of, 304; purchase of Reinhart's outfit by, 305
faro: played by miners, 33
Farewell Bend, Snake River: Reinhart at, 224
Farrel, Jack: 261–262, 263
*Fauntleroy:* and boundary survey, 139 n
Fauntleroy, Captain: and *Santa Cruz,* 144
Fay, [miner]: and shooting of Indian, 30, 32
Feather River: Cornelius J. Hills on, 10, 10 n; Oregon wagon train passes, 21–22, 22 n
Ferrel, Seth: and Clearwater gold discoveries, 184 n
Fetterman, George: and Reinhart, 57, 58–59, 59–63; prospecting by, 60–61; dishonesty of, 62, 64–65, 65–66, 67–68, 75–76
Fetterman, Captain William J.: and Fetterman disaster, 261 n
Fetterman, Fort: See Fort Fetterman, Wyoming
Fillmore, Utah: finding of Reinhart's horses at, 284–288; Reinhart in, 286, 309; vigilantes in, 288
firemen's benefit: riot at, 204, 205–206
Fish, Charles: rides with Reinhart, 272
Fitzgerald, Major Edward H.: 164 n
Fitzpatrick, Thomas: as Indian agent, 7 n
Flanagan, John: 220
Flannery, [homesteader]: See Flournoy, H. B.
Flint, Addison R.: and Winchester, 51 n
Floras Creek, Oregon: miners from, 82
Florence, Idaho: 202, 202 n, 203, 223
Flournoy [Flannery], H. B.: 57, 57 n
Fond du Lac, Wisconsin: 163, 307
food: prices of, 33, 35, 58, 65, 66, 131, 134, 248 n, 300
Foot, Commodore: See Sloat, Commodore John D.
Ford, Pat: gold claim by, 34–35
Forster, [unidentified]: 113
Fort Benton, Montana: freight hauled to, 252, 309; mention of, 252 n; shipping to, 253, 267; passengers hauled to, 254, 255, 256, 265; Indians at, 255; drowning of Governor Meagher at, 266–267
Fort Benton [no. 2]: location of, 254 n
Fort Bridger, Wyoming: 304, 309
Fort Des Moines, Iowa: See Des Moines, Iowa
Fort Douglas, Utah: 273, 273 n
Fort Ellis, Montana: soldiers at, 263 n
Fort Fetterman, Wyoming: 261, 261 n
Fort Hall, Idaho: 9, 10 n, 248
Fort Hayes, Idaho: See Pioneer City, Idaho
Fort Hill, Illinois: Joel Sherman from, 164
Fort Hope, British Columbia: mention of, 106, 125, 130; location of, 106 n, 138 n; Reinhart arrives at, 137–138
Fort Jones, California: 164, 164 n
Fort Kamloops, British Columbia: See Fort Thompson, British Columbia
Fort Kearney, Nebraska: 196 n, 261 n
Fort Langley, British Columbia: 138, 138 n
Fort Lapwai, Idaho: 190 n

Fort Laramie, Wyoming: Oregon wagon train reaches, 7; Indians at, 7; mention of, 260 n, 261 n
Fort Leavenworth, Kansas: See Leavenworth, Kansas
Fort Lemhi, Idaho: 269, 269 n
Fort Lewis, Montana: See Fort Benton, Montana
Fort Okanogan, Washington: mention of, 120, 125; Fraser River company arrives at, 124
Fort Phil Kearney, Wyoming: 261, 261 n
Fort Randall [on Missouri River]: 253 n, 254 n
Fortress Monroe, Oregon: 4 n
Fort Scott, Kansas: Reinhart at, 310
Fort Simcoe, Washington: Fraser River Company at, 114, 117, 118; establishment of, 115 n; Reinhart arrives at, 117–118
Fort Thompson, British Columbia: mention of, 127, 134; Fraser River company at, 128
Fort Victoria, British Columbia: 141 n
Fort Walla Walla, Washington: William Reinhart at, 102; mention, 114 n, 125, 186, 194–195; wagon stolen from, 214
Fort Yale, British Columbia: 136 n, 241
Foster, Colonel: 264
Foster, John: 264
Foster, Philip: 182, 182 n
Fountain, William [Dancing Bill]: mention of, 61–62; and Reinhart's saloons, 64, 65, 70; and George Fetterman, 67
Fountain, The: See The Fountain, British Columbia
Fountain Creek, British Columbia: 130 n
Fowler, [unidentified]: and Gold Hill Mining Company, 174
Fox Indians: at Omaha, Nebraska, 3
Frankfort, Oregon: Three Sisters near, 80 n
Fraser, Paul: 128 n
Fraser Falls: 136 n
Fraser River, British Columbia: gold discoveries on, xxi, 108, 115, 128, 131; Daniels and Hill on, 105–106; Reinhart on, 106, 131–132, 133, 138–140, 309; Fort Hope on, 106 n; mention of, 125, 129 n, 133 n, 140, 152, 171, 184; Indians on, 130, 135, 136; Fraser River company reaches, 130; travel on, 133 n, 135; and Thompson River, 134; and Anderson River, 135 n; law enforcement on, 143–144, 143 n–144 n; the Standifers on, 241
Fraser River George: and Ferd Patterson, 241–242
Fraser River miners: and Indians, 109 n, 114
Fraser River mines: evaluations of, 112,
133 n, 137; profits from, 132, 133
Fraser River trip: organization of, 118–119; and Indians, 119, 120–124, 125–126, 126–127, 128, 129, 130, 131, 135; difficulties of, 119–120; costs of, 138
Frazer, Jobe: and Reinhart, 27
Free, Judge J. N.: naming of Freetown for, 27, 27 n
Freetown, California: Reinhart at, 27, 28; justice in, 30
Freez, Joe: and Charley Reinhart, 97; as deserter from army, 99, 155
freeze-out poker: playing of, 67, 150, 170, 173–174
freight hauling: by Reinhart, 218–221, 228, 231–235, 248–251, 264–265, 265–266, 268, 269, 271, 297; conditions of, 219, 220; profits from, 228; rates for, 228, 248, 248 n–249 n; and Charley Reinhart, 229; losses from, 236
Frémont, John C.: 1845 expedition of, 19 n; and Lost River, 22 n; mention of, 85; and Mexican War, 189
Fremont Center, Illinois: Reinhart at, 308
Fremont's Peak: Oregon wagon train passes, 18
French: in Chinook jargon, 56; spoken by Reinhart's neighbors, 80; spoken by Bartholomew, 108
French Bar, California: 27 n, 97
French blacksmith: and Indians, 122–123
French Prairie, Oregon: mention of, 81; account of, 81 n, 182 n; Reinhart passes through, 108, 112, 182
Fritz [butcher's helper]: pigs bought by, 224–225
*Frontier Index:* publishing of, 301 n
frontier justice: Reinhart on, xxii: at Freetown, California, 30–32; at Jacksonville, Oregon, 36–39; at Cow Creek, Oregon, 47–48; at Nevada, Montana, 52; at Althouse Creek, Oregon, 70, 72–75; at Port Orford, Oregon, 91; at Indian Creek, California, 91–93; in British Columbia, 143–144, 143 n–144 n; at Roseburg, Oregon, 155, 156–157, 180; at Florence, Idaho, 207–208; in Auburn, Oregon, 211–212; at Boise Mines, 222
Frost, [railroad paymaster]: 299
Fruid, Thomas: background of, 252; passengers hauled by, 254; winters with Reinhart, 258–262; mention of, 262
Fulkerson, Peter: prospecting by, 45, 46; gray horse of, 49
Fuller, [unidentified]: rides with Reinhart, 272
Fuller, Alexander: 57, 58, 66–67, 79, 172
Fullerton, Sheriff John: and McPherson, 157
furs, prices of, 87

Gabestadt, Prussia: Reinhart's family from, xx, 308
Gaiety Theater: in Walla Walla, 204 n
Gaines, J. P.: as Indian commissioner, 34 n
Gallagher [Gallecher], Jack: and Plummer gang, 240 n
Gallatin City, Montana: Reinhart passes through, 258
Gallatin River, Montana: emigrants to, 195; gold discoveries on, 211 n; Reinhart on, 258–262, 264, 309; soldiers on, 263, 263 n
Gallecher, Jack: *See* Gallagher, Jack
Gallus Creek, Oregon: and Rogue River War, 97
gambling: in saloons, 27, 55, 67, 116; of the Reinhart Brothers, 60, 77, 83, 99–100, 104–105, 148, 170–171, 173–174, 194
*Ganges*: at Victoria, 142 n
Gans Brothers: Reinhart hauls for, 231, 234
Gardiner City, Oregon: 53, 150–151
Garnett, Major R. S.: and Indians, 114 n
Garrison Creek, Washington: Reinhart crosses, 186
Gaston, John A.: 267, 267 n, 293
Gaston & Simpson: 267
Gates, Colonel: in Illinois Valley, 78; in Yamhill County, 78 n
Gates, Lafayette: in Illinois Valley, 78; in Jackson County, 78 n; public house kept by, 146
Gates, N. H.: 78 n
Gates, Wallace: in Jackson County, 78 n
Gazeley, Ab: *See* Gazley, James F.
Gazley [Gazeley], James F.: 161
Geiselle family: murder of, 94
*General Sumner*: naming of, 144 n
George [butcher]: 46, 48
George [Englishman]: and Reinhart's hauling trip, 218–221
George, Abel: account of, 58, 167 n, 222; killing of Hough McCaslin by, 164, 167, 168
George, Clubfoot: *See* Clubfoot George
George, Fraser River: *See* Fraser River George
George, John: and trial of Abel George, 168
Georgia: John M. Bozeman from, 259 n
Georgia, Gulf of: *See* Gulf of Georgia, British Columbia
German: spoken by Bartholomew, 108
Germany: Reinhart born in, xx, 308; Charles Pope from, 204 n
Gestler, Charlie: *See* Getzler, Charlie
Getzler [Gestler], Charlie: 204, 204 n

Gibbs, A. C.: and trial of Joseph Knott, 47 n
Gifford, Clarence: and S. M. Charles revolver, 117; Reinhart works for, 192
Gilbert, Frank B.: and Green River, 295 n
Gilbert, Sy: gold claims of, 152
Gillette, W. C.: road built by, 254 n
Gillmore, Jack: horse of, 297
Girton, Ward: and Salmon River gold, 270 n
Glasgow, Clement: businesses of, 43, 50, 56; Reinhart's trip with, 51–56; gold claims of, 150, 152; ranch of, 152
Gloria Mundy apples: in Oregon, 176
Gold Beach, Oregon: mining at, xxi, 50, 75, 81; Benjamin Wright at, 22 n; mention of, 76, 85, 90, 118, 145, 172, 177, 199, 220, 257; Reinhart at, 78–79, 309; trail to, 80; and Rogue River War, 94
Gold Beach massacre: *See* Rogue River War
gold claims: prices of, 60, 102; handling of, 77; types of, 101; of Reinhart, 60–62, 65, 66, 90, 97, 99, 101–102, 103, 106, 108
Gold Hill, Oregon: gold discoveries at, 40, 69 n, 175, 207, 278
Gold Hill Mining Company: 174–175
Goldsten, [hotel owner]: 197
Goldsten, Dan: 192, 193, 197
Goldsten, Perry: 192, 192–193, 193, 197
Good, [auctioneer]: and government cattle, 196
Goodale's Hotel: murder at, 164
Goose Lake: Oregon wagon train passes, 20 n
Gordon, Captain Samuel: 162
Graham, Pat: rides with Reinhart, 225
Grande Ronde Valley, Oregon: 195, 209, 211 n, 219, 236
Grand Harbor, Michigan: Edward Ryan from, 97
Grant County, Oregon: Canyon City in, 211 n
Grappy, Joe: 92
Grass Flat, Oregon: 70, 71
Grasshopper, Montana: gold discoveries at, 211 n
Grave Creek, Oregon: Reinhart on, 41, 41 n–42 n; mention of, 154, 169. *See also* Woodpile Creek, Oregon
Grave Creek Hotel: Reinhart stops at, 63, 111, 148
Grave Creek Indians: Warty as member of, 41, 42; and settlers, 47–48
Graves, William: and Plummer gang, 240, 240 n
Gray, [miner]: rides with Reinhart, 225

Green, [teamster]: hired by Reinhart, 233–235
Green, Major: as southern sympathizer, 195
Green, George: killing of George Stout by, 179–180
Green, Harvey: mining activities of, 28
Green Brier County, Virginia: William Craig from, 190 n
Greenhorn, California: 164
Green River, Wyoming: mention of, 9 n, 267 n; Indian battle on, 28 n; Reinhart at, 293, 295, 310; description of, 295; passengers hauled from, 296, 297
Griffin Creek, Oregon: gold discoveries at, 211 n
Grimes, [partner of Brown]: in Browntown, 63
Grimes, Governor James W.: and Des Moines, Iowa, 2 n
Grimes Basin: See Boise Basin, Idaho
Grimes' Creek, Idaho: Pioneer City on, 221 n
grist mill: at Oakland, 175
Grizzly Gulch, Montana: Reinhart in, 252
Grosluis, John: 80–81, 83, 108
Grosluis, Peter: 80–81, 83, 108, 112
Grover, Caleb: Reinhart works for, 186–187; cattle of, 187, 200–201, 210; mention of, 214
Groville, [Frenchman]: 266
Gulf of Georgia, British Columbia: mention of, 130, 131, 138 n; Reinhart's adventures on, 139–140
Gunn, Captain [sea-captain]: gambling by, 99–100
Guthings Hotel: Reinhart stops at, 146

*Hackstaff, W. L.*: wreck of, 10 n
Hadley, Massachusetts: Joseph Hooker from, 159 n
Hall, [contractor]: and Union Pacific Railroad, 301
Hall, Anna: in Oregon wagon train, 4 n
Hall, Fort: See Fort Hall, Idaho
Halle, Germany: Reinhart in, 308
Hall's: Reinhart stops at, 271
Hamburg, California: rivers near, 28 n
Hamilton, Nevada: gold discoveries near, 294 n
Hammer, Elisha: in Oregon wagon train, 3; as Charley Reinhart's partner, 35, 39, 41
Ham's Fork, Wyoming: 293
Happy Camp, California: mention of, 92, 105, 150, 152, 184; Reinhart at, 93, 309; Charley Reinhart at, 95
Harding, [lawyer]: and trial of Joseph Knott, 47 n

Harding, Dr. [Indian victim]: 26
Harding, Benjamin F.: 26 n
Harding, Dr. John R.: 26 n
Harding, L. L. D.: 26 n
Hard Scrabble, Oregon: 40 n
Harkness, McDonough: hotel of, 65, 148, 169; Reinhart buys whipsaw from, 170
Harper, [contractor]: hires Reinhart, 298–299
Harper, John: mention of, 210; Reinhart stays with, 236
Harper & Reed: Reinhart works for, 301
Harris and Marks: Reinhart hauls freight for, 248
Harrison County, Ohio: Thomas Mercer from, 11 n
Harrison Lake, British Columbia: 130 n, 133 n
Harrison-Lilloet route: 130 n, 133, 133 n
Harrison River, British Columbia: mention of, 130, 130 n, 133 n, 136; dangers on, 135
Hart, Thom: and shooting of Indian, 30
Hart, Tom: and Charley Johnson, 28
Harter, George: marriage of, 266
Harter and Keeler: 266
Hartford, Connecticut: Henry Switzer from, 274
Hartman, [hotelkeeper]: murder of, 240
Hart's Bluffs, Iowa: renaming of 2 n
Harvey, Hogue, and Solomon Jacobs: 176
Haskin Brothers: 57, 58, 66–67
Haskins, [miner]: 79
Hasting Cutoff: 10 n
Havard Brothers: 222, 223
Hawkins, Sam: Reinhart's trip with, 51–56; hotel kept by, 235
Hayes, James: and Gold Hill Mining Company, 175 n
Hayes, Fort: See Pioneer City, Idaho
Hay Press Ranch: 233
Hazard Powder Company: wagon of, 265
Headspath Cutoff: See Hudspeth Cutoff
Heaps, Hugh: 91
Heinige, Louis: in Helena, 249, 252
Helena, Montana: gold discoveries near, xxi, 248, 252 n; founding of, 211 n; mention of, 251, 254, 256, 257, 263, 265, 267, 271, 294, 306; vigilantes at, 251, 263; location of, 252 n; description of, 252; soldiers from, 263; Reinhart at, 265, 309
Helena Accommodation Line: 255
Hell Gate, Montana: hanging of Bill Bunton at, 52 n
Hellmuth, Joe: brewery run by, 204; mention of, 216, 221
Helm, Boone: mention of, 207; hanging of, 208, 208 n; and Plummer gang, 240

*Index* 331

Helm, James [Old Tex]: mention of, 207; death of, 208, 208 n
Helms, Colonel: in Oregon wagon train, 3, 9, 9–10, 10
Helms, Thomas Benton: 10
Hendershott, James: and trial of Abel George, 167, 168–169; in Grande Ronde Valley, 219; as state senator, 219 n
Hendershott, Sydney: 219, 219 n
Hendricks, C.: marriage of, 19 n
Henley, California: 26 n
Henry [baker]: as fireman, 205
Herman, James: 117 n
Heron, Green: chase of horse by, 50
Herzog, [barber]: hotel kept by, 220–221, 233, 235
Hibbard, Samuel: 208
Hickman, William A.: and Reinhart's horses, 279–280; mention of, 284 n
Hickman Brothers: 280 n
Hicks, Justice Richard: in British Columbia, 143–144, 144 n
High Rock Canyon, Nevada: Oregon wagon train passes, 18 n, 20 n; mention of, 129
Hill, [gambler]: rides with Reinhart, 271
Hill, Bill: prospecting by, 46
Hill, Flem: and Bill Hill, 46; and Joseph Knott, 46–47, 47 n; killing by, 47
Hill, Guss: Reinhart gambles with, 105; on Fraser River, 105–106, 137, 138; mention of, 143; on Umatilla River, 184
Hill, James: killing of, 47 n
Hill, Fort: *See* Fort Hill, Illinois
Hills, Alvin: Reinhart's meeting with, 27
Hills, Cornelius Joel: as leader of Oregon wagon train, 3, 4 n, 12; mention of, 4 n, 7, 10, 20–21, 27, 57; and Indians, 14–15, 15, 18–19, 23; and Samuel B. Briggs, 20
Hills, Elijah: in Oregon wagon train, 3, 10 n; and Indians, 14
Hills, Elijah C.: as son of C. J. Hills, 4 n
Hills, Erastus: in Oregon wagon train, 3, 10 n; and Indians, 14
Hills, Jasper: as son of C. J. Hills, 10 n
Hills, Mrs. Sophronia Briggs: marriage of, 4 n, 10 n; and Indians, 14–17
Hill's Bar, British Columbia: account of, 106, 136, 137 n; mention of, 135, 135 n, 241; Hill and Daniels on, 137
Hills Creek, Oregon: 10 n
Hinks, Samuel: on prospecting trip, 98–99; as partner of Reinhart, 99; mention of, 107, 113
Hiskey, [mormon]: 278, 291
Hoboken, New Jersey: Pat Graham from, 225

Hoge, Enos D.: 280 n
Hogem, Idaho: as name for Pioneer City, 221 n, 309
Hogem Creek, Idaho: 221, 228
Hogue, Charles P.: 176, 181–182
Hogue, Henry: in Portland, 176, 182
Hogue, Ida: 176
Hogue, James P.: threshing machine of, 176; Reinhart works for, 176–179
Hogue, Montgomery: 176
Holmes, [convict]: and Hough O'Niell, 91
Holt, James: on Dry Creek, 194; and Reinhart's land claim, 207
Holton, Dr.: removal of Reinhart's tooth by, 89
Holton, Dr. David S.: hotel of, 78; account of, 78 n
Hood, Mount: *See* Mount Hood, Oregon
Hook, H. M.: and Green River, 295 n
Hooker, Colonel Joseph: at Round Prairie, 159, 160; and Molly Burnett, 159, 160; description of, 159, 160; background of, 159 n, 160; and bed bugs, 160
Hope, Jimmy: 101, 211
Hope, Fort: *See* Fort Hope, British Columbia
horse racing: at The Dalles, 116
horses: of Charley Reinhart, 40, 225, 228; chase of, 49; of Reinhart, 1–2, 49–50, 78, 79, 117, 119, 120, 124, 136, 162, 166, 173–174, 183, 188, 193, 196, 197, 214, 225, 227, 231, 233, 247, 251–252, 256, 257, 258–262, 264, 265, 266, 274–277, 278–288, 291, 294, 295–296; prices of, 196, 196 n, 231, 258
Horseshoe Bend [Payette River]: Reinhart at, 224, 231
Horseshoe Mountains, Idaho: Reinhart crosses, 220, 232
Hostetter's Stomach Bitters: Reinhart's use of, 277
Hot Spring House [near Idaho City]: 243
hot springs: uses for, 21
House, J. E.: of Union Pacific Railroad, 273 n
Howard, [fighter]: 91, 93
Howard, Daniel: and murder of Magruder, 238–240
Howard, Virginia: marriage of, 204 n
Howe, [miner]: killing of Indian by, 30–32; gold discovery by, 34, 46
Howell, Jeff: mention of, 57, 58, 67, 79, 172, 173; and Reinhart, 172; success of, 173
Howie, Captain Neil: in command of soldiers, 263, 263 n
Howlins, the: 252

How-tome Creek, Oregon: and Umatilla Indian Reservation, 185 n
Hubbard, [soldier]: murder of, 206, 223, 240
Hubbel, Harry: mention of, 277 n
huckleberries: on Fraser River, 136
Hudson Bay Company: operations of, 56–57; and Chinook jargon, 56–57; area of operations of, 56–57, 142; and American Fur Company, 57; and U.S. Government, 57; and French Prairie, 81 n, 112, 182 n; and Fort Hope, 106, 138; Dr. McLaughlin in, 125; and Indians, 128, 129, 131; and Fraser gold rush, 133 n; ships belonging to, 137 n; land claims of, 212, 212 n–213 n; and Marcus Whitman, 213; and Fort Benton, 254
Hudspeth, Benoni M.: route blazed by, 9 n
Hudspeth Cutoff: 9, 9 n
Hulin, Lester: as guide, 19 n
Humboldt, Kansas: Reinhart at, 310
Humboldt Bay, Oregon: *Sea Gull* sinks in, 83 n
Humboldt Mountains, Oregon: near Pacific Ranch, 80
Humboldt River, Nevada: Oregon wagon train on, 10, 10 n; Indians near, 15–16; and Lassen's Road, 22 n; mention of, 66
Humboldt Sink: Cornelius J. Hills at, 10, 10 n
Humbug City, California: Reinhart at, 27, 29, 164; Sunday in, 29; mention of, 41
Humbug Creek, California: Reinhart on, 27, 29, 33, 35, 164; miners on, 29, 37, 39; Minersville on, 36; mention of, 61, 70, 81, 118, 163
Humbug War: 98 n
Hunt, Jim: 95, 96
hunting: from Oregon wagon train, 11, 22; of sea otters, 87
Huntsman, Peter: 284 n
Hurd, Samuel: death of, 135–136, 137
Hurst, [miner]: shooting of Indian by, 30, 32
Hurt, George: rides with Reinhart, 272
Hussy, Stephen: in Oregon wagon train, 4 n
Huston, Billy: and Jeff Standifer, 298 n
Hutchinson, Edward: in Victoria, 144; saloon run by, 221; rides with Reinhart, 271

Idaho: fish in, 54–55; gold discoveries in, 211; hauling in, 218–222, 224–225, 226, 228, 231–236, 249–250, 269–270, 271–272; Reinhart in, 220–222, 224–225, 226–227, 228, 231–235, 249–250, 269–270, 271–272; vigilantes in, 227, 244, 250
Idaho City, Idaho: mention of, 213, 221, 221 n, 228, 232, 243–244, 274; description of, 221; Reinhart at, 221, 231–235, 309; vigilantes in, 244
Idaho Springs, Idaho: Reinhart passes through, 272
Illinois: Reinhart in, xxi, 63, 308, 309, 310; people from, 3, 4 n, 33–34, 37, 61, 99, 152, 159 n, 164, 165, 176, 219 n, 264; mention of, 18
Illinois River, Oregon: mines on, 58, 66, 67; and Sucker Creek, 68–69; mention of, 78, 107, 109, 113; and Rogue River War, 97; Reinhart's trip to, 146
Independence, Missouri: Oatman's leave from, 166 n
Independence Rock: Oregon wagon train passes, 18 n
Indiana: mention of, 18; people from, 28, 52, 52 n, 102, 261
Indian affairs, superintendent of: Joel Palmer as, 48 n
Indian Agencies: at Omaha, Nebraska, 3; at Fort Laramie, 7 n
Indian Agents: Judge Snelling as, 31; Alonzo A. Skinner as, 34
Indian commissioner: Beverly S. Allen as, 34 n; J. P. Gaines as, 34 n; A. A. Skinner as, 34 n
Indian Creek, California: mention, 61, 91, 92, 118, 144, 150, 151, 212, 219, 220, 221, 257, 271, 298; mining on, xxi, 90; location of, 90 n; Happy Camp on, 93; Reinhart at, 93, 97, 99, 164, 309; Indians near, 97–98
Indian Creek City, California: Reinhart at, 93, 97; Eldorado Saloon and Bakery at, 93, 97, 99, 144, 309
Indian Creek saloon: *See* Eldorado Saloon and Bakery
Indian department, California: Indian scouts of, 24
Indian language: in Chinook jargon, 56; Reinhart learns, 84
Indians: conference of, 7; and Reinhart, 7, 14–17, 36, 54–55, 80, 84, 85, 87, 88, 90, 147, 148; and Oregon wagon train, 9, 10–11, 12, 13, 14–17, 18, 19, 20, 22–23, 23–24; and Captain Cornelius J. Hills, 14–15, 15, 18–19, 23; and John B. Welch, 16–17; injury of Ann Davis by, 19 n; and Captain Ben Wright, 22–23, 22 n, 94, 95, 96, 220; massacres by, 28, 94–97, 213, 213 n, 261–262; thefts by, 30; on Klamath River, 35; and Charley Reinhart, 41, 42; adultery among, 45; and settlers, 47–48, 88; honesty of, 80,

## Index

82; Yreka Creek, 80; on Mussel Creek, 80; mention of, 87, 89; and Rogue River War, 94–98; and Fraser River miners, 109, 109 n, 114, 130, 135, 136; and Colonel Edward J. Steptoe, 114; and Major Robinson, 114, 115, 119, 121; near The Dalles, 114; and General Newton S. Clarke, 114 n; and Fraser River trip, 119, 120–124, 125–126, 126–127, 128, 129, 130, 135; thefts from, 125–126; killing of, 30, 47–48, 126–127; and Hudson Bay Company, 129, 131; at Victoria, 142–143; handwork of, 143; and gold discoveries, 184 n; on Umatilla Indian Reservation, 185 n; killing of Whitman by, 213, 213 n; and Missouri River steamers, 253 n, 254 n; and Fort Benton, 255; and Bozeman trail, 260
Indian school: at Bellevue, Nebraska, 3 n
Indian scouts, and Oregon wagon train, 24
Indian superintendents: James W. Nesmith as, 54 n; General Palmer as, 69; A. A. Skinner as, 69
Iowa: Reinhart in, 2, 309, 310; people from, 3, 4 n, 57, 63, 92, 184, 194, 219 n, 277, 296; mention of, 18, 61, 83
Iowa City, Iowa: as state capital, 2, 2 n
Irish: working on railroad, 296
Irish Jimmy: 170
Ish, George: 166 n, 175 n
Ish, Jacob, 166 n
Ishe, William: 166
isinglass: *See* mica
Italian: spoken by Bartholomew, 108
Ives, George: and Plummer gang, 240, 246; and Freelon Schnebley, 246; hanging of, 258


Jack [horse]: 233, 247, 289
Jack, Captain: *See* Captain Jack's Company
Jack, Crossroad: *See* Elliot, Jack
Jack, Little: *See* Little Jack
Jackson, Dave: death of, 23, 28; and Reinhart, 29
Jackson County, Oregon: gold discoveries in, 34 n; mention, 78 n; and trial of Abel George, 168; Wolf Creek in, 169; Gold Hill in, 175
Jackson Creek, Oregon: mines on, 34, 40
Jackson Lake: 9 n
Jackson's Hill, Oregon: gold discoveries at, 34
Jacksonville, Oregon: mining at, xxi, 34, 34 n, 40; James Lingenfelter at, 4 n; Indians near, 23 n; founding of, 34 n; Charley Reinhart in, 35, 39, 162; Reinhart's trip to, 35, 39, 40, 309; mention of, 41, 48, 58, 60, 79, 149, 152, 160, 166, 167, 171, 174, 183, 278; miners from, 45, 46; winter of 1852 in, 58; description of, 162; trial of Abel George at, 164; Charley Adams at, 194; Thomas Fruid from, 252
Jacobs, Cyrus: 227, 228
Jacobs, Dick: 164, 248, 249
Jacobs, John M.: and John M. Bozeman, 260 n
Jacobs, Richard: at Browntown saloon, 65, 69; mention of, 227
Jane, Captain: 41
Janesville, Wisconsin: people from, 3, 4, 16
Jasper, Oregon: Cornelius J. Hills in, 10 n
Jefferson, Mount: *See* Mount Jefferson, Oregon
Jefferson Basin, Montana: gold discoveries in, 211 n
Jefferson Bridge and Toll Road: 264–265
Jefferson River, Montana: mention of, 250; Reinhart passes, 251, 265, 309
Jena, Saxony: as Reinhart's birthplace, xx, 308
Jeromes, the: 252
Jesus Maria mines: gold from, 109
Jim [butcher]: Reinhart's poker game with, 93
Jim [horse]: 247
Jim [Scotchman]: 258
Jim [half-breed]: 71–72
Jim, Long: *See* Long Jim
Jim, Spanish: *See* Spanish Jim
Jimmy, Irish: *See* Irish Jimmy
Jinks Hotel: Reinhart at, 305
Jocelyn, [contractor]: Reinhart works for, 295–296
Joe [Mexican]: shooting by, 234–235
Joe, Old: See Old Joe [Indian]
John [horse]: 247, 265
John, Captain [Indian]: See Captain John [Indian]
John, Dutch: *See* Schertz, John
John, Long: *See* Long John
John, Old: *See* Old John
John, Squire: *See* Pringle, John
John, Yankee: *See* Pringle, John
John Day River, Oregon: mention of, 178, 199; Reinhart crosses, 183; gold discoveries on, 211, 211 n
Johnson, [Swede]: 192
Johnson, Andrew: 269
Johnson, Charley: Reinhart employed by, 27–28; and Charley Reinhart, 217
Johnson, Coarse Gold: gold discovery of, 82
Johnson, James: pack train of, 85–86; store kept by, 96
Johnson, James: breaking of horses by,

179; drives horses to The Dalles, 181–183
Johnson, Levi: works for Reinhart, 269
Johnson, Samuel: cattle purchase of, 210
Johnson Creek, Oregon: berries on, 82
Johnson Diggings: 91
Johnson Mountain, Oregon: 82 n
Jones, C. C.: hanging of, 302 n
Jones, T. C.: as Collector of Internal Revenue, 255; in Chanute, Kansas, 255 n
Jones, Fort: See Fort Jones, California
Jordan River, Utah: 285
Josephine County, Oregon: creeks in, 41 n, 57 n, 155; mention of, 61, 70 n, 173, 238; towns in, 78, 164, 219; trial of Abel George in, 168; Reinhart in, 309
Josephites: 272 n–273 n
Joshua Indians: and Rogue River War, 94, 94 n
Juan de Fuca Straits: 138 n
Jullian: See Neuschwander, Julien
Jump-off-Joe Creek, Oregon: Reinhart at, 41
Junction House: Reinhart at, 227
justice, frontier: See frontier justice

Kamloops, Fort: See Fort Thompson, British Columbia
Kamloops Lake, British Columbia: gold discoveries at, 108, 112, 115; mention of, 120, 127, 134; Fraser River company on, 129; Reinhart at, 129, 309
Kanesville, Iowa: Mormons at, 2
Kansas: Reinhart in, xx, 310
Kansas City, Missouri: Reinhart at, 310
Karpville, Utah: See Case Ward, Utah
Kavanaugh, Thomas: and James Wilson, 167, 174, 207, 278; theft of horse from, 167, 174, 207, 278; and Gold Hill Mining Company, 174–175
Kearney, Fort: See Fort Kearney, Nebraska
Kearney, General Stephen F.: in Mexican War, 189 n
Keller, George: on Applegate Road, 10 n
Kelley, Andrew: on prospecting trip, 98–99; as Reinhart's partner, 99; as deserter from Army, 99, 155; mention of, 107, 113
Kelly, Judge John: and Reinhart's lawsuit, 216–217
Kenley, William: and government cattle, 196
keno game: Reinhart wins at, 83
Kentucky, Old: See Old Kentuck
Kentucky: mention of, 4 n; people from, 43 n, 177, 266, 267, 268, 269 n
Kerbyville, Oregon: mention of, 78 n, 107, 113, 183, 192, 196, 219, 221, 241–242; Reinhart's suit at, 101–102, 164, 172;

Reinhart at, 108, 109, 162; Charley Reinhart at, 146; naming of, 147 n; sawmill near, 148; trial of Abel George at, 167–168
Kernan, Peter: 103
Kijus Indians: See Kuitsh Indians
Kimball, [aerialist]: performances of, 283, 295; rides with Reinhart, 291
King, [unidentified]: drowning of, 80
King, [road builder]: 254 n
King, Nathan: and King's Valley, 54 n
King and Gillette: and Mullan Road, 254
Kingland Bakery, New York: Reinhart's father works for, 308
"King Lear": performance of, 283
King, Severe, and Marshall: on Strawberry Island, 136 n–137 n
King's Valley, Oregon: Reinhart passes through, 54
Kinyen, [emigrant]: on Reinhart's farm, 237
Kinyen, Mrs.: reputation of, 238
Kinyen, Cynthia: Reinhart wants to marry, 237; marriage of, 237–238; reputation of, 238
Kippel, Henry; and Jacksonville, 34 n
Kirby, James: and Kerbyville, 147 n
Kirkpatrick, [schoolteacher]: and Charley Johnson, 28
Kitchen, [unidentified]: chase of horse by, 50
Kitchen, Eph: prospecting of, 46
Kitchen, and Co. Canyon John: mine of, 44
Klamath County, California: 72
Klamath County, Oregon: Lost River in, 22 n
Klamath Indians: and Oregon wagon train, 23, 24; in Freetown, 30–31, 35; funeral rites of, 31
Klamath Lake, Oregon: 24, 32
Klamath River: Oregon wagon train passes, 24, 26; and Scott River, 28 n; prospecting on, 31; Reinhart at, 35, 36, 163, 309; ferry on, 36; and Shasta River, 36; mention of, 85, 93, 98, 146; Indians on, 90; Happy Camp on, 92, 93
Klickitat Indians: and Oregon wagon train, 23, 23 n; theft of Indian horses by, 115–116
Klippel, John: and Gold Hill Mining Company, 175 n
Knickerbocker and Argenta Mining and Smelting Company: 276 n
Knott, Joseph: sawmill of, 42, 46–47; hotel of, 42, 43, 46–47, 147; account of, 42 n; and Indians, 46; and Flem Hill, 46–47; killing of James Hill by, 47 n
Kraft, Charley: 225, 248, 249

# Index

Kroney, [miner]: mention of, 145, 146, 147, 221
Kroney, Henry: Reinhart works for, 203; and Emil Myers, 207; at Idaho City, 213
Kuitsh Indians: and Oregon wagon train, 23, 23 n

LaBarge, Captain Joseph B.: and the *Octavia*, 267 n
Labenee, Joe: killing of, 211–212
labor: cost of, 66
Ladd, John: 175, 208–209
Ladd, Sarah: and John Story, 208–209
Lafayette, Oregon: Reinhart visits, 177
Lager Beer Joe: *See* Hellmuth, Joe
La Grande, Oregon: Reinhart passes through, 219
Lake Chelan, Washington: Fraser River company crosses, 120
Lake County, Illinois: Reinhart in, xxi, 63, 88, 308; people from, 61, 99
LaMar, General: 230
LaMar, James: and stage robbery, 230; killing of, 230–231
land: price of, 58
land claims: of Charley Reinhart, 35, 39, 40, 43–44; of H. F. Reinhart, 40, 43–44, 53, 63, 187–188, 193, 207, 209–210; size of, 44
Lander Cutoff: 9
Lane, George: and Plummer gang, 240 n–241 n
Lane, General Joseph: and Rogue River War, 48; farm of, 52, 111; background of, 52, 52 n, 160; political activities of, 69, 70, 72; and Joseph Hooker, 159; and bedbugs, 160; the Buntons near, 185
Lane County, Oregon: 4 n, 9; settlers in, 10 n, 53 n
Langell Valley, Oregon: Lost River in, 22 n
Langley, Fort: *See* Fort Langley, British Columbia
languages: Chinook jargon as, 56; spoken by Bartholomew, 108
La Perle Creek, Wyoming: 261 n
Lapwai, Fort: *See* Fort Lapwai, Idaho
Lapwai Mission, Idaho: 184 n, 190 n
Laramie, Wyoming: vigilantes at, 302
Laramie, Fort: *See* Fort Laramie, Wyoming
Laramie River, Wyoming: Indians camped on, 7
Laramie's Fork: and Platte River, 7 n
Larkin, John: 171, 173, 174, 175
Lassater, [lawyer]: and Reinhart's lawsuit, 217

Lassen Road: Cornelius J. Hills on, 10, 10 n; and Humboldt River, 22 n
Lassen's Meadows: 22 n
Last Chance Creek, Montana: Helena on, 252
Last Chance Gulch, Montana: *See* Helena, Montana
Latham, William [Boston Bill]: as partner of Reinhart, 28, 29, 30, 33, 34–35; on Fraser River Trip, 118, 127
Latzenheiser, John: character of, 205; murder of, 206, 223, 240
Lava Beds: killing of General Canby at, 23
law enforcement: *See* frontier justice
Lawrence, J. W.: 277 n
Lawson Road: *See* Lassen Road
Learn, Richard: account of, 188 n
Leavenworth, Kansas: 95 n, 267, 310
Leby, [hurdy-gurdy-house keeper]: in Idaho City, 235
Lee, [carpenter]: and Sucker Creek bowling alley, 71, 75
Lee, Al: 68
Lee County, Iowa: Sophronia Briggs in, 10 n
Leeman, Robert: in Idaho City, 221; Reinhart hauls freight for, 235. *See also* Leeman and Sallaro
Leeman and Sallaro: Reinhart hauls for, 231, 232, 235
Leesburg, Idaho: 270, 270 n, 271
Lee's Encampment, Oregon: 185 n, 233 n
Leighton, Colonel Davis: and James P. Hogue, 176
Leighton, Mrs. Emma: 176
Leighton, John: 176
Leipzig, Germany: Reinhart's father in, 308
Leland Creek: *See* Grave Creek, Oregon
Lemhi, Fort: *See* Fort Lemhi, Idaho
Lemhi County, Idaho: gold discoveries in, 269 n
Lemhi River, Idaho: gold discoveries on, 269 n; Reinhart passes, 269, 309
Levan, Utah: 289 n
Levens, Dan: store kept by, 169, 170; Reinhart stops with, 173
Lewellen Brothers: 214
Lewiston, Idaho: founding of, 181, 181 n; Reinhart at, 190, 309; description of, 190–191, 290 n–291 n; mention of, 202, 223, 238, 239 n; vigilantes in, 239
lice: race of, 56
Lightning Creek, British Columbia: gold discoveries on, 137 n
Lillooet, British Columbia: 130 n, 131 n, 133 n

Lillooet Lake, British Columbia: travel on, 133 n
Lincoln, Abraham: 159 n, 226, 242, 269, 295 n
Lincoln & Brush: 217
Lindley, Bowen: in Oregon wagon train, 3, 14
Lindsey, Lavollette: Reinhart's poker game with, 93
Lindsey Brothers: as partners of Reinhart, 97
Lingenfelter, James W.: in Oregon wagon train, 4, 4 n
Linn, David: in furniture business, 171 n
Linn County, Oregon: Thomas Benton Helms in, 10; Brownsville in, 60 n; Albany in, 111, 182; Peoria in, 175; suit against Wheeler in, 178
liquor: finding of, 20–21; price of, 29, 33, 35, 64, 65, 66, 131, 203
Little, Feramorz: hotel run by, 273, 273 n; and Brigham Young, 273, 273 n
Little Cotton Canyon, Utah: silver discovered in, 276 n
Little Jack: and stage robbery, 229–231
Little Luckiamute River: See Luckiamute River, Little
Little Prickley Pear Toll Road: 254 n
Little Shasta River, California: Reinhart crosses, 26; mention of, 165
Lockhart, [Indian agent]: killing of, 23
Lockridge, John: in Oregon wagon train, 4, 16–17, 19, 24; and Indians, 16–17, 19; in Humbug City, 29
Lodi, California: 28 n
Logan, David: as candidate for Congress, 159, 173
Long, Hiram: saloon run by, 170–171
Long Jim: winters with Reinhart, 261–262
Long John: mention of, 92; and Gold Hill Mining Company, 175 n
Long's Ferry: 109 n
long tom: use of, 33, 40; description of, 33 n
Looking Glass Valley, Oregon: 179
Lorrilard Brothers: 252
Lost River, Oregon: Oregon wagon train passes, 22; natural bridge on, 23, 24
Louis, Dutch: See Dutch Louis
Loup River, Nebraska: Oregon wagon train on, 5
Love, Thomas: gold claims of, 150, 152; death of, 152
Lowery, Christopher: and murder of Magruder, 238, 239 n
Lucas, Governor Robert: founding of Iowa City by, 2 n

Luckiamute River, Big: Reinhart crosses, 54
Luckiamute River, Little: Reinhart crosses, 54
lumber: price of, 64, 71, 78, 101, 170, 248 n; Reinhart hauls, 217, 298–299
Luna [Lunar] House: in Lewiston, 191 n; and Robert Talbot, 223
Lunar House: See Luna House
Lynch, [blacksmith]: 249
Lyons, Haze: and Plummer gang, 240
Lytton, [unidentified]: in Iowa, 61

Mackinoo Indians: See Mikono tunne Indians
Madison River, Montana: Reinhart on, 263, 264, 309
Magin, Tom: 91
Magnolia Saloon: in Centerville, 221
Magnolia Saloon: in Placerville, 221 n
Magruder, Floyd: murder of, 238–239, 239 n
Magruder, Mrs. Floyd: 239
Maguire, [prizefighter]: saloon and bowling alley of, 90
Maguire's Opera House, San Francisco: 204 n
mail: methods of handling, 81–82
Maine: 4 n
Malad City, Idaho: 272, 272 n
Malheur River, Idaho: 202 n
*Maria*: on Fraser River route, 137 n
Marion County, Oregon: mention of, 81 n; Salem in, 112, 182; George Eads in, 177
Marsden Hotel: Reinhart stops at, 272
Marshall, [hotelkeeper]: Reinhart stays with, 184, 185; mention of, 207
Marshall, John: 269 n
Marshall, Steve: See Marshland, Steven
Marshelburg, Germany: Reinhart in, 308
Marshland [Marshall], Steven: and Plummer gang, 240, 240 n–241 n
Marsh Valley, Idaho: Reinhart passes through, 272
Martin, James: Reinhart works for, 175–176; account of, 176, 259; rides with Reinhart, 266
Mary, Spanish: See Spanish Mary
Maryland: C. Miller from, 261
Marysville, California: See Sacramento, California
Marysville, Oregon: See Corvallis, Oregon
Mason, Colonel: 95 n
Masons: and trial of Abel George, 164, 167; and Reinhart's lawsuit, 216; and George Porter trial, 240
Massachusetts: mention of, 4 n; people from, 28, 159 n, 195 n, 213, 215, 274, 282, 297

# Index

Mayflower Saloon & Bakery & Bowling Alley: 99, 103, 104–105, 147
McBride, [hotelkeeper]: and recovery of Reinhart's horses, 286, 287–288
McCandless, [teamster]: death of, 123–124
McCaslin, Hough: murder of, 164, 167, 222
McClain, Jake: 212
McClancy, John: background of, 154; and death of Andrew Robertson, 154–155; gold claims of, 169, 171; and John Larkin, 171; stabbing of, 173; mention of, 205
McClellan, George B.: as candidate for president, 226
McCloud, [gambler]: and vigilantes, 72–75; whipping of, 73, 74, 172, 173; mention of, 75
McCormick reaper: Reinhart works on, 176
McCoy, [boy]: and vigilantes, 72–74; whipping of, 73, 74, 172, 173; mention of, 75
McCoy, William: 175
McCrady River: See Lost River, Oregon
McCurdy, Solomon P.: 280 n
McDermit, Captain Charles: mining activities of, 28; background of, 28 n
McDonald, [bowling-alley keeper]: and killing of Indians, 30
McGill, Lucinda: in Oregon wagon train, 4 n
McGowan, Ned: on Fraser River, 106, 143; account of, 106 n, 143 n
McGuire, Jimmy: and Jeff Standifer, 298 n
McJanes, [Scotchman]: murder of, 92–93
McKay, Deputy: and McPherson, 157
McKay, Mrs. Alexander: marriage of, 125 n
McKay, Jerry: store kept by, 150; gold claims of, 150; in Helena, 265–266
McKay, William C.: and Umatilla Indian Reservation, 185 n
McKenzie River, British Columbia: 19 n
McKinlay, Archibald: as leader of a Fraser River Company, 109 n
McKinley, Mount: See Mount McKinley, Alaska
McKokety, Iowa: Sol Norton from, 277
McLancy, John: departure of, 175
McLaughlin, Captain David: as leader of Fraser River company, 109 n, 124–125, 125 n; Reinhart's comments on, 125
McLaughlin, Dr. John: account of, 125; marriage of, 125 n; and Gold Hill Mining Company, 175 n; and Oregon City, 183 n

McLaughlin Canyon, Washington: Indian fight in, 124–125, 125 n
McLean, Captain Donald: influence of, 128, 131; mention of, 128 n; at Fort Hope, 138, 138 n
McPhearson, [Irishman]: as deserter from Army, 99, 155; murder of Brad Robinson by, 155–157
Meacham, Alfred B.: wounding of, 23, 23 n; mention of, 215 n; and Mountain House, 233 n
Meacham, Harvey J.: mention of, 215 n; and Mountain House, 233 n
Meacham toll road: 215, 215 n, 218
Meadow Creek, Idaho: Florence near, 202 n
Meagher, Thomas Francis: See Meagher, Thomas J.
Meagher, Thomas J. [Francis]: as governor of Montana, 260, 260 n, 264; and Indian fighting, 260, 260 n, 264; account of, 264, 264 n; drowning of, 266–267, 267 n, 269; mention of, 274
Meagher, Mrs. Thomas: freight hauled for, 269
meals: prices of, 58, 87, 197
Meamber, Gus: 28 n
Medford, Oregon: Indian battle near, 48 n
Medicine Rock, Montana: Reinhart crosses, 254
medicines: use of, 98
Mellon [saloon owner]: 90
Mercer, Asa S.: bringing of women to Washington by, 111 n
Mercer, Thomas: 109–111, 111 n
Mexicans: in gold mines, 27, 29, 122, 251
Mexican War: Captain McDermit in, 28 n; Joseph Lane in, 52, 160; William H. Tichenor in, 83 n; Joseph Hooker in, 159 n, 160; Edward Baker in, 160; James Brady in, 189; in California, 189 n; Charles Russell in, 195 n; Green Clay Smith in, 269 n; Patrick E. Connor in, 276 n
mica: discovery of, 21
Michigan: people from, 51 n, 56 n, 97, 106
Middleton, Jack: in Oregon wagon train, 4, 4 n
Mikono tunne Indians: Reinhart and, 85; in Tututni tribe, 85 n; and Rogue River War, 94, 94 n
Military Academy, United States: 114 n, 159
Millard, Ezra: rides with Reinhart, 256–257
Mill Creek, Oregon: mention of, 114 n, 214, 217; Whitman mission on, 213 n
Miller, C.: winters with Reinhart, 261–262
Miller, Charley: 72

Miller's Hollow, Iowa: renaming of, 2 n
Miller's Hotel: Reinhart passes, 220
Milner, Moses: killing of Abel George by, 222
Miluk, Indians: 83 n
Milwaleta: as chief of Grave Creek Indians, 47 n
Milwaukee, Wisconsin: Reinhart at, 310
mine claims: *See* gold claims
miners: and Oregon wagon train, 18; card games of, 33; and Indians, 94–98, 109, 114, 130, 135, 136
Miner's National Bank: in Salt Lake City, 271 n
Minersville, California: A. Dejarlais at, 27, 27 n, 36; mining activities near, 27; Reinhart's bakery at, 28; George Rogers' store in, 33; on Humbug Creek, 37
mines: operation of, 33; profits from, 33; techniques of, 44, 46, 50, 65
Mission Ferry: Reinhart crosses on, 263
Missouri: people from, 3, 18, 29, 172, 239 n; Reinhart in, 310
Missouri Bottom, Oregon: founding of, 47 n; Reinhart passes through, 56
Missouri Indians: and Oregon wagon trains, 3 n
Missouri River: Oregon wagon train crosses, 3, 5 n; mention of, 250; Fort Benton on, 252; navigation on, 253, 253 n; Reinhart crosses, 258; drowning of Governor Meagher in, 267
Missouri Roan [horse]: in horse race, 297
Mitchell, Colonel D. D.: as head of Indian Agencies, 3 n
Mitchell, Mike: fate of, 189–190
Mix, Albert: and Charley Russell, 194–195; and government cattle, 196
Mix, James D.: 195 n
Mix, W. A.: 195 n
Modoc Indians: and Bloody Point massacre, 22 n, 28; and Oregon wagon train, 23, 24; use of Chinook jargon with, 57
Modoc War: 233 n
money, paper: devaluation of, 203
Monmouth, Illinois: James P. Hogue from, 176
Monroe, Jack: 214
Monroe, Fortress: *See* Fortress Monroe, Oregon
Monson, [Swede]: 192
Montana: vigilantes in, 52, 240; emigrants to, 195; gold discoveries in, 211, 248; Reinhart in, 250–269; 271–272
Montana Bar, Montana: mining at, 256
monte bank: played by miners, 33, 72, 73, 93; Reinhart runs, 105
Monterey, California: James Brady at, 189

Moore's Creek, Idaho: Idaho City on, 274
Morford, R. B.: 168
Morley, R.: Reinhart works for, 308
Mormons: in Iowa, 2, 2 n; and polygamy, 272 n–273 n, 274, 282; in Idaho, 269 n; working on railroad, 296, 301
Morrisina, New York: Thomas Fruid in, 252
Morrison, Charley: 264
Morse, Samuel: store of, 258, 264
Morse Brothers Store: at Gallatin City, 258, 264
Moss's Bar, California: 33
Mountaineers, The: kept by Jeff Standifer, 298 n
Mountain House: founding of, 53 n; Reinhart stops at, 233
Mountain Meadows Massacre: 280, 280 n
Mount Hood, Oregon: 113, 163, 183
Mount Jefferson, Oregon: 113
Mount McKinley, Alaska: 113
Mount Rainier, Washington: 113, 183
Mount St. Helen, Oregon: 113, 183
Mount Shasta, California: mention of, 22 n, 163; view of, 25, 36
Moylen, Tom: beating of, 302 n
Muddy Bridge: 293
Mud Lake: Oregon wagon train passes, 21 n, 22 n
mules: prices of, 196 n
Mulkey, Bill: 199
Mulkey, Cy: 199
Mulkey, Elijah: and Salmon River gold, 270 n
Mulkey, John: 199–200
Mullan, Lieutenant John: and Mullan Road, 254 n
Mullan Road: 254, 254 n
Mullen Mill: Reinhart hauls to, 217
Multnomah County, Oregon: Eagle Creek in, 182 n
Murain, Antoine: as Reinhart's neighbor, 80–81; mention of, 83, 108
Murian [Frenchman]: 90
Murphy & Stevenson: Reinhart hauls freight for, 263
Musgat, Emma: as Reinhart's sister, 148; Reinhart visits, 307
Musgat, John: 163, 238
Mussel Creek, Oregon: Indians on, 80
Mussel Creek Indians: mention of, 80; and Reinhart, 85; in Takelma tribe, 85 n
Myers, Emil: Reinhart works for, 203; and Henry Kroney, 207; and Charley Reinhart, 208, 213; in Idaho City, 213; and Reinhart's lawsuit, 216
Myers, John: Reinhart's poker game with, 93
Myers, John J.: as Oregon trail guide, 9 n
Myers Brewery and Bakery: 212

# Index

Myrtle Creek, Oregon: Reinhart at, 51, 56, 111, 149, 175; mention of, 158 n, 228

Nappias Creek, Idaho: gold discovery on, 270 n
Nash, Reverend John A.: on Des Moines, 2 n
National Prohibition Party: and Green Clay Smith, 269 n
Natural Bridge: on Lost River, 23, 24
Nauvoo, Illinois: Mormon's expelled from, 2 n
Nauvoo Legion: 280 n
Navy, English: See English Navy
Nebraska: Reinhart passes through, 309
Nelly [horse]: 247
*Nelly Tichenor*: mention of, 83: and Rogue River War, 96
Nesmith, James W.: 54, 54 n
Neuschwander, Julien: 165–166
Nevada: Captain McDermit in, 28 n; gold discovered in, 180, 294; Reinhart passes through, 309
Nevada, Montana: vigilantes at, 52, 52 n; founding of, 211 n
Newark, New Jersey: William H. Tichenor from, 83 n; Reinhart in, 308, 310
New Berlin, New York: Delazon Smith from, 159 n
New Chicago, Kansas: See Chanute, Kansas
New Jersey: William H. Tichenor from, 83 n
New Mexico: 283
New York: Reinhart in, xx, 259, 308, 310; people from, 3, 10 n, 29, 34, 53 n, 99–100, 105, 159 n, 161, 188, 197, 204 n, 225, 250, 252, 256, 261
New York Gulch, Montana: 258, 263
New York Gulch Quartz Mining Company: 256
Nez Percé Indian Agency: 188 n, 190
Nez Percé Indians: and Fraser River miners, 114; and Colonel Steptoe, 114 n; and gold discoveries, 184 n
Nez Percé mines: discussion of, 181, 181 n, 202 n–203 n; mention of, 186, 197, 202; Reinhart starts for, 188, 309
Niagara Falls: Reinhart visits, 307, 310
Nicholas [Indian chief]: authority of, 127, 128; and Fraser River company, 129, 131
Nicholls, Samuel: pack trains of, 86
Nichols, I. B.: and Charley Reinhart, 43, 59; and Ed Northcut, 59
Nicommen, British Columbia: gold discoveries near, 108 n
Niecely, Ab: hired by Reinhart, 64, 65; and George Fetterman, 67, 76
Nimrod, Dr.: and emigrant train, 196
Nitsell, Joe: as partner of Reinhart, 29
Nixon, A. J.: and Umatilla House, 183 n
Nobles, [saloon keeper]: in Humbug City, 29; movements of, 41
Nollan, [sailor]: marriage of, 83; and Hough O'Niell, 91
North Carolina: General Joseph Lane from, 52 n, 160
Northcut, Ed J.: 59, 59 n, 60, 74
North Umpqua River: See Umpqua River, North
Norton, Ben: Reinhart's suit with, 169, 172
Norton, Sol: at South Pass City, 294; stays with Reinhart, 277; hunts for Reinhart's horses, 279, 281, 282, 283, 284; at Green River, 295; works for Reinhart, 296; leaves for east, 296
Noumburg, Germany: Reinhart in, 308
Nounnan, Colonel Joseph F.: Reinhart hauls freight for, 271, 301, 304; account of, 271 n; and Union Pacific Railroad, 301 n
Nounnan & Kiskadin: Reinhart hauls freight for, 271, 273; and Union Pacific Railroad, 301
Nugent, Edward: See Nugent, John
Nugent, John [Edward]: as American consul for British Columbia, 141–142, 142 n; duties of, 142, 145
No. 8 Hotel: 79

Oakland, Oregon: Reinhart passes through, 55, 111, 175, 309; mention of, 208
Oatman, Harrison B.: 166 n
Oatman, Lorenzo: hotel run by, 166; in Indian attack, 166 n
Oatman, Mary Ann: and Indians, 166 n
Oatman, Olive: hotel run by, 166; and Indians, 166 n
Oatman, Royce: killing of, 166 n
Oberlin College Institute: Delazon Smith at, 159 n
O'Brien, John: and Three Sisters Hotel, 80; and Hough O'Niell, 91; and Rogue River War, 96
*Octavia*: in Fort Benton, 267–268
Ogden, Peter: 138 n
Ogden, Utah: and railroads, 273, 273 n; Reinhart passes through, 273, 309; mention of, 305
Ohio: mention of, 4 n; people from, 34 n, 11 n, 176
Okanogan, Fort: See Fort Okanogan, Washington
Okanogan River, Washington: mention

of, 120, 135; Fraser River company crosses, 124
Old Joe [Indian]: taken to San Francisco, 48
Old John: 92
Old Kentuck: 92
Old Tex: *See* Helm, James
Olds, Rubin: ferry of, 220
Oliver Stage Company, J. C.: Reinhart buys coach from, 272; account of, 272 n
Olney, Nathan: ferry run by, 183 n
Omaha, Nebraska: Reinhart at, 3, 310; Ezra Millard from, 256
Omaha Indians: at Omaha, Nebraska, 3; and Oregon wagon trains, 3 n
Oneida County, Idaho: Malad City as county seat of, 272 n
O'Neil, [railroad contractor]: mention of, 297; and Bear River City riot, 304
O'Neil, Jack: hanging of, 302, 302 n
O'Neil, James: works for Reinhart, 296
O'Neil, Tim: works for Reinhart, 296
O'Niell, Hough: and Three Sisters Hotel, 80, 91; and Harry Shannon, 91; at Port Orford, 91; fight of, with Con Oram, 257–258
Oram, Con: fight of, with Hough O'Niell, 257–258; rides with Reinhart, 258
O'Regan, Peter: mine of, 79
Oregon: mention, 1, 3, 9, 10; Indian department of, 24; boundary of, 25; A. A. Skinner in, 34 n; senators from, 52; governors of, 52; timber in, 54; farming in, 54; fish in, 54–55; Chinook jargon in, 56–57; Hudson Bay Company in, 56–57; Reinhart in, 24, 37–78, 79–91, 98–99, 100–107, 146–162, 166–185, 187, 193–194, 207, 208, 209–210, 212, 214, 215–216, 217, 225, 237, 247, 306, 309
Oregon City, Oregon: mention of, 52 n, 125; General Joseph Lane at, 52 n; water power at, 112; Reinhart passes through, 112, 182
Oregon Springs, Utah: Reinhart passes through, 273
Oregon trail: Reinhart Brothers on, 1–25; Indians on, 9; cholera on, 12; graves on, 12
Oregon wagon train: members of, 3, 4 n; at Missouri River, 3; practices of, 6, 7–8, 12; and Indians, 7, 10, 11, 12, 13, 14–17, 18, 19, 20, 22–23, 23–24; teams in, 7; splitting of, 9–10; hunting from, 11, 22; miners met by, 18
Orford, Port: *See* Port Orford, Oregon
Oro Fino, Idaho: founding of, 211 n
Orofino, Idaho: mining at, xxi; mention of, 186, 191 n, 197, 202; description of, 191–192, 192 n

Orofino River, Idaho: gold discoveries on, 181, 184; Reinhart reaches, 191
Osage Mission, Kansas: Reinhart at, 310
Oswego, New York: David Ransom from, 161
otter, sea: *See* sea otter
Ottoe Indians: and Oregon wagon trains, 3 n
Overland Hotel: in Boise, Idaho, 227
Owyhee, Idaho: gold discoveries at, 211 n, 214 n; mention of, 229, 231, 243
Owyhee, Nevada: Indian battle near, 28 n
oxen: in Oregon wagon train, 7; price of, 9

*Pacific:* transportation of miners on, 142
Pacific Hotel: attempt to rob, 91
Pacific Ranch: description of, 80; timber on, 80; Reinhart's stay at, 80–89, 309; operation of, 81; mention of, 83, 91, 108, 167; food at, 87; view from, 87; and Rogue River War, 94, 220; Reinhart sails past, 145
Paddy [teamster]: in Oregon wagon train, 4
Page, William: and murder of Magruder, 238–239, 239 n
Pahranagut, Nevada: 283
Painter, Elizabeth: *See* Reinhart, Elizabeth
Palmer, Joel: mention of, 21 n, 69; Indian scouts of, 23–24; as Indian superintendent, 24 n, 69 n; and Rogue River War, 48 n; and Richard Jacobs, 69; as leader of a Fraser River company, 109 n, 114 n–115 n; and Umatilla Indian Reservation, 185 n
Palouse Indians: and Colonel Steptoe, 114 n
parfleshes: use of, 51
Paris, France: sale of furs in, 87
Parish, Frank: and Plummer gang, 240; 240–241 n
Parker, [blacksmith]: on Oregon trail, 1
Parks, Samuel C.: and murder of Magruder, 239
Parranegut: *See* Pahranagut, Nevada
Pataha River, Washington: 188, 190
Patten, William: and Summerville, 236
Patterson, Ferdinand J.: and William Terry, 241, 243; and Preston Standifer, 241; and Fraser River George, 241–242; nature of, 242, 243, 244, 298; killing of Captain Staples by, 242; as southern sympathizer, 242, 243; and Thomas Donnahough, 243, 244–245; in Idaho City, 243–244; killing of Captain Pinkham by, 243–244
Payette River, Idaho: mention of, 202 n, 226, 227; Reinhart at, 220, 224, 231
Pease, [schoolteacher]: on Oregon trail, 1

## Index

Peck, Henry: hotel run by, 272, 272 n
pedro: playing of, 67
Pemberton portage : travel via, 133 n
Pend l'Oreille, Idaho: 186
Pennsylvania: people from, 28 n, 57, 71, 205, 215, 221–222; mention of, 102
Peoples, William: robbery by, 240 n
Peoria, Oregon: naming of, 175, 176
Perkins, Joel: 109 n
Perkins' Ferry: on Rogue River, 41
Pettijohn, Issac: on Applegate Road, 10 n
Philadelphia, Pennsylvania: people from, 221–222, 261; Reinhart at, 310
Phillips, [unidentified]: murder by, 91–93, 97
Phillips, William: murder of, 239 n
Pickett, [miner]: Reinhart gambles with, 105
Piegan Indians: at Fort Benton, 255
Pierce, Elias Davidson: and Pierce City, 184, 184 n, 191 n
Pierce, President Franklin: and Joel Palmer, 24 n
Pierce City, Idaho: naming of, 184; mention of, 186, 191 n, 211 n; Reinhart reaches, 191; population of, 191 n
pigs: hauled by Reinhart, 223–225, 228; prices of, 224–225, 226
Pike, [miner]: 121
Pike County, Illinois: John Brown from, 37
Pilcher: [wagon driver]: hired by Reinhart, 228
Pilcher, Joshua: founding of Fort Benton [no. 2]: by, 254 n
Pine Creek, Oregon: 185, 210, 212, 124
Piney Creek, Big: 261 n
Piney Creek, Little: 261 n
Pinto [horse]: 278
Pinkham, Captain: murder of, 243–244
Pioneer City, Idaho: mention of, 221; reputation of, 221 n; Reinhart hauls to, 228, 309
Pioneer Livery Feed & Sale Stable: Reinhart owns, 307, 310
Pioneerville, Idaho: See Pioneer City, Idaho
Pipestem Creek Montana: Reinhart passes, 251
pitcher, silver water: adventures of, 86
pitfalls: Indians' use of, 84–85
Pitney, [unidentified]: 40
Pitt, Billy: and Alvin Hills, 27
Pitt River, California: Oregon wagon train passes, 22; Indians on, 22, 57; Indian massacre on, 28
Pitt River Indians: and Oregon wagon train, 23, 24
Pitts, Dr.: 60

Pittsburgh, Pennsylvania: Reinhart at, 310
Placerville, Idaho: gold discoveries at, 211 n; Reinhart at, 220–221, 228, 233 n, 235, 309; mention of, 221 n, 237; pigs sold at, 224, 226; Christmas in, 235
Plano, Illinois: 272 n–273 n
Platte River: mention of, 5, 6, 7 n; dangers of, 5, 6, 9; Chimney Rock on, 18
Platte River, North: 260 n, 261 n, 299
Pleasands, Wesley: and trial of Abel George, 168
*Pleiades*: at Victoria, 142 n
Plum Creek: 19 n
Plum Creek Mountain: Indians at, 18; Oregon wagon train passes, 18 n, 20 n
Plummer, Henry: gang of, xxii, 240, 246, 258, 264; hanging of, 240 n
*Plumper*: at Victoria, 142 n
Plymouth, Massachusetts: William Latham from, 28
Point Blanco, Oregon: gold discoveries at, 77, 82; mention of, 82, 145
Point Roberts, Washington: Reinhart's attempts to reach, 139–140; location of, 139 n; account of, 140
Point of Rock, Wyoming: mention of, 280; Union Pacific terminus at, 296
poker: playing of, 33, 67, 72; played by Reinhart, 77, 83, 105; played by Charley Reinhart, 148
Poker Bill: See Willowby, William
pole cat: See skunk
politics: Reinhart on, xxii; at Althouse Creek, 69
Polk, President Franklin: and General Joseph Lane, 52 n
Polk County, Oregon: Dallas in, 177
Pollards, the: 192
polygamy: of Mormons, 272 n–273 n, 274, 282
Pool, John R.: pack trains of, 33–34; and Jacksonville, 34 n; mine of, 40
Pope, Charles: theater built in Walla Walla by, 204; account of, 204 n
Pope, Charley: mining activities of, 28
Pope, Jack: and Gold Hill Mining Company, 174
Port Douglas, British Columbia: 130, 130 n, 133, 133 n
Porter, George: as southern sympathizer, 205; at firemen's benefit riot, 205–206; murder of Hartman by, 240
Portland, Oregon: Joseph Knott in, 42 n; supplies shipped from, 66, 81; ships from, 87; Reinhart at, 111, 112, 182, 309; mention of, 116, 144, 158, 186, 238; Henry Hogue in, 176, 182; death of Mike Mitchell in, 189–190; Ferd Pat-

terson in, 242; Thomas Donnahough in, 243
Portneuf Canyon, Idaho: Reinhart passes through, 272
Port Orford, Oregon: gold discoveries near, 77; goods shipped from, 80, 81; trail to, 80; description of, 82; Spanish Mary in, 83; founding of, 83 n; mention of, 86, 144, 145, 166–167, 257; Hough O'Niell leaves, 91; and Rogue River War, 97; Bill Shepherd at, 158; Reinhart at, 309
Portuguese: spoken by Bartholomew, 108
post master: Reinhart as, 56
powder horns: making of, 21
Powder River, Oregon: gold discoveries at, 211, 211 n; Reinhart passes, 220; mention of, 228, 233; Baker City in, 236
Powder River, Montana: Indians on, 260
Powers, Jed: and James P. Hogue, 176, 178, 179; drives horses to The Dalles, 181–183
Powers, Jimmy: hanging of, 302 n
Pratt, F. H.: and Prattsville, 79 n; and Rogue River War, 96
Pratt, O. C.: and trial of Joseph Knott, 47 n
Prattsville, Oregon: Reinhart at, 79, 309; trail to, 80; Charley Reinhart at, 81, 86, 89, 95; mention of, 85, 86, 88, 89, 90, 93, 145, 177, 199, 220; and Rogue River War, 94, 95–96; location of, 97 n
Prattsville bakery and saloon: 81, 86, 95
Pre-emption claim: See land claim
Preston, General: and Reinhart, Wilson and Company lawsuit, 101
Prevost, Captain James C.: 139 n, 142 n
Prickly Pear Valley, Montana. Reinhart passes through, 251, 254
Prince [horse]: 247, 265
Princeton, Illinois: Thomas Mercer starts from, 111 n
Pringle, John [Squire John, Yankee John]: 92–93, 147, 149
Promontory Point: railroads joined at, 273 n
Provo, Utah: District Court at, 285; Reinhart passes through, 290, 309
Prussia: Reinhart from, xx, 308; Guss Hill from, 105
Pudding River, Oregon: French Prairie near, 81 n
Puget Sound, Washington: 140, 144
Purier, Allix: as Reinhart's neighbor, 80–81; mention of, 83, 108
Purier, Frank: as Reinhart's neighbor, 80–81; mention of, 83, 108
Pyburn, [unidentified]: and Mrs. Wheeler, 177–178

Pyburn, Mrs.: and Mrs. Wheeler, 177–178
Pyle, James M,: and Pyle Canyon, 219 n
Pyle Canyon, Oregon: Reinhart passes through, 219

Quartz mills: on Indian Creek, 164
Queen Victoria: See Victoria, Queen
quicksilver: use of, in mining, 50

Race, John: Reinhart buys wagon from, 298, 299
Racine, Wisconsin: George Ives from, 240
racing, horse: See horse racing
Racoon River, Iowa: Des Moines on, 2 n
railroads: Reinhart works on, 295–296, 299
Randall, Fort: See Fort Randall [on Missouri River]
Ranier, Mount: See Mount Ranier, Washington
Ransom, David: and Reinhart, 150, 171; gold claims of, 150; store kept by, 150; indiscretions of, 151; about Elizabeth Reinhart, 153–154; and John McClancy, 154; on Ab Gazley, 161; on Sarah Ladd, 209
Ransom & McKay: 265
Rapp, Joseph: and Salmon River gold, 270 n
Redding, California: 25 n, 118
Red Face Bill: See Willowby, William
Red Rock, Montana: Reinhart passes through, 250, 269
Red Wolf [Indian chief]: 190 n
Reed, Jimmy: 303 n
Reed, S. B.: Reinhart works for, 299; and Union Pacific Railroad, 299 n
Reedall, H. S.: 277 n
Reinhart, Carl Heinrich: as father of H. F. Reinhart, xx
Reinhart, Charley: finances of, 1; on Oregon trail, 1–25; in water, 5; and Indians, 7, 41, 42; and H. F. Reinhart, 9, 24, 161–162, 216, 217, 223, 226, 233, 236, 247, 306; and William Riddle, 9, 24, 43, 59, 63; at Jacksonville, 35, 39; land claims of, 35, 40, 43, 63; horses of, 40, 225, 228; and Elish Hammer, 41; mining by, 44–45, 46, 97, 147; houses of, 46, 49, 213; and I. B. Nichols, 43, 59; at Althouse Creek, 62–63, 78; at Browntown, 63; and Webfoot Brown, 63–64; at Sucker Creek, 70, 71, 75, 77, 77–78, 99, 108, 155; gambling by, 77, 105, 148; and Pacific Ranch, 80–81; at Prattsville, 81, 82, 86, 89; and John Waddell's pack train, 86; mention of, 90; and Eldorado Saloon and Bakery, 93, 97, 99–

# Index

100; and Rogue River War, 96; whipsawing by, 97, 99; drinking habits of, 100–101; to Crescent City, 103; chickens raised by, 104; at Kerbyville, 147; at Roseburg, 147–148, 149, 151; marriage of, 151; illness of, 151; and Brad Robinson, 157; farming by, 158–161, 162, 169, 175, 217; operation of threshing machine by, 160–161, 162; and Colonel Burnett, 161; and Captain Gordon, 162, 196; son of, 173; in Walla Walla, 196–197, 213; daughter of, 197; and Emil Myers, 208; haying operations of, 212–213; and stolen wagon, 214; livestock of, 217; and Reinhart's loads of pigs, 223, 226; and freight hauling, 229, 233, 236, 247; bills of, 247; death of, 249 n

Reinhart, Elizabeth: marriage of, 151; Reinhart's comments on, 151, 154, 161; and joke played upon Coom, 153–154; and Colonel Burnett, 161; mention of, 175, 233; in The Dalles, 196–197; in Walla Walla, 197; and Sarah Ladd, 208–209; and Cynthia Kinyen, 237

Reinhart, George Ellwood: birth of, 173; mention of, 175, 196–197

Reinhart, Maria Concordia: as mother of H. F. Reinhart, xx

Reinhart, Herman Francis: birth of, xx; schooling of, xx; character of, xx, xxi–xxii; death of, xx, 310; immigration of, xxi; and Brigham Young, xxii, 280, 290; finances of, 1, 249, 282, 306; horses of, 1–2, 49–50, 78, 79, 117, 119, 120, 124, 136, 162, 166, 173–174, 183, 188, 193, 196, 197, 214, 225, 227, 231, 233, 247, 251–252, 256, 257, 258–262, 264, 265, 266, 274–277, 278–288, 291, 294, 295–296; on Oregon trail, 1–25; in water, 6, 35–36; and Indians, 7, 14–17, 36, 47, 54–55, 80, 84–85, 85, 87, 88, 90, 120–121, 132, 147, 148; and Charley Reinhart, 9, 24, 161–162, 216, 217, 223, 226, 233, 236, 247, 306; cattle of, 9, 84, 214, 216; buckskin coat of, 17; loss of belongings by, 17, 19; theft of money from, 19; in Yreka, 26–27; at Minersville, 28; mining activities of, 28, 29, 33, 34–35, 44–45, 46, 60–61, 67, 68, 90, 97, 99, 101–102, 103, 108, 133; and William Latham, 30, 33; trip of, to Jacksonville, 35; illnesses of, 36, 89, 98, 170, 176–177, 218, 219, 277, 299, 304; land claims of, 40, 43–44, 53, 63, 187–188, 193, 207, 209; horses broken by, 50, 178–179; trip of, to Northern Oregon, 51–56; and stealing of pig, 55; killing of calf by, 59; and William Riddle, 59; trip of, to Althouse Creek, 59–60, 78; and George Fetterman, 59–63; gambling of, 60, 77, 83, 99–100, 104–105, 170–171, 173–174, 194; and Webfoot Brown, 63–64; Browntown bakery and saloon of, 63–66; and Sucker Creek bowling alley, 75; trip of, to Gold Beach, 78–79; at Pacific Ranch, 80–89; reading by, 33, 83, 158, 259, 276; and Indian boy, Skeesy, 84; and Indian language, 84; and John Waddell's pack train, 86; trip of, to Siskiyou, 90, 91, 93–94; and Enos, 95; and Eldorado Saloon and Bakery, 97; law suits of, 101–102, 161, 171–172, 216–217, 223; drinking habits of, 100–101; and gunpowder accident, 106; whipsawing by, 108, 150, 170, 173, 309; keepsake gold of, 109; trips of, to The Dalles, 109–113, 181–183; trip of, to the Fraser River, 117–141; boat trips of, 138–140, 141, 145; devil-fish killed by, 141; at Coffee Creek, 149–155; employment of, 150, 165, 176–179, 186–187, 192, 195, 196, 203; and Elizabeth Reinhart, 151, 154, 161; as deputy postmaster, 158–159; farming by, 158–161, 169, 209–210, 212, 214, 215–216; operation of threshing machine by, 160–161, 176, 195, 309; and Colonel Burnett, 161; and Ben Norton, 171–172; church attendance of, 176; trip of, to Walla Walla, 183–186; log house built by, 193; and Charley Russell, 195, 196; at Walla Walla Hotel, 197; haying operations of, 212–213; and stolen wagon, 214; freight hauling by, 214, 217, 218–221, 223–225, 226, 228, 231–235, 236, 248–251, 257, 258, 263, 264–265, 265–266, 268, 269, 271, 297, 298–299; lumber hauled by, 214, 217, 298–299; and John Ding, 216–217; and Peter Scheible, 218–221, 223; passengers hauled by, 225, 248–251, 254, 255, 256, 257–258, 265, 266, 269, 271, 272, 277–278, 291, 296, 297; and Cynthia Kinyen, 237; loss of horses by, 251–252, 264, 278–288, 294; mules of, 256, 257; works on railroad, 295–296, 299, 300; return of, to east, 306, 307; marriage of, 307; in Chanute, Kansas, 310; owns Bon Ton Restaurant, 310; owns Pioneer Livery Feed & Sale Stable, 310. *See also* Reinhart, Wilson and Company [second]

Reinhart, William: account of, 102. *See also* Reinhart, Wilson and Company [first]

Reinhart, Wilson and Company [first]: lawsuit of, 101–102

Reinhart, Wilson and Company [second]: lawsuit of, 101–102

Rents: 197
revolver, S. M. Charles: purchase of, 109; defect in, 113; Reinhart's sale of, 117; fate of, 192
Reynolds, [miner]: death of, 135
Reynolds, [timekeeper]: and Reinhart's time, 300
Reynolds, Jack: gold claims of, 150, 152; mention of, 155
Reynolds, Jackson: hotel bought by, 147 n
Rhett Lake: *See* Tule Lake, California
Rhodes, Bill: mines of, 191
Rhodes Creek, Idaho: gold discoveries on, 184, 191, 191 n; Pierce City on, 191; Reinhart at, 309
Richards, Dick: 249
Richards, Francis: as Reinhart's neighbor, 80; and Indians, 85
Richardson, Clark: in Oregon wagon train, 4, 24; in Humbug City, 29
Rich Diggings, Oregon: farms on, 163
Rich Gulch, Oregon: 34, 40 n, 162
Richmond, Tom: 183, 187
Richmond, William: rides with Reinhart, 272
Richmond, Kentucky: Green Clay Smith from, 269 n
Richmond, Oregon: 199
Rickett, Harry: ferry operated by, 272 n
Rickmoth, Charles: rides with Reinhart, 256
Rickreall Creek, Oregon: Senator Nesmith on, 54
Riddle, Abner: in Oregon wagon train, 4 n
Riddle, F. Stilley: in Oregon wagon train, 4 n
Riddle, George W.: in Oregon wagon train, 4 n
Riddle, Isabella: in Oregon wagon train, 4 n; marriage of, 43
Riddle, John Bouseman: in Oregon wagon train, 4 n
Riddle, Maximilla Bouseman: in Oregon wagon train, 4 n
Riddle, William R.: in Oregon wagon train, 3, 4 n; and Charley Reinhart, 9, 24, 43, 59, 63; and Bramson Brothers, 40 n; marriage of daughters of, 147
Riddle, William H. Jr.: in Oregon wagon train, 4 n
Riddle, Oregon: 4 n
Riddle Station, Kentucky: 4 n
Rio de Janeiro, Brazil: 138, 184
Rippley, Robert: rides with Reinhart, 272
River Jordan, Utah: 285
Rivers' Hotel: Reinhart stays at, 235
Roades, Elmer James: and McCoy and McCloud, 74

Roads, [teamster]: works for Reinhart, 268
Roberts, Captain Henry: 139 n
Roberts, Jesse: hotel kept by, 147; on Coffee Creek mines, 149; mention of, 169
Roberts, Joseph: hotel bought by, 147 n
Roberts, Point: *See* Point Roberts, Washington
Robertson, Major: *See* Robinson, Major
Robertson, Andrew: death of, 154–155
Robinson, Major: and Indians, 114–115, 121, 129; as leader of Fraser River company, 115 n, 117, 118–119; identification of, 131 n
Robinson, Brad: hotel kept by, 149; and Charley Reinhart, 149, 151; and joke played on Coom, 153–154; murder of, 156–157
Robinson, Mrs. Brad: and Charley Reinhart's marriage, 151; and joke played on Coom, 153–154; as widow, 158
Robinson, Clara: 204 n
Robinson, Fred: 158, 204
Robinson, Dr. J. King: hack operated by, 255; killing of, 255, 255 n, 290
Robinson, Dr. Jesse: 149 n
Robinson, Joseph B.: 204 n
Robinson, Susan: marriage of, 204; benefit for, 204, 205; mention of, 204 n
Robinson, William: 204 n
Robinson Family: in Walla Walla, 204; in San Francisco, 204 n
Robinson's Bar, British Columbia: 131 n
Robnette, John: marriage of, 207
Rochester, New York: people from, 3, 204 n
rocker: use of, 33 n, 40
Rockwell, Orrin Porter [Porter C.]: and Reinhart's lost horses, 280; and Destroying Angels, 280, 284; account of, 280 n
Rockwell, Porter C.: *See* Rockwell, Orrin Porter
Rocky Bar, Oregon: Reinhart in mining partnership on, 29–30
Rocky Mountains: Reinhart crosses, 250, 272; mention of, 254; Bozeman trail across, 259; South Pass City in, 293; ties hauled in, 299
Rogers, George: at Minersville, 33
Rogue River, Oregon: wreck of *W. L. Hackstaff* near, 10 n; murder of Benjamin Wright on, 22 n; Oregon wagon train on, 24; Indians on, 23 n, 26, 34, 48, 48 n, 85, 94–97; mention of, 29, 69, 86, 145, 158, 173; gold discoveries on, 34, 75, 77; Perkins' Ferry on, 41; miners from, 45; Abel George at, 58; Reinhart at, 79, 109, 166; description of, 162

Rogue River Indians: A. A. Skinner as agent to, 34
Rogue River War: mention of, 23 n, 48; account of, 94-97; end of, 98
Romain, [bookseller]: See Roman, Anton A.
Romain, James P.: and murder of Magruder, 163 n, 238-240
Roman [Romain], Anton A.: 163, 163 n
Rose, Aaron: in Roseburg, 51 n, 56; background of, 56 n; daughters of, 160
Rose, Emma: 160
Rose, Lissy: 160
Roseburg, Oregon: Reinhart in, 51, 55, 111, 151, 175, 309; as county seat of Douglas County, 51 n; mention of, 53, 153, 158, 158 n, 160, 189, 265, 278; Charley Reinhart in, 147-148, 149, 151, 158; as source of supplies for Coffee Creek, 150; McPherson at, 155; James Wilson at, 166
Rosenthal, [jeweler]: on trip to Montana, 249-251
Rosenthal, Mrs.: rides with Reinhart, 256-257
Ross, Colonel George: livery stable kept by, 149, 151; Louisa Bowen working for, 155; as son of John E. Ross, 174
Ross, Colonel John E.: and Gold Hill Mining Company, 174-175, 175 n
rounce: playing of, 33, 67
Round Prairie, Oregon: Reinhart at, 56, 158-161, 175; location of, 158 n; mention of, 179
Round Tent: in Browntown, 78
Round Valley, Utah: Reinhart passes through, 286, 289
Ruby City, Idaho: 211, 229, 231
Ruckle, J. S.: as stage operator, 224 n, 230 n; Thomas Road built by, 224 n
Rushman, John: works for Reinhart, 296
Rush Valley, Utah: silver mining in, 276 n
Russell, Charley: activities of, 194-195, 196; Reinhart works for, 195, 196; account of, 195 n
Russell's Creek, Oregon: 194
Russian: spoken by Bartholomew, 108
Ryan, Edward: and Eldorado Saloon and Bakery, 97; whipsawing by, 97; on prospecting trip, 98-99; as partner of Reinhart, 99; drinking of, 99; as deserter from army, 99, 155; death of, 106; mention of, 107
Ryan, Jimmy: shooting of, 302
Ryan, Tom: gold discovery by, 277 n
Ryder, M.: hotel built for, 147 n

Sacramento, California: 27, 186, 189

Sacramento River, California: and Pitt River, 22 n
Sacs Indians: at Omaha, Nebraska, 3
saddles: prices of, 116
Sailor Diggings, Oregon: mention of, 57, 58, 90, 219; Reinhart stops at, 146; Ferd Patterson in, 241-242. See also Waldo, Oregon
St. Helens, Mount: See Mount St. Helens, Oregon
St. Joseph, Missouri: 19 n
St. Louis, Missouri: Charles Ebert from, 197; shipping from, 253
St. Louis Restaurant and Hotel: in Helena, 252
St. Paul, Nebraska: 5 n
St. Petersburg, Russia: sale of furs in, 87
Salem, Oregon: French Prairie near, 81 n; Reinhart passes through, 112, 177, 182
saleratus: 21
Sallaro, [liquor dealer]: in Idaho City, 221. See also Leeman and Sallaro
salmon: in northern streams, 54-55
Salmon City, Idaho: gold discoveries at, 269; Reinhart hauls freight to, 269-270, 309
Salmon Creek, Oregon: 182
Salmon River, Idaho: mining on, xxi, 18, 202, 269 n; mention of, 172, 199, 215, 294; Henry Kroney at, 203; pack trains to, 210
Salmon River mines: 203 n, 222
saloons, Browntown: 63-64, 65, 66, 67-68, 75, 77-78
saloons, Eldorado: 93, 97, 99, 144, 309
saloons, Mayflower: 99, 103, 104-105, 147
saloons, Prattsville: 81, 86, 95
saloons, Sucker Creek: 70, 71, 75, 77-78, 97
Salt Lake City, Utah: mention of, 3, 9, 10, 248, 273; Reinhart's trip to, 271-273; description of, 273; Reinhart at, 273-288, 309
Salt Lake City Hotel: 273, 273 n
Salt Lake road: 12
Salt Lake Theater: building of, 283
Sands, John L.: Reinhart's poker game with, 93
Sandwich Islands: cattle ship from, 144; Jeff Standifer in, 241
San Francisco, California: Indians taken to, 48; shipment of supplies from, 53, 71, 81, 87; mention of, 72-73, 90, 105, 117, 144, 184, 186; sale of furs in, 87; vigilantes in, 143
Sangamon County, Illinois: William Riddle from, 3, 4 n
San Quinton Penitentiary: 91
*Santa Cruz:* Reinhart travels on, 144-145

Sarpy, Peter A.: ferry run by, 3 n
Satellite, H.M.S.: and boundary survey, 139 n; at Victoria, 142, 142 n
Saunders, James: as partner of Reinhart, 29
Savage, Dr.: gold claims of, 70
Savage, James: 70 n
Saxony: Reinhart born in, 308
Scheible, Phillip: Reinhart hauls for, 217, 218–221; mention of, 222; nature of, 223; Reinhart's lawsuit against, 223
Schertz, John [Dutch John]: mention of, 61, 92, 148; as partner of Reinhart, 99, 103, 108, 133; salting of mine by, 147; at sawmill, 148; in Kerbyville, 168; as Charley Reinhart's partner, 169, 173
Schnebley, Freelon [Stubbs, Stump]: Reinhart stops with, 188; account of, 188 n; and George Ives, 240, 246; theft of mules by, 245–246; killing of, 246
schutler wagon: 266, 272, 299
Scott, [stockman]: murder of, 239
Scott, John W.: gold strike of, 28 n
Scott, Levi: as founder of Scottsburg, 53 n
Scott, Nelson: robbery by, 240 n
Scott, Fort: See Fort Scott, Kansas
Scott Bar, California: lithograph of 27 n; O. Dejarlais at, 27 n, 36, 36 n; Brown's store at, 33; mention of, 97
Scott County, Kentucky: James K. Duke from, 268
Scott River, California: gold in, 18; miners on, 28; mention of, 51, 97
Scott Valley, California: Indian Agent in, 31; Reinhart goes to, 164
Scotts Bluff: Oregon wagon train passes, 18
Scottsburg, Oregon: location of, 53; supplies shipped from, 55, 66, 150–151; mention of, 73, 144, 145
scouts, Indian: See Indian scouts
scurvy: Reinhart's case of, 36
Sea Gull: wreck of, 83, 83 n
sea otter: hunting of, 87; price of furs of, 87
Seattle, Washington: mention of, 109; Thomas Mercer as founder of, 111 n
Seavey, [storekeeper]: mine of, 79
Sena, Saxony: See Jena, Saxony
Sessions, Peregrine: founding of Sessions by, 293 n
Sessions, Utah: Reinhart passes through, 273, 293
Seton Lake, British Columbia: 133 n
Seven Mile Spring, Montana: Reinhart passes, 256
seven-up: playing of, 33, 67
Sevier Lake, Utah: 287
Sevier River, Utah: 287

Shaffer, [teamster]: in Oregon wagon train, 4
Shaffer, [hotelkeeper]: and Rogue River War, 96, 220; hotel of, 232, 233
Shaffer Canyon, Idaho: 221
Shaffer Hill: See Horseshoe Mountains, Idaho
Shannon, Harry: as fighter, 91
Sharkey, Frank B.: and Salmon River gold, 270 n
Sharp, John: and Union Pacific Railroad, 301 n
Shasta, California: gold at, 18
Shasta, Mount: See Mount Shasta, California
Shasta Butte City, California: See Yreka, California
Shasta County, California: 250 n
Shasta River, California: Reinhart's crossing of, 26, 35–36
Shasta River, Little: See Little Shasta River, California
Shasta Valley, California: 165
Shaw, Robert: ball alley of, 66, 69, 72, 73
Shaw Brothers: Reinhart hires horses from, 280, 291
Shears, George: and Plummer gang, 240
Sheil, [lawyer]: and trial of Joseph Knott, 47 n
Shepherd, Billy: mention of, 83; in minstrel troupe, 158
Shepherd, George: as Reinhart's partner, 170, 173
Shepherd, Mollie: Reinhart hauls freight for, 297
Sherman, Joel: account of, 164
Shively, [miner]: mine of, 40
Shuswap Lake, British Columbia: See Kamloops Lake, British Columbia
Sickle, Van: See Van Sickle, [miner]
Sierra Nevada Mountains: Oregon wagon train passes, 18 n, 20 n
Silcott, Washington: 190 n
Silver Canyon, Utah: Reinhart winters in, 273–277; description of, 274; mention of, 278, 283, 306
silver deposits: at Argenta, Montana, 264 n; at Stockton, Utah, 276
Sim, [lawyer]: and trial of Joseph Knott, 47 n
Simcoe, Fort: See Fort Simcoe, Washington
Similkameen River, British Columbia: gold discoveries on, 115; Fraser River Company splits at, 125; mention of, 135, 184
Simmonds, [miner]: See Simonds, [miner]

*Index* 347

Simmonds' Bend, Oregon: 79, 80
Simmons, [miner]: *See* Simonds, [miner]
Simonds, [miner]: 57, 58, 79
Simons, [miner]: *See* Simonds, [miner]
Simpson, [merchant]: 267
Singleton, [unidentified]: and McPherson, 156
Sioux City, Iowa: shipping from, 253 n
Sioux Indians: and Oregon wagon train, 7; on Bozeman trail, 260
Siskiyou County, California: Reinhart's mining in, 26–39; Scott River in, 28 n
Siskiyou Mountains: Reinhart crosses, 25, 36, 39, 90, 163; Oregon wagon train reaches, 29
Siuslaw River, Oregon: 53 n
Sixes River, Oregon, 77, 82
Skeesy [Indian boy]: and Reinhart, 84
Skene, Peter: 138 n
Skinner, Alonzo A.: as Indian agent, 34, 34 n; background of, 34 n; political activities of, 69, 72
Skinner, Cyrus [John]: and Plummer gang, 240, 240 n
Skinner, James: discovery of Rich Gulch by, 34, 40, 252
Skinner, John: *See* Skinner, Cyrus
skunk: Charley's capture of, 2
Skunk River, Iowa: Reinhart passes, 2
Slate Creek, Oregon: and Rogue River War, 97; mention of, 172, 173
Slaterville, Idaho: 191, 191 n
Sloat [Foot], Commodore John D.: in Mexican War, 189 n
sluices: use of, 40
Smith, [paymaster]: and Reinhart's pay, 299, 305
Smith, [company president]: rides with Reinhart, 256–257
Smith, Judge: and recovery of Reinhart's horses, 287, 290, 291
Smith, Dr. A. M. C.: [Judge]: mining activities of, 28
Smith, Bullhead: and William Terry, 241, 243
Smith, Delazon: 159, 159 n, 160
Smith, Elias: account of, 287 n, 291 n–292 n
Smith, Frank: saloon kept by, 151
Smith, Green Clay: as governor of Montana, 267–268; anecdote about, 267–268; Reinhart hauls freight for, 268; account of, 268–269, 269 n
Smith, John: as partner of Reinhart, 29
Smith, Joseph Jr.: 272 n–273 n
Smith, Pegleg: 92
Smith, General Persifer Frazer: 94–95, 95 n
Smith, Sam: 190 n

Smith, Thomas: ferry kept by, 43, 152; store of, 50
Smith, Thomas: and Bear River City riot, 303
Smith, William: and Salmon River gold, 270 n
Smith's Fork: 293
Smith's River, California: 78
Snake Indians: and Fraser River miners, 114
Snake River: mention of, 9 n, 114 n, 188, 188 n, 202, 202 n, 236; Indians on, 114; ferry on, 190; gold discoveries on, 181; Reinhart reaches, 190, 220, 224, 272
Snelling, Judge R. B.: as Indian Agent, 31
Soda Springs, Idaho: 9, 9 n
Sonoma, California: Joseph Hooker at, 159 n
South America: Hill and Daniels go to, 138, 184
Southard, Joe: gold claims of, 150, 152
Southern Restaurant: in Bear River City, 298, 303; shooting at, 302
southern sympathizers: 195, 204–207, 242, 243
South Pass, Wyoming: 9 n
South Pass City, Wyoming: mention of, 267 n, 277 n, 278, 285, 291; gold discoveries at, 271; Reinhart in, 293, 309; location of, 293
South Umpqua River: *See* Umpqua River, South
South Water County, Illinois: Reinhart in, 308
Spalding, Rev. Henry H.: 190 n, 213 n
Spanish: in Chinook jargon: 56, spoken by Bartholomew, 108
Spanish Fork, Utah: Reinhart passes through, 289
Spanish Jim: and Reinhart's horses, 251–252
Spanish Mary: account of, 83; and Hough O'Niell, 91; Bill Shepherd working for, 158
Spanish Monta Banking game: *See* monte bank
Spears, Horace: and James Cranston, 165
Spokane, Washington, 125
Spokane Indians: and Fraser River miners, 114; and Colonel Edward J. Steptoe, 114, 114 n
Sprague, [lawyer]: and Reinhart, Wilson and Company lawsuit, 102
Sprague, Davis and Associates: 299
Springfield, Illinois: William Riddle from, 3, 4 n
Spuzzum, British Columbia: 136 n
Squire John: *See* Pringle, John

Standifer, Jeff: trial of, 241; in Bear River City, 298; saloon kept by, 298 n
Standifer, Preston: and Ferd Patterson, 241; and Jeff Standifer's trial, 241
Staples, Captain: killing of, 242
Stark, James: in Salt Lake City, 283
Statler, [unidentified]: and Gold Hill Mining Company, 174
steers: Reinhart's care of, 84
Steinberger, Colonel Justus: 195 n
Steptoe, Colonel Edward J.: and Charles Russell, 195 n; and Indians, 114; account of, 114 n
Steptoe, Washington: See Walla Walla, Washington
Sterling, Oregon: gold discoveries at, 40
Stevens, Colonel: See Stevenson, Colonel Jonathan D.
Stevens, Robert L.: 225
Stevenson, Colonel Jonathan D.: in Mexican War, 189 n
Steward, Robert: 170–171, 173–174
Stine, Fred: 248, 249
Stinking Water, Montana: gold discoveries at, 211 n
Stinkwater River, Montana: Reinhart on, 251, 265, 309
Stinson, Buck: and Plummer gang, 240
Stockton, Commodore Robert F.: in Mexican War, 189 n
Stockton, Utah: Reinhart winters near, 273–277, 309; history of, 273 n, 276; description of, 274; silver mines near, 276; James Wilson near, 278; mention of, 306
Stone, Harry: as partner of Reinhart, 97, 98
Stonebraker, Isaac: at Umatilla River, 184; and Reinhart, 207, 208, 209–210
Storey, Silas: and James P. Hogue, 176
Stout, Ben: 152, 154–155, 180
Stout, George: mention of, 152, 154–155; murder of, 179–180
Stout, Colonel Lansing: and General Joseph Lane, 159; election of, 159 n; and Jefferson Howell, 173
Stratton, R. E.: and trial of Joseph Knott, 47 n
Strausburg, France: fancy woman from, 29
Strawberry Island, British Columbia: 136 n
Straw Ranch: Reinhart passes, 220; mention of, 228; stage robbery near, 229
Strickland, [mail driver]: 232
Stubbs: See Schnebley, Freelon
Stump: See Schnebley, Freelon
Sublette, William: and Fort Laramie, 7 n; route blazed by, 9 n; and William Craig, 190 n

Sublette Cutoff: 9 n
Sucker Creek, Oregon: mention of, 61, 79, 82, 105, 113, 135, 144, 149, 150, 151 155, 164, 168, 177, 184, 188, 192, 211, 221, 222; gold discoveries on, 68, 98; farms on, 68–69; Reinhart at, 71–78, 98, 99, 148, 309; Charley Reinhart on, 77; lumber on, 101; Chinese miners on, 104
Sucker Creek Bowling Alley and Bakery [second]: See Mayflower Saloon & Bakery & Bowling Alley
Sucker Creek saloon and bowling alley: building of, 70, 71, 75; costs of, 75; selling of, 77–78; burning of, 97
Sulphur Creek, Wyoming: 293, 297
Summers, James: mention of, 61–62; and Browntown saloon, 64
Summerville, Oregon: mention of, 236; location of, 236
Summit, Montana: founding of, 211 n
Sumners, Charley: winters with Reinhart, 215
Sun River, Montana: Reinhart crosses, 254, 256, 269
*Surprise:* on Fraser River route, 137 n
Surprise Valley: Oregon wagon train passes, 18 n, 20 n
Sutherlin, Oregon: Joseph Knott at, 42 n
Swan, [guard]: 256–257
Swedish: spoken by Bartholomew, 108
Sweetwater Mines: mention of, 285, 291; Reinhart arrives at, 293; nature of, 294 n
Sweetwater River, Wyoming: mining on, xxi, 271, 277 n; Oregon wagon train passes, 18; mention of, 267 n, 278; Reinhart at, 309
Swill [Indian]: shooting of, 31
Switzer, Major Henry: works for Reinhart, 271, 274; mention of, 276–277
Sykes [unknown]: 34 n
Sykes, Ben: and Judge Young, 69–70; and gold discovery, 69 n; mention of, 72
Syracuse, New York: B. D. Cartwright from, 53 n

Tabernacle: in Salt Lake City, 273
Table Rock, Oregon: Indians near, 23 n, 48, 48 n
Takelma Indians: Mussel Creek Indians as, 85 n
Talbot, Robert [Cherokee Bob]: as southern sympathizer, 205; at firemen's benefit riot, 206; death of, 206 n, 223; career of, 223
Talbotte, Henry: See Talbot, Robert
Tar Head Indians: See Klamath Indians
Tate, Hough: 241
Taylor, [hotel keeper]: daughter of, 217
Taylor, Judge: and trial of Phillips, 92

## Index

Taylor, James Maddison [Matt]: bridge built by, 272 n
Taylor, Steve: 86
Taylor's Ferry: Reinhart uses, 272
ten-dice game: playing of, 194
Ten Mile Creek, Montana: 252
Tennessee: James D. Burnett from, 56 n
Ten Pin Pool: playing of, 104–105
Terry, William: and Ferd Patterson, 241, 243; and Bullhead Smith, 241; as southern sympathizer, 243
Teton River, Montana: and Missouri River, 252 n
Tex, Old: *See* Helm, James
Texas: people from, 29, 43 n
Texas Bar, British Columbia: 136, 136 n
Thayer, Kansas: Reinhart at, 310
The Dalles, Oregon: mention of, 78 n, 99, 135, 148, 149, 155, 171, 175, 188, 198, 241; Fraser River miners leave from, 108; Reinhart at, 113, 117, 183, 309; description of, 113, 116, 183; Indians near, 114; stolen Indian horses at, 116; horses driven to, 181–183; Eightmile Creek near, 185; Mike Mitchell in, 189; Charley Reinhart in, 196–197
The Fountain, British Columbia: Fraser River company at, 130; prices at, 133
Thomas, Reverend Eleazar: killing of, 23 n
Thomas, George F.: as stage operator, 224 n, 230 n; road built by, 224 n; and James Lamar, 230
Thomas Road: Reinhart travels on, 224
Thompson, [Englishman]: 99, 155
Thompson, Bill: 106
Thompson, Clay: rides with Reinhart, 272, 273
Thompson, David: 127 n
Thompson, Fort: *See* Fort Thompson, British Columbia
Thompson City, British Columbia: mention of, 134 n, 135 n; Reinhart passes through, 135
Thompson River, British Columbia: gold discoveries on, 108, 108 n, 112, 115; mention of, 120, 127, 129 n; Fraser River company on, 129; and Fraser River, 134; dangers on, 135
Thompson's, Idaho: Reinhart sells pigs at, 224
Thompson's Dry Diggings, California: *See* Yreka, California
Thoroughman, Colonel Thomas: in command of volunteers, 263 n
Thorp, Russell, Sr.: in vigilantes, 302 n
Thousand Spring Valley, Nevada: Indian attack in, 14–17
Three Sisters [mountains], Oregon: Reinhart's sight of, 113

Three Sisters, Oregon: mention of, 88, 91, 257; in Rogue River War, 94, 96
Three Sisters Hotel: 80, 88
threshing machines: manufacture of, 152–153; operated by the Reinhart Brothers, 160–161, 162, 176, 195, 309
Throgs Neck, New York: Thomas Fruid in, 252
Thurber, John K.: 289 n
Tichenor, Captain William H.: account of, 79 n, 83, 83 n; and Rogue River War, 96
ties: hauling of, 299
timber: in northwest, 54; on Pacific Ranch, 80; on Dry Creek, 193
Tincture, William: saloon run by, 221
Tioga County, New York: Dave Ransom from, 161
Tirrell, A.: as partner of Reinhart, 29
Titus, John: 280 n
tobacco: price of, 65
Tobias [dog]: on East Gallatin River, 259
Todd, William: and Caleb Grover, 187
toll bridges: operation of, 148
toll roads: charges of, 221; Reinhart's use of, 221, 250, 251, 265, 272, 293
Tooele, Utah: Reinhart winters near, 273–277; founding of, 273 n; New Years Day in, 274–276; James Wilson near, 278
tools: prices of, 35
Toottootenay Indians: *See* Tututunne Indians
Touchet River, Washington: Reinhart crosses, 188; mention of, 245
Towhee Creek, British Columbia: gold discoveries on, 137 n
Townsend, Peter: 269, 269 n
Townsend, Elizabeth: marriage of, 269 n
Transylvania University: Green Clay Smith at, 269 n
traps: pitfalls as, 84–85
Trask, [hotelkeeper]: 79
Travis Brothers' stable: 257, 266, 270
Treasure Hill, Nevada: gold discoveries at, 294 n
*Tribune:* at Victoria, 142 n
Trinidad: 28 n
Trinity River, California: gold at, 18; mention of, 69, 72, 90 n
Tripp, Robert: on prospecting trip, 68
Trumbles, Henry: Reinhart stays with, 236
Tucannon River, Washington: Reinhart crosses, 188; mention of, 189
Tule Lake, California: Oregon wagon train passes, 20 n; Indian massacre on, 22 n
Tumlum River, Washington: Reinhart crosses, 185; mention of, 216

Turner, John: and Plummer gang, 240, 240 n–241 n
Tututni Indians: Mikono tunne Indians as, 85 n; and Rogue River War, 94 n
Tututunne Indians: and Rogue River War, 94, 94 n, 220
Tutt & Donald's: Reinhart hauls freight for, 265
Twogood, James: hotel of, 65, 148; mention of, 111; and Reinhart, 170, 173
Tygh Hotel: Reinhart stops at, 113
Tygh Valley, Oregon: Reinhart passes through, 113; stolen Indian horses at, 116
Tyson, J.: and Henry H. Brown, 60 n

*Umatilla:* on Fraser River route, 133 n, 137 n, 138 n
Umatilla County, Oregon: Reinhart's farm in, 53, 187, 309
Umatilla House: 183
Umatilla Indian Reservation: location of, 185, 185 n; Indians on, 185 n; mention of, 219
Umatilla Landing, Oregon: mention of, 229, 236; Reinhart at, 231
Umatilla Reserve Road: 215, 215 n, 218
Umatilla River, Oregon: farms on, 184; Reinhart at, 184, 219, 231; Umatilla Landing on, 229 n
Umpqua Canyon: *See* Canyon Creek Canyon, Oregon
Umpqua County, Oregon: 83 n
Umpqua Indians: Grave Creek Indians as, 47; and death of Warty, 48
Umpqua River, Oregon: Indians near, 23 n; Scottsburg on, 53; game on, 55; mention of, 60, 62, 65, 66, 111, 145, 147, 154; the Buntons on, 185
Umpqua River, North: Reinhart crosses, 51, 111, 175; mention of, 52; lost horse on, 52; gold mining on, 53; Winchester on, 55
Umpqua River, South: Charley Reinhart's land on, 35, 39; and Canyon Creek, 41 n–42 n; crossing of, 51, 62–63, 111, 147–148, 309; mention of, 75, 94, 180; farms along, 149; gold discoveries near, 149; Round Prairie on, 185 n
Union, Oregon: *See* Uniontown, Oregon
Union County, Oregon: Summerville in, 236
Union Pacific Railroad: at Ogden, Utah, 273, 273 n; Reinhart works for, 295–296, 299, 300; terminus of, 296; mention of, 299; and government, 305; payment of employees by, 305
Uniontown, Oregon: 219 n, 236
United States-Canadian boundary survey: 138–139, 138 n,–139 n

United States Government: and Hudson Bay Company, 57; and Union Pacific Railroad, 305
United States Military Academy: 114 n, 159
Updike, Dave: hanging of, 250, 250 n
Upper Platte and Arkansas Indian Agency: and Indian conference, 7 n
Upton, Gus: pack train of, 85–86; store kept by, 96
Upton, John: pack train of, 85–86; store kept by, 96
Utah: Reinhart in, 273–294, 309
Utah War: 280 n

Vancouver, Captain George: 139 n
Vancouver, British Columbia: ships from, 87; mention of, 106, 117; Reinhart at, 141–145, 309
Vancouver, Washington: 155
Vanderbilts, the: 252
Van Dyke, [unidentified]: and government horses, 196
Vannoy, James N.: operation of ferry by, 109, 109 n, 173
Van Sickle, [miner]: prospecting by, 45, 46
Venango County, Pennsylvania: George Fetterman from, 57
Vera Cruz, Mexico: 28 n
Victoria, Queen: 131
Victoria, British Columbia: ships from, 87; mention of, 106, 117, 130, 137, 221, 241; Reinhart at, 141–145, 309; description of, 141, 142, 144; Indians at, 142–143; miners at, 142, 143
Victoria, Fort: *See* Fort Victoria, British Columbia
Vigilance Committees: *See* vigilantes
vigilantes: hanging of John Brown by, 36–39; and Flem Hill, 47; killing of Warty by, 47–48; and Bill Bunton, 52; and McCoy and McCloud, 73–75; and Hough O'Niell, 91; in California, 105; and Ned McGowan, 106 n; in San Francisco, 143; near Corvallis, 177–178; at Virginia City, Montana, 208; in Boise, 227, 250; and James LaMar, 230–231; in Walla Walla, 230–231; and Dave English, 239; and Plummer gang, 240; in Montana, 208, 240, 251, 263; and Ferd Patterson, 244; in Idaho City, 244; and Dave Updike, 250; at Helena, 251, 263; at Fillmore, 288; in Bear River City, 301, 302, 302 n; at Laramie, 302; and Jack O'Neil, 302, 302 n; and Jimmy Powers, 302 n; and C. C. Jones, 302 n
Vining, George T.: hotel built by, 147, 147 n

# Index

Virginia: mention of, 102; people from, 114 n, 190 n
Virginia City, Montana: emigrants to, 195; gold discoveries at, 211, 238; founding of, 211 n; mention of, 239 n, 248, 252 n, 256, 260 n, 274, 294, 306; vigilantes in, 240, 258; location of, 248 n; Reinhart at, 251, 263, 268, 309; soldiers from, 263; governor at, 264; as state capital, 267
Virginia City, Nevada: 21, 180

Waddell, John: pack trains of, 82, 86
Waddingham, [New Yorker]: 197
Waddingham, Wilson: 197
Wade, Rob: horse of, 297
Waggoner, George: *See* Wagner, George
Wagner [Waggoner], George: and Plummer gang, 240 n–241 n
Wagner, John: 46, 50
wagons: mishaps to, 5, 6; abandoning of, 9, 12
Waiilatpu Mission: founding of, 213 n
Waldo, Oregon: Reinhart at, 78, 91, 146; snow near, 90; mention of, 92, 93, 219, 257; Indians near, 97–98; Ferd Patterson at, 241–242. *See also* Sailor Diggings, Oregon
Walker, Squire: 72
Walker, Charles: 70, 91
Walker, Joseph R.: 190 n
Wallace, [unidentified]: stabbing of John McClancy by, 173
Walla Walla, Washington: mention of, 53, 69, 114, 117, 148, 180, 202 n, 214, 217, 219, 220, 221, 227, 236, 254, 265, 278; Reinhart at, 53, 183–186, 197–201, 309; description of, 186; James A. Buckley as sheriff of, 188, 188 n; emigrants to, 195; bad winter in, 197–201, 203; Robinson Family in 204; Thomas Donnahough in, 243, 244–245; ruffians in, 240; Ferd Patterson in, 244, 245
Walla Walla, Fort: *See* Fort Walla Walla, Washington
Walla Walla Hotel: Reinhart Brothers work for, 197
Walla Walla Indians: and Umatilla Indian Reservation, 185 n
Walla Walla River: the Bunton Family on, 53, 194; Reinhart reaches, 185; farms along, 185; John Broughton on, 194
Walling, A. G.: 59 n
Walling's Ranch: 74 n
Wallula, Washington: 186, 197, 239
Walters, Christopher: 265
Ward, [merchant]: 210
Ward, Hester L.: marriage of 111 n

Ward, James: works for Reinhart, 296, 297, 298
Wards Gulch, Idaho: gold discovery in, 270 n
Warner & Whitman: Reinhart hauls freight for, 297
Warren, James: gold discovery by, 202 n
Warren's Diggings, Idaho: 202, 211
Warta-hoo [Indian]: as father of Warty, 48 n
Warty [Indian]: and Charley Reinhart, 41, 42; killing of, 47–48
Warwick, Ike: mine of, 79
Wasatch, Utah: 300, 300 n, 306
Wasatch Mountains: ties hauled in, 299
Wasatchor Mountains: *See* Wasatch Mountains
Wasco County, Oregon: The Dalles in, 113, 116, 148; Eightmile Creek in, 185
Washington: farming in, 54; timber in, 54; fish in, 54–55; Hudson Bay Company, 56–57; Chinook jargon in 56–57
Washington, D.C.: Reinhart at, 310
Watkins, Dr.: 78, 90
Watson, Old Eph: 158, 189
Weaver, Anna: 47 n
Weaver, Daniel: and William Reinhart, 102
Weaver: Harriet: 47 n
Weaver, James B.: 47 n, 56 n
Weaver, John W.: 47 n
Weaver, William: and Graves Creek Indians, 47; mention of, 56, 94
Weaver & Bailey's Saloon: 302 n
Weber Canyon, Utah: Reinhart passes through, 293
Webster & Davis: Reinhart hauls freight for, 266
Weewich River, Washington: 114, 120
Weimar, Germany: Charles Pope from, 204 n
Weiser River, Idaho: Reinhart crosses, 220, 235
Weiser River House: Reinhart stays at, 235
Welch, Jerry: in Oregon wagon train, 3; death of, 4–5, 5 n
Welch, John B.: in Oregon wagon train, 3; shot by Indians, 16–17, 19
Wells, Fargo and Company Express: and Charlie Getzler, 204; mention of, 227, 294 n; robberies of, 229, 272; Reinhart competes with, 297
Wenatchee River, Washington: 114
Wenatchee Indians: and Fraser River miners, 114
Westfall, C.: as partner of Reinhart, 29
Weston, [miner]: in Captain Jack's Company, 28

Whales' Head Hotel: Reinhart at, 79
Whannell, P. B.: as magistrate at Fort Yale, 144 n
Whatcom County, Washington: Point Roberts in, 139 n
Wheeler, [school teacher]: troubles of, 177–178
Wheeler, Mrs.: infidelity of, 177–178
whipsaw: price of, 170
whipsawing: by Charley Reinhart, 97, 99; by Reinhart, 99, 108, 150, 170, 173, 309
Whirlpool [on Fraser River]: dangers of, 137
Whiskey Creek, Washington: 188, 194 n
White, [miner]: 126
White, Fred: and stolen Indian horses, 116
White, Captain William: as captain of *W. L. Hackstaff*, 10 n
Whiteman, [unidentified]: and government horses, 196
White Pines Mines, Nevada: gold discoveries at, 294, 294 n
White Tail Deer Creek, Montana: freight hauling to, 264
Whitey [horse]: 278
Whitman, Marcus: account of, 213, 213 n
Whitman College: founding of, 213 n
Whitman Mission: near Reinhart's farm, 213
Whitman Seminary: founding of, 213 n
Whitt, John: ball alley of, 66, 69, 72, 73, 77–78; and Dupuy, 75
Wienhanover, Germany: Reinhart in, 308
Wiggans, Henry: 100
Wild Horse Creek, Oregon: on Umatilla Indian Reservation, 185; mention of, 197; Reinhart passes, 219
Wild Horse Prairie: Reinhart passes through, 269
Willamette River, Oregon: ferries across, 42 n; navigation on, 54; French Prairie on, 81 n; Portland on, 112
Willamette Valley, Oregon: mention of, 10, 49, 53, 58, 75, 81, 167, 177, 198, 212, 222; Reinhart in, 51–56, 57, 111, 175, 182, 309; game in, 55; winter of 1852 in, 58
Willey, [carpenter]: rides with Reinhart, 272
William Creek, British Columbia: gold discoveries on, 137 n
Williams, [miner]: gold claims of, 150
Williams, Alex: rides with Reinhart, 272
Williams, Captain Bob: and McCloud, 74; as captain of volunteers, 172; and Jack Driscoll, 172

Williams, Charles: and Gold Hill Mining Company, 174, 175 n
Williams, George. adventures of, 41–43; mine claims of, 44–45, 46
Williams, Jacob D.: and Robert Talbot, 223
Williams, John: works for Reinhart, 296
Williams, Captain M. M.: public house kept by, 146; sawmill operated by, 148
Williamson, Lieutenant R. S.: and Pitt River, 22 n
Willowby, William [Red Face Bill, Poker Bill]: and Robert Talbot, 223; killing of, 223
Willow Creek, Oregon: Reinhart crosses, 183; cattle on, 187; mention of, 199
Willow Springs, Oregon: Indian ambush near, 26 n
Wilson, [boardinghouse keeper]: 27
Wilson, H.: gold claims of, 150
Wilson, Henry: *See* Reinhart, Wilson and Company [first]
Wilson, Henry: *See* Wilson, James
Wilson, James [Henry]: theft of Tom Kavanaugh's horse by, 174, 175; discovery of gold by, 174–175; account of, 166–167; mention of, 207; in Utah, 278; theft of Reinhart's horses by, 286–291
Wilson, Jim: gold claims of, 152
Wilson, Thomas: and Charley Reinhart, 99; as partner of Reinhart, 101–102; chickens raised by, 104; mention of, 192. *See also* Reinhart, Wilson and Company [second]
Winchester, Oregon: Flem Hill in, 46; as county seat of Douglas County, 51; Reinhart passes through, 51–52, 55, 111, 175, 309; founding of, 51 n; General Joseph Lane near, 52; the Buntons near, 185
Wind Mill Flour Store: Reinhart works for, 308
Wind River Mountains: 9 n
Wingate and Humphrey's bar: 94
Winn, William: helps drive horses, 181–182
Winters, [paymaster]: 299
Wisconsin: mention, 1; people from, 3, 4, 4 n, 16, 70, 94, 240; John Musgat in, 163; governor of, 210; Reinhart in, 310
Wolf Creek, Oregon: gold discoveries on, 169; sawmill on, 170; mention of, 172, 173, 173 n, 175, 222; Reinhart at, 309
Wolff, [hunter]: 131 n
wood: scarcity of, 9, 11–12, 200; price of, 200; hauled by Reinhart, 214
Woodbury, George: works for Reinhart, 271, 273, 274, 296; hunts for Reinhart's

*Index*

horses, 279, 281, 282, 284; in Wyoming, 293–294, 295; leaves for east, 296
Woodpile Creek, Oregon: Reinhart at, 41. *See also* Grave Creek, Oregon
Wood River, Nebraska: Oregon wagon train crosses, 6
Woods, Bill: as partner of Reinhart, 29
Woolf Brothers: Reinhart hauls freight for, 248–251; mention of, 252, 306
Woolsey, [sugar refiner]: Thomas Fruid works for, 252
Woosley, William: gold claims of, 169, 171; mention of, 222
Worcester, Massachusetts: George Woodbury from, 274, 297
Wright, Captain Ben: and Indians, 22–23; murder of, 22 n, 94, 95, 96, 220; and Howe, Hurst, and Fay, 32–33
Wright, Colonel George: and Indians, 114 n; building of Fort Walla Walla by, 186 n
Wright, Lazarus: mention of, 56; Reinhart stops with, 111, 149; in Grande Ronde Valley, 228
Wyoming: Reinhart in, 293–306, 309

Yakima Indians: and Fraser River miners, 114; and Colonel Steptoe, 114 n
Yakima River, Washington: mention of, 114; gold discoveries on, 115; Fraser River company on, 118
Yak-to-in Creek, Idaho: 188 n
Yale, James M.: 138 n
Yale, Fort: *See* Fort Yale, British Columbia
Yale's Bar, British Columbia: mention of, 135; location of, 135 n; gold at, 136; Reinhart passes, 136; prices at, 136
Yamhill County, Oregon: mention of, 78 n, 177; Reinhart in, 177
Yankee John: *See* Pringle, John
Yellow Creek, Wyoming: 293
Yellowhawk River, Idaho: 186, 214

Yellowstone River: Fort Benton [no. 2] on, 254 n; mention of, 259; Indians on, 260
Yewka Indians: *See* Yukwitee Indians
Yocum, James: gold claims of, 150
Yokum, Samuel: in Oregon wagon train, 4 n; mention of, 63 n
Yokum, Sandy: in Oregon wagon train, 4 n; mention of, 63 n
Yokum Family: land claim of, 45; ferry operated by, 62–63, 75; toll bridge operated by, 148
Yoncalla Creek, Oregon: Reinhart passes, 53, 111
Young, Judge: and Ben Sykes, 69–70, 72; and McCoy and McCloud, 73
Young, Brigham: Reinhart's meeting with, xxii, 280, 290; visit of, to Idaho, 269 n; ownership of hotel by, 273 n; and Brigham Hamilton Young, 281; and Salt Lake Theater, 283; and Union Pacific Railroad, 301 n
Young, Brigham Hamilton: and Brigham Young, 281; and Reinhart, 281, 284, 291; wives of, 282
Young, John [Pony]: in Victoria, 144
Young, Joseph A.: and Union Pacific Railroad, 301 n
Yreka, California: mining at, xxi, 18; Ben Wright from, 22; history of, 22 n; road to, 24; General Palmer at, 24; mention of, 29, 39, 41, 60 n, 80, 82, 83, 88, 90 n, 117, 118, 122, 135, 146, 160, 163, 199, 238, 241, 259, 309; Reinhart in, 26, 165; Dejarlais' store at, 27 n, 36; Captain McDermit in, 28 n; District Court in, 32; description of, 163
Yreka Indians: 80, 85
Yukwitee Indians: and Rogue River War, 94, 94 n

Zackery [Zakeria], Robert: and Plummer gang, 240 n–241 n
Zakeria, Robert: *See* Zackery, Robert

www.ingramcontent.com/pod-product-compliance
Lightning Source LLC
Chambersburg PA
CBHW021141240426
43661CB00075B/1601